New Insights in Medical Mycology

New Insights in Medical Mycology

Edited by

Kevin Kavanagh

Department of Biology,
National University of Ireland Maynooth,
Co. Kildare,
Ireland

 Springer

A C.I.P. Catalogue record for this book is available from the Library of Congress.

ISBN 978-1-4020-6396-1 (HB)
ISBN 978-1-4020-6397-8 (e-book)

Published by Springer,
P.O. Box 17, 3300 AA Dordrecht, The Netherlands.

www.springer.com

Printed on acid-free paper

Contents

Preface

Although the means of diagnosing and treating fungal infections have improved greatly over the last decade, fungi still represent a serious threat to the health of immunocompromised and immunodeficient patients. In addition to the more commonly encountered fungi, recent years have also seen the emergence of life-threatening infections due to fungi that had previously been seen only rarely in clinical practice. Many of these fungi are difficult to detect and to treat and their emergence as serious agents of disease among specific patient cohorts presents new challenges to the delivery of safe and effective antifungal therapy.

Recent developments in antifungal drug development have led to the welcome introduction of the Echinocandins and various azole derivatives into routine clinical use. Diagnosis of fungal infections has been improved with the utilization of PCR and immunoassay techniques. Despite these advances in diagnosis and therapy, fungi remain serious threats to the health of susceptible patients and we must strive to fully understand the fungi responsible for these infections and their interactions with the host's immune system if improved means of dealing with these infections are to be developed in the future.

The aim of this book is to give an in-depth assessment of our current understanding of the Biology of the main fungal pathogens and how they interact with the host. Each chapter focuses on a specific fungal pathogen or group of pathogens and examines their biology and the factors that allow the fungus colonize and disseminate within the host. In addition each chapter gives an indication of the challenges that remain to be tackled over the next 5–10 years in increasing our understanding of fungal pathogenicity. Each chapter is written by internationally recognized experts and this has ensured that the book is as comprehensive and authorative a text as is possible to assemble.

Chapter 1 gives a detailed description of the immune response of humans to pathogenic fungi and illustrates how an in-depth understanding of the host immune response can be utilized to better challenge fungal infection. Chapter 2 describes various *in vivo* models used to study the virulence of fungal pathogens. Chapter 3 presents an examination of the possibility of using 'alternative' animal models and demonstrates how the structural and functional similarities between the innate immune response of mammals and the insect immune response can be utilized to allow insect be used in place of mammals for the routine screening of fungal mutants. Chapter 4 describes recent developments in the antifungal therapy and highlights the possibility of vaccines being used to prevent fungal infections.

Chapter 5 is the first chapter that deals with a specific pathogen and in this case the pathogen is *Candida albicans*, which has been the subject of much molecular examination in recent years to elucidate its virulence factors and this chapter also describes the current state of our knowledge and highlights the challenges that remain. Chapter 6 describes the biology of *Cryptococcus neoformans* and examines how this fungus interacts with the immune response. The Zygomycetes have emerged in recent years as serious etiological agents of disease in immunocompromised patients. In Chapter 7 the epidemiology, diagnosis, and treatment of zygomycosis is described. Recent developments in our understanding of the pathogenicity of *Aspergillus fumigatus* are described in Chapter 8, and Chapter 9 describes the factors affecting the virulence of *Penicillium marneffei*, which is a serious cause of disease in Southeast Asia among AIDS patients. Dermatophytic infections are one of the most widely encountered of all fungal infections and Chapter 10 describes how dermatophytes interact with the immune system to colonize areas of the body (skin, hair) that would normally be considered extremely hostile. Chapter 11 describes the biology of *Paracoccidioides brasiliensis*, which is the agent of the human systemic disease paracoccidioidomycosis that affects individuals in endemic areas extending from Argentina to Central America. Finally, Chapter 12 describes the occurrence and biology of *Fusarium* spp. and *Scedosporium* spp., which have recently emerged as important fungal pathogens causing significant morbidity and mortality especially in immunocompromised patients.

As well as providing a comprehensive assessment of our current understanding of the biology and pathogenicity of the principal fungal pathogens, each chapter provides an indication of the main challenges that remain to be tackled over the next 5–10 years in our efforts to improve patient recovery. It is the hope of all the contributors that this book will facilitate increased research into the interaction of pathogenic fungi with the immune response and allow the development of new and improved means of diagnosing and treating fungal infections.

Kevin Kavanagh

Contributors

Dr. Khaled H. Abu-Elteen, Department of Biology and Biotechnology, Faculty of Science, Hashemite University, Jordan

Professor Alex Andrianopoulos, Department of Genetics, University of Melbourne, Victoria, Australia

Dr. David M. Arana, Departamento de Microbiología II, Facultad de Farmacia, Universidad Complutense de Madrid, Plaza de Ramón y Cajal s/n, E-28040 Madrid, Spain

Dr. David Cánovas, Department of Genetics, University of Melbourne, Victoria, Australia

Professor Arturo Casadevall, Albert Einstein College of Medicine, Bronx, New York, USA

Dr. Eric Dannaoui, Centre National de Référence Mycologie et Antifongiques, Unité de Mycologie Moléculaire, Institut Pasteur, 75724 Paris Cedex 15, France

Dr. John Dotis, Fellow, 3rd Department of Pediatrics, Aristotle University, Thessaloniki, Greece

Dr. Helene C. Eisenman, Albert Einstein College of Medicine, Bronx, New York, USA

Dr. Susanne Gola, Departamento de Microbiología II, Facultad de Farmacia, Universidad Complutense de Madrid, Plaza de Ramón y Cajal s/n, E-28040 Madrid, Spain

Dr. Gustavo Goldman, Unidade de Oncologia Experimental, Universidade Federal de São Paulo, Rua Botucatu 862, 8 andar, São Paulo, SP 04023-062, Brazil, Spain

Dr. Mawieh M. Hamad, Taif University School of Medicine, Taif, Saudi Arabia

Professor Roderick J. Hay, School of Medicine and Dentistry, Queens University Belfast, BT9 7BL, Northern Ireland, UK

Dr. Dea Garcia-Hermoso, Centre National de Référence Mycologie et Anti-fongiques, Unité de Mycologie Moléculaire, Institut Pasteur, 75724 Paris Cedex 15, France

Dr. Aspasia Katragkou, Fellow, 3rd Department of Pediatrics, Aristotle University, Thessaloniki, Greece

Dr. Kevin Kavanagh, Department of Biology, National University of Ireland Maynooth, Co. Kildare, Ireland

Dr. Donna MacCallum, Aberdeen Fungal Group, School of Medical Sciences, Institute of Medical Sciences, University of Aberdeen, Aberdeen AB25 2ZD, UK

Professor Erin E. McClelland, Albert Einstein College of Medicine, Bronx, New York, USA

Dr. Rebeca Alonso-Monge, Departamento de Microbiología II, Facultad de Farmacia, Universidad Complutense de Madrid, Plaza de Ramón y Cajal s/n, E-28040 Madrid, Spain

Dr. Nir Osherov, Department of Human Microbiology, Sackler School of Medicine, Tel Aviv University, Ramat Aviv, Tel-Aviv 69978, Israel

Professor Jesús Pla, Departamento de Microbiología II, Facultad de Farmacia, Universidad Complutense de Madrid, Plaza de Ramón y Cajal s/n, E-28040 Madrid, Spain

Dr. Rosana Puccia, Unidade de Oncologia Experimental, Universidade Federal de São Paulo, Rua Botucatu 862, 8 andar, São Paulo, SP 04023-062, Brazil

Dr. Julie Renwick, Department of Biology, National University of Ireland Maynooth, Co. Kildare, Ireland

Professor Emmanuel Roilides, 3rd Department of Pediatrics, Aristotle University, Thessaloniki, Greece

Dr. Elvira Román, Departamento de Microbiología II, Facultad de Farmacia, Universidad Complutense de Madrid, Plaza de Ramón y Cajal s/n, E-28040 Madrid, Spain

Professor Luigina Romani, Department of Experimental Medicine and Biochemical Sciences, University of Perugia, Via del Giochetto, 06122 Perugia, Italy

Dr. Carlos P. Taborda, Unidade de Oncologia Experimental, Universidade Federal de São Paulo, Rua Botucatu 862, 8 andar, São Paulo, SP 04023-062, Brazil

Professor Luiz R. Travassos, Unidade de Oncologia Experimental, Universidade Federal de São Paulo, Rua Botucatu 862, 8 andar, São Paulo, SP 04023-062, Brazil

Chapter 1
Immunity to fungi

Luigina Romani

Abbreviations CMC – chronic mucocutaneous candidiasis; CMI – cell-mediated immunity; CR – complement receptor; CR3 – complement receptor 3; DC – dendritic cell; FcR – Fc receptor; IDO – indoleamine 2,3-dioxygenase; IL – interleukin; MBL – mannose-binding lectin; MR – mannose receptor; MyD88 – *Drosophila* myeloid differentiation primary response gene 88; PRR – pattern recognition receptor; TGF-β – transforming growth factor-β; Th, helper T cell; TLR – Toll-like receptor; Treg cell – regulatory T cell

Introduction

The kingdom of fungi consists of a number of species that are associated with a wide spectrum of diseases in humans and animals, ranging from allergy and autoimmunity to life-threatening infections. Most fungi (such as *Histoplasma capsulatum, Paracoccidioides brasiliensis, Coccidioides immitis, Blastomyces dermatitidis, Cryptococcus neoformans, Aspergillus fumigatus*, and *Pneumocystis jirovecii*) are ubiquitous in the environment. Some, including *Candida albicans*, establish lifelong commensalism on human body surfaces. Not surprisingly, therefore, human beings are constantly exposed to fungi, primarily through inhalation or traumatic implantation of fungal elements. The most common of the human diseases caused by fungi are the opportunistic fungal infections that occur in patients with defective immunity.

The switch of emphasis from morbidity to mortality has made the study of fungi a research priority. Because fungal pathogens are eukaryotes and therefore share many of their biological processes with humans, most antifungal drugs are associated with severe toxicity. No standardized vaccines exist for preventing any of the fungal infections of humans, a situation attributed both to the complexity of the pathogens involved and their sophisticated strategies for surviving in the host and evading immune responses (Dan & Levitz, 2006). However, provided that immunotherapy be tailored to specific immunocompromised states (Segal et al., 2006), the proper manipulation of the immune system is the challenge for future strategies that will prevent or treat fungal infections in susceptible patients.

Although not unique among infectious agents, fungi possess complex and unusual relationships with the vertebrate immune system, partly due to some

1

K. Kavanagh (ed.), *New Insights in Medical Mycology.*
© Springer 2007

prominent features. Among these, their ability to exist in different forms and to reversibly switch from one to the other in infection (Nemecek et al., 2006). Examples are the dimorphic fungi (*H. capsulatum, P. brasiliensis, C. immitis*, and *B. dermatitidis*) which transform from saprobic filamentous molds to unicellular yeasts in the host, the filamentous fungi (such as *Aspergillus* spp.) that, inhaled as unicellular conidia, may transform into a multicellular mycelium, and some species of *Candida*, capable of growing in different forms such as yeasts, blastospores, pseudohyphae, and hyphae. This implicates the existence of a multitude of recognition and effector mechanisms to oppose fungal infectivity at the different body sites. For commensals, two prominent features are also important, the highly effective strategies of immune evasion they must have evolved to survive in the host environment and the prolonged antigenic stimulation of the host that can have profound immunoregulatory consequences. Thus, in the context of the antagonistic relationships that characterize the host–pathogen interactions, the strategies used by the host to limit fungal infectivity are necessarily disparate (Romani, 2004b) and, in retaliation, fungi have developed their own elaborate tactics to evade or overcome these defenses (Romani, 2001; Huffnagle & Deepe, 2003; Shoham & Levitz, 2005; Hohl et al., 2006). This may have resulted in an expanded repertoire of cross-regulatory and overlapping antifungal host responses that makes it extremely difficult to define the relative contribution of individual components of the immune system in antifungal defense.

Within the limitations imposed by these considerations, this chapter is an advanced attempt to analyze the role of innate and adaptive immunity in resistance to pathogenic fungi. Through the involvement of different pattern recognition receptors, cells of the innate immune system not only discriminate between the different forms of fungi, but also contribute to discrimination between self and pathogens at the level of the adaptive T helper (Th) immunity (Romani, 2004b and references therein). Thus, the traditional dichotomy between the functions of innate and adaptive immunity in response to fungi has been recently challenged by the concept of an integrated immune response to fungi (Romani, 2004b and references therein).

The Immune Response to Fungi: From Microbe Sensing to Host Defensing

Most pathogenic fungi need a stable host–parasite interaction characterized by an immune response strong enough to allow host survival without elimination of the pathogen, thereby establishing commensalisms and latency. Therefore, the balance of pro-inflammatory and anti-inflammatory signaling is a prerequisite for successful host–fungus interaction. In light of these considerations, the responsibilities for virulence is shared by the host and the fungus at the pathogen–host interface, regardless the mode of its generation and maintenance. Studies with *C. albicans* have provided a paradigm that incorporates contributions from both the fungus and the host to explain the theme of the origin and maintenance of virulence for pathogens and commensals (Romani, 2004a). Through a high degree of flexibility, the model accommodates the concept of virulence as an important component of fungus fitness *in vivo* within the plasticity of immune responses orchestrated by dendritic cells (DC). Conceptually, this implies that the qualitative development of adaptive response to a fungus may not primarily depend on the nature of the fungal form being presented but rather on the type of cell signaling initiated by the ligand–receptor interaction in DC. Therefore, the functional plasticity of DC at the pathogen–host interface may offer new interpretative clues to fungal virulence.

The host defense mechanisms against fungi are numerous, and range from protective mechanisms that appeared early in the evolution of multicellular organisms (referred to, collectively, as 'innate immunity') to sophisticated adaptive mechanisms, which are specifically induced during infection and disease ('adaptive immunity'). The innate mechanisms appeared early in the evolution of multicellular organisms and act early after the infection. Innate defense strategies are designed to detect broad and conserved patterns which differ between pathogenic organisms and their multicellular hosts. Most of the innate mechanisms are inducible upon infection and their activation requires specific recognition of invariant evolutionarily conserved molecular structures shared by large groups of pathogens by a set of pattern recognition receptors (PRR), including Toll-like receptors (TLR) (Akira & Takeda, 2004). In vertebrates, however, if the infectious organism can breach these early lines of defense an adaptive immune response will ensue, with generation of antigen-specific T helper (Th) effector and B cells that specifically target the pathogen and memory cells that prevent subsequent infection with the same microorganism. The two systems are intimately linked and controlled by sets of molecules and receptors that act to generate a highly coordinated and unitary process for protection against fungal pathogens. The dichotomous Th cell model has proven to be a useful construct that sheds light on the general principle that diverse effector functions are required for eradication of different fungal infections (Romani, 1999). The paradigm has greatly contributed to a better understanding of the host immune response to fungi from a regulatory perspective and has been helpful to accommodate clinical findings in a conceptual framework amenable to strategies of immune-interventions.

The Innate Immunity: The Art of Microbe Sensing and Shaping of Specific Immunity

The innate immune system distinguishes self from nonself and activates adaptive immune mechanisms by provision of specific signals. The constitutive mechanisms of defense are present at sites of continuous interaction with fungi and include the barrier function of body surfaces and the mucosal epithelial surfaces of the respiratory, gastrointestinal, and genitourinary tracts. Microbial antagonism (lactobacilli and bifidobacteria have shown efficacy in the biotherapy of candidiasis, i.e. probiotics), defensins, and collectins comprise the major constitutive mechanisms of fungal immunity (Romani, 2004a).

Antigen-independent recognition of fungi by the innate immune system leads to the immediate mobilization of immune effector and regulatory mechanisms that provide the host with three crucial survival advantages: (i) rapid initiation of the immune response and creation of the inflammatory and co-stimulatory environment for antigen recognition; (ii) establishment of a first line of defense, which holds the pathogen in check during the maturation of the adaptive immune response; and (iii) steering of the adaptive immune response towards the cellular or humoral elements that are most appropriate for protection against the specific pathogen. Therefore, in order to achieve optimal activation of antigen-specific adaptive immunity, it is first necessary to activate the pathogen-detection mechanisms of the innate immune response.

The bulwark of the mammalian innate antifungal defense system is built upon effector mechanisms mediated by cells, cellular receptors, and a number of humoral factors (Romani, 2004a; Mansour & Levitz, 2002). The professional phagocytes, consisting of polymorphonuclear leukocytes (neutrophils), mononuclear leuko-cytes (monocytes and macrophages) and DC play an essential role. The antifungal effector functions of phagocytes include fungicidal and growth-inhibiting mecha-nisms, as well as processes to resist fungal infectivity, including inhibitory effects on dimorphism and promotion of phenotypic switching. The optimal restriction of fungal growth occurs via a combination of oxidative and complementary nonoxida-tive mechanisms, the latter consisting of intracellular or extracellular release of effector molecules, defensins, neutrophil cationic peptides, and iron sequestration. Enzymes such as the nicotinamide adenine dinucleotide phosphate (NADPH) oxidase and inducible nitric oxide synthase initiate the oxidative pathways known as the respiratory burst. Myeloperoxidase, a lysosomal hemoprotein found in azurophilic granules of neutrophils and monocytes, but not macrophages, is also a mediator in the oxygen-dependent killing of fungi. Patients with inherited X-linked chronic granulomatous disease (CGD), resulting from a deficiency in formation of activated oxygen radicals due to an NADPH-oxidase deficiency, have increased susceptibility to aspergillosis (Segal et al., 2000). However, transplantation of bone marrow cells transfected with the NADPH-oxidase gene has been shown to restore fungicidal activity of CGD patients (Barese et al., 2004). Myeloperoxidase defi-ciency predisposes to pulmonary candidiasis and aspergillosis, although it has not been shown to play an isolated role in fungal host defense in the absence of the NADPH oxidase.

The fact that both quantitative and qualitative defects of neutrophils are associated with an undue susceptibility to major disseminated fungal infections points to the important role that neutrophils play in the protective immunity to fungal diseases (Romani, 2004a; Fradin et al., 2006). Their functions may well go beyond microbicidal activity, and also include an immunoregulatory role in adaptive immunity (Romani et al., 1996). Myeloid suppressor cells are responsible for the immunosuppression observed in pathologies as dissimilar as tumor growth, immunosuppression, overwhelming infections, graft-versus-host disease, and pregnancy.

The reciprocal relationship of neutrophils and T lymphocytes further implies that the immune resistance to fungi is a highly coordinated and unitary process.

Macrophages are a heterogeneous population of tissue resident cells possessing the machinery for antigen presentation; however, their main contribution to antifungal defense is through phagocytosis and killing of fungi and immunomodulation (Vazquez-Torres & Balish, 1997; Cortez et al., 2006). Not surprisingly, therefore, fungi have exploited a variety of mechanisms or putative virulence factors to evade phagocytosis, escape destruction and survive inside macrophages (Woods, 2003; Alvarez & Casadevall, 2006). Macrophages serve as a protected environment in which the dimorphic fungi multiply and disseminate from the lung to other organs. *H. capsulatum* is a teaching example of a successful intracellular pathogen of mammalian macrophages.

Humoral factors contribute to and enhance the innate defense mechanisms. Mannose-binding protein or lectins (MBL), collectins, complement (a group of proteins activated in cascading fashion) and antibodies promote binding (opsonization) of the fungal organism and represent a recognition mechanism carried out by a variety of receptors and PRR that have a hierarchical organization. The collectin pentraxin 3 has shown a nonredundant role in antifungal resistance to *A. fumigatus* by promoting conidial recognition and phagocytosis, as well as activation of effector phagocytes (Garlanda et al., 2002). The specific biological activities of the complement system and antibodies, which contribute to host resistance are multifaceted and interdependent (Romani & Kaufmann, 1998 and references therein). For example, antibodies greatly contribute to the activation of the complement system by fungi and complement is essential for antibody-mediated protection. Each receptor on phagocytes not only mediates distinct downstream intracellular events related to clearance, but it also participates in complex and disparate functions related to immunomodulation and activation of immunity, depending on cell type. The receptors below are teaching examples. Engagement of CR3 (also known as CD11b/CD18) is one most efficient means of engulfing opsonized fungi, but it also has the remarkable characteristic of a broad recognition capacity of diverse fungal ligands. The multiplicity of binding sites and the existence of different activation states enable CR3 to engage in disparate (positive and negative) effector activities against fungi. Thus, because signaling through CR3 may not lead to phagocyte activation without the concomitant engagement of receptors for the Fc portion of immunoglobulins (FcR), this may contribute to intracellular fungal parasitism. It is of interest, therefore, that *H. capsulatum* uses this receptor for entry into macrophages, where it survives, and not into DC, where it is rapidly degraded. Likewise, *Candida* exploits entry through CR3 to survive inside DC. In contrast, ligation of FcR is usually sufficient to trigger phagocytosis, a vigorous oxidative burst, and the generation of pro-inflammatory signals. Ultimately, recognition of antibody-opsonized particles represents a high-level threat.

The absence of an association between deficits in antibodies and susceptibility to fungal infections and the presence of specific antibodies in patients with progressive fungal infections have been the main arguments against a protective role of antibodies in fungal infections. Recent advances have demonstrated that both protective and nonprotective antibodies against fungi can be demonstrated, the relative composition and proportion of which may vary greatly in infections (Cassone et al., 2005). As a matter of fact, antibodies to HSP90 are associated with recovery from *C. albicans* infections, protection against disseminated disease in patients with AIDS, and synergize with antifungal chemotherapy (Pachl et al., 2006). Complement, antibodies and collectins not only fulfill the requirement of a first line of defense against fungi, but have also an impact on the inflammatory and adaptive immune responses, through several mechanisms, including regulation of cytokine secretion by and costimulatory molecule expression on phagocytes. The local release of these effector molecules regulates cell trafficking in various types of leukocytes, thus initiating an inflammatory response, activates phagocytic cells to a microbicidal state, and directs Th/Treg-cell development (Romani, 2004b).

Sensing Fungi: The TLR and Non-TLR Recognition System

TLR, which are broadly distributed on cells of the immune system, are arguably the best-studied immune sensors of invading pathogens, and the signaling pathways that are triggered by pathogen detection initiate innate immunity and help to strengthen adaptive immunity. TLR belong to the TIR (Toll/interleukin-1 (IL-1) receptor) superfamily, which is divided into two main subgroups: the IL-1 receptors and the TLR. All members of this superfamily signal in a similar manner owing to the presence of a conserved TIR domain in the cytosolic region, which activates common signaling pathways, most notably those leading to the activation of the transcription factor nuclear factor-κB (NF-κB) and stress-activated protein kinases that activate the transcription of the inflammatory and adaptive immune responses. The common signal pathways utilized by IL-1R and TLR involve the recruitment of several adapter proteins, including MyD88 (*Drosophila* myeloid differentiation primary response gene 88), that activates, in turn, a series of kinases that are crucially involved in innate immunity (Akira, 2003). Evidence suggests that individual members of the TLR family or other PRR interact with each other and cumulative effects of these interactions instruct the nature and outcome of the immune response to the provoking pathogen (Mukhopadhyay et al., 2004). TLR activation is a double-edged sword. It is essential for provoking the innate response and enhancing adaptive immunity against pathogens. However, members of the TLR family are also involved in the pathogenesis of autoimmune, chronic inflammatory inflammatory disorders, such as asthma, rheumatoid arthritis, and infectious diseases. Thus, by hyperinduction of pro-inflammatory cytokines, by facilitating tissue damage or by impaired protective immunity, TLR might also promote the pathogenesis of infections.

A number of cell wall components of fungi may act as TLR activators (Levitz, 2004; Netea et al., 2006). The different impact of TLR on the occurrence of the innate and adaptive Th immune response to fungi is consistent with the ability of each individual TLR to activate specialized antifungal effector functions on phagocytes and DC. Although not directly affecting phagocytosis, TLR influence specific antifungal programs of phagocytes, such as the respiratory burst, degranulation, and production of chemokines and cytokines (Bellocchio et al., 2004a, b). As the quantity and specificity of delivery of toxic neutrophil products ultimately determine the relative efficiency of fungicidal activity versus inflammatory cytotoxicity to host cells (Bellocchio et al., 2004b), this implicates that TLR may contribute to protection and immunopathology against fungi. Although the simultaneous engagement of multiple TLR, as well as TLR cooperativity *in vivo*, makes it difficult to gauge the relative contribution of each single fungal morphotype in TLR activation and functioning, the emerging picture calls for: (i) the essential requirement for the MyD88-dependent pathway in the innate and Th1-mediated resistance to fungi (Bellocchio et al., 2004a; Biondo et al., 2005; Rivera et al., 2006); (ii) the crucial involvement, although not essential, of the TLR4/MyD88 pathway in recognition of and resistance to *A. fumigatus* (Netea et al., 2003; Bellocchio et al., 2004a); (iii) the beneficial effect of TLR9 stimulation on immune-mediated resistance to fungal pneumonia (Bozza et al., 2002; Edwards et al., 2004); (iv) the dependency of Treg induction on selected TLR (Netea et al., 2004); (v) the exploitation of TLR as a mechanism to divert and subvert host immune responses (Netea et al., 2004), and (vi) the association of selected TLR polymorphisms with susceptibility to fungal infections (Kesh et al., 2005).

C-type lectin receptors (i.e. Dectin-1 and 2, DC-SIGN, and the galectin family) are major mammalian PRR for several fungal components and are the prototype of innate non-TLR signaling pathway for innate antifungal sensing (Brown, 2006). The finding that N-linked mannosyl residues on fungal cells are bound by the MR, and O-linked mannosyl residues are bound by TLR4 (Netea et al., 2006) provide mechanistic insights into the cooperative signaling between TLR and non-TLR for full innate immune cell activation (Steele et al., 2005; Gross et al., 2006; Sato et al., 2006; Taylor et al., 2006; Saijo et al., 2006; Gersuk et al., 2006).

Tuning the Adaptive Immune Responses:
The Instructive Role of DC

As DC are equipped with several TLR, they are the main connectors of the innate and adaptive immune systems. DC are bone marrow-derived cells of both lymphoid and myeloid stem cell origins that populate all lymphoid organs, as well as nearly all nonlymphoid tissues and organs. The dual activation/tolerization function of DC is mediated by their capacity to change the context of antigen presentation and to communicate to T cells the nature of the antigens they are presenting. This process exemplifies the importance of TLR not only in direct early immune responses, but also in activation of adaptive immunity. The DC system consists of a network of different subpopulations (Romani & Puccetti, 2006a). The ability of a given DC subset to respond with flexible activating programs to the different stimuli, as well as the ability of different subsets to convert into each others confers unexpected plasticity to the DC system. DC are uniquely adept at decoding the fungus-associated information and translating it in qualitatively different adaptive T-cell immune responses (Romani & Puccetti, 2006a). PRR (such as CR, FcR, C-type lectins (such as DC-SIGN and dectin-1), MR, and TLR determine the functional plasticity of DC in response to fungi and contribute to the discriminative recognition of the different fungal morphotypes. DC (both human and murine) are now known to recognize and internalize a number of fungi, including *A. fumigatus, C. albicans, C. neoformans, H. capsulatum, Malassezia furfur*, and *Saccharomyces cerevisiae* and fungi and fungal products may affect DC functioning as well (Romani & Puccetti, 2006a; Buentke & Scheynius, 2003). DC are also known to cross-present exogenous fungal antigens through uptake of apoptotic macrophage-associated fungal antigens (Lin et al., 2005). Profiling gene expression on DC by microarray technologies has revealed that both shared response and a pathogen-specific gene expression program were induced upon the exposure to bacteria, viruses and fungi. Additional studies with *S. cerevisiae* have shown that recombinant yeast could represent an effective vaccine for the generation of broad-based cellular immune responses. It seems, therefore, that DC are uniquely able at decoding the fungus-associated information at the host–fungus interface. *Candida* and *Aspergillus* proved to be useful pathogen models to dissect events occurring at the fungus–DC interface. Murine and human DC internalize *Candida* yeasts, *Aspergillus* conidia and hyphae of both. The uptake of the different fungal elements occurred through different receptors and forms of phagocytosis. Transmission electronic microscopy indicated that internalization of yeasts and conidia occurred predominantly by coiling phagocytosis, characterized by the presence of overlapping bilateral pseudopods, which led to a pseudopodal stack before transforming into a phagosome wall. In contrast, entry of hyphae occurred by a more conventional zipper-type phagocytosis, characterized by the presence of symmetrical pseudopods which strictly followed the contour of the hyphae before fusion. Recognition and internalization of unopsonized yeasts and conidia occurred through the engagement of MR of different sugar specificity, DC-SIGN, dectin-1, and partly, CR3 (Claudia et al., 2002; Mansour et al., 2006). In contrast, entry of hyphae occurred by a more conventional, zipper-type phagocytosis and involved the cooperative action of FcγR II and III and CR3. Phagocytosis does not require TLR/MyD88. Consistent with the findings that signals from protein kinase C (PKC) and/or protein tyrosine kinases are required for phagocytosis in a variety of systems, the PKC inhibitor staurosporine was required for CR- and FcγR-mediated phagocytosis, while FcγR- and, to a lesser extent, MR-mediated phagocytosis required signaling through protein tyrosine kinases (Claudia et al., 2002). The results are consistent with the view that fungi have exploited common pathways for entry into DC, which may include a lectin-like pathway for unicellular forms and opsono-dependent pathways for filamentous forms.

The engagement of distinct receptors by distinct fungal morphotypes translates into downstream signaling events, ultimately regulating cytokine production and costimulation, an event greatly influenced by fungal opsonins, such as MBL, C3, and/or antibodies (Romani et al., 2004; Romani et al., 2002). Entry through MR and dectin-1 resulted in the production of pro-inflammatory cytokines, including IL-12, upregulation of costimulatory molecules and histocompatibility Class II antigens. IL-12 production by DC required the MyD88 pathway with the implication of distinct TLR. These events were all suppressed upon entry through CR3. In contrast, coligation of CR3 with FcγR, as in the phagocytosis of hyphae, resulted in the production of IL-4/IL-10 and upregulation of costimulatory molecules and histocompatibility Class II antigens. The production of IL-10 was largely MyD88-independent. Therefore, TLR collaborate with other innate immune receptors in the activation of DC against fungi through MyD88-dependent and MyD88-independent pathways (Romani & Puccetti, 2006a).

A remarkable and important feature of DC is their capacity to produce IL-10 in response to fungi. These IL-10-producing DC activate CD4$^+$ CD25$^+$ Treg cells that are essential components of antifungal resistance (see below). Thus, by subverting the morphotype-specific program of activation of DC, opsonins, antibodies, and other environmental factors may qualitatively affect DC functioning and Th/Treg selection *in vivo*, ultimately impacting on fungal virulence. In this scenario, the qualitative development of the Th cell response to a fungus may not primarily depend on the nature of the fungal form being phagocytosed and presented. Rather, the nature of the cell response is strongly affected by the type of cell signaling initiated by the ligand–receptor interaction in DC. For *Candida*, the paradigm would predict that dimorphism per se can no longer be considered as the single most important factor in determining commensalism versus infection, nor can specific forms of the fungus be regarded as absolutely indicative of saprophytism or infection at a given site. The selective exploitation of receptor-mediated entry of fungi into DC could explain the full range of host immune–parasite relationships, including saprophytism and infection. Importantly, as both fungal morphotypes, but particularly hyphae, activate gut DC for the local induction of Treg cells and because the morphogenesis of *C. albicans* is activated *in vivo* by a wide range of signals, it appears that the discriminative response towards Treg cell function is of potential teleological meaning. It could indeed allow for fungal persistence in the absence of the pathological consequences of an exaggerated immunity and possible autoimmunity, a condition which represents the very essence of fungal commensalism. Therefore, in addition to the induction of phase-specific products enhancing fungal survival within the host, transition to the hyphal phase of the fungus could implicate the induction of immunoregulatory events that will benefit the host.

Fungus-pulsed DC translated fungus-associated information to Th1, Th2, and Treg cells, *in vitro* and *in vivo* (Romani & Puccetti, 2006a). *In vivo*, the balance among the different DC subsets determined whether protective or nonprotective antifungal cell-mediated immune responses developed. Fungus-pulsed DC activated different CD4$^+$ Th cells upon adoptive transfer in a murine model of allogeneic bone marrow transplantation (Bozza et al., 2005; Romani et al., 2006). The ability of fungus-pulsed DC to prime for Th1 and Th2 cell activation upon adoptive transfer *in vivo* correlated with the occurrence of resistance and susceptibility to the infections. Recent data have shown that the infusion of fungus-pulsed

DC of the different subsets accelerated the recovery of peripheral antifungal Th1 immunity and increased resistance to fungal infections in a murine model of allogeneic bone marrow transplantation (Romani et al., 2006). However, only the co-infusion of DC of both subsets resulted in: (i) induction of T reg cells capable of a fine control over the inflammatory pathology; (ii) tolerization toward alloanti- gens; and (iii) diversion from alloantigen-specific to antigen-specific Tcell responses in the presence of donor T lymphocytes. Thus, the adoptive transfer of DC may restore antifungal immunocompetence in hematopoietic transplantation by contributing to the educational program of T cells through the combined action of activating and tolerizing DC. These results, along with the finding that fungus- pulsed DC could reverse T-cells anergy of patients with fungal diseases (Romani & Puccetti, 2006a), may suggest the utility of DC for fungal vaccines and vaccination (Bozza et al., 2004; Lam et al., 2005).

The Adaptive Immunity: Th1, Th2, and Th17 Cells

Serological and skin reactivity surveys indicate that fungal infections are common, but clinical disease is rare, consistent with the development of acquired immunity. Underlying acquired immunity to *C. albicans*, such as the expression of a positive delayed type hypersensitivity, is demonstrable in adult immunocompetent individuals, and is presumed to prevent mucosal colonization from progression to symptomatic infection (Puccetti et al., 1995). Lymphocytes from healthy subjects show strong proliferative responses after stimulation with fungal antigens and produce a number of different cytokines (Romani, 2004b). For many fungal pathogens, the effective tissue response to invasion is granulomatous inflammation, a hallmark of cell-mediated immunity (CMI). There is extensive plasticity in the T-cell response to fungi (Romani & Puccetti, 2006a). The heterogeneity of the $CD4^+$ and $CD8^+$ T cell repertoire may account for the multiplicity and redundancy of effector mechanisms through which T lymphocytes participate in the control of fungal infections. The flexible program of T lymphocytes also implicates the production of a number of mediators, including cytokines. Due to their action on circulating leukocytes, the cytokines produced by fungus-specific T cells are instrumental in mobilizing and activating antifungal effectors, thus providing prompt and effective control of infectivity once the fungus has established itself in tissues or spread to internal organs. Therefore, host resistance to fungi appears to be dependent upon the induction of cellular immunity, mediated by T lymphocytes, cytokines, and a number of effector phagocytes (Romani, 2004b).

The clinical circumstances in which fungal infections occur definitely suggest an association with impaired CMI. AIDS and severe hematological malignancies are examples of acquired defects in T-cell function that predispose to severe fungal infections. Interestingly, however, defective CMI may also be a consequence of fungal virulence (Fischer et al., 1978; Yauch et al., 2006). Furthermore, the occur- rence of severe disseminated infections by filamentous fungi in non-granulocytopenic patients, as well as in concomitance with the onset of graft-versus-host disease in bone marrow transplant recipients are compelling evidence of the pathogenic role of T-cell dysreactivity in infection. In endemic mycosis, the severity of the disease correlates with the degree of impairment of CMI, associated with elevated levels of antibodies (Romani, 2004b).

Generation of a dominant Th1 response driven by IL-12 is essentially required for the expression of protective immunity to fungi. Through the production of the signature cytokine IFN-γ and help for opsonizing antibodies, the activation of Th1 cells is instrumental in the optimal activation of phagocytes at sites of infection. Therefore, the failure to deliver activating signals to effector phagocytes may predispose patients to overwhelming infections, limit the therapeutic efficacy of antifungals and antibodies, and favor persistency and/or commensalism. Immunological studies in patients with polar forms of paracoccidioidomycosis demonstrate an association between Th1-biased reactivity and the asymptomatic and mild forms of the infection, as opposed to the positive correlation of Th2 responses with the severity of the disease. Not surprisingly, therefore, patients with disseminated infection show defective production of IFN-γ and DTH anergy, associated with elevated levels of type 2 cytokines (IL-4 and IL-5), IgE, IgG4 and IgA, and eosinophilia, which is a marker of poor prognosis in endemic mycoses (Romani & Kaufmann, 1998 and references therein). In patients with defective IL-12/IFN-γ pathway, such as those with hyperimmunoglobulinemia E syndrome, fungal infections, and allergy are both observed (Romani, 2004b). Deficient IFN-γ receptor-mediated signaling occurs in neonates and may predispose to fungal infections. IL-4 is one major discriminative factor of susceptibility and resistance in most fungal infections. The most important mechanism underlying the inhibitory activity of IL-4 in infections relies on its ability to act as the most potent proximal signal for commitment to Th2 reactivity that dampens protective Th1 responses and favors fungal allergy. In atopic subjects, the suppressed DTH response to fungi is associated with elevated levels of antifungal IgE, IgA, and IgG. However, susceptibility to fungal infections may not always be associated with an overt production of IL-4. For instance, although an association between chronic disseminated candidiasis and genetic variants of IL-4 has been recently described (Romani, 2004b), IL-4 or IL-5 are not always increased in patients with chronic mucocutaneous candidiasis (CMC), despite a defective type 1 cytokine production (Lilic et al., 2003).

Recent studies have suggested a greater diversification of the CD4[+] T cell effector repertoire than that encompassed by the Th1/Th2 paradigm (Dong, 2006). Th17 cells are now thought to be a separate lineage of effector Th cells contributing to immune pathogenesis previously attributed to the Th1 lineage. Although the developmental pathways leading to Th17 differentiation *in vitro* are still unclear, IL-23 is a critical cytokine for the generation and maintenance of this lineage. IL-12 and IL-23 are members of a small family of pro-inflammatory heterodimeric cytokines that share a common p40 subunit linked to the IL-12p35 chain or the IL-23p19 chain. IL-23 functions through a receptor complex composed of the IL-12Rβ1 subunit and a unique component, the IL-23R chain. Both cytokines induce IFN-γ expression in CD4[+] T cells, though only IL-23 facilitates a T-helper state marked by production of the pro-inflammatory cytokine, IL-17. Despite these similarities,

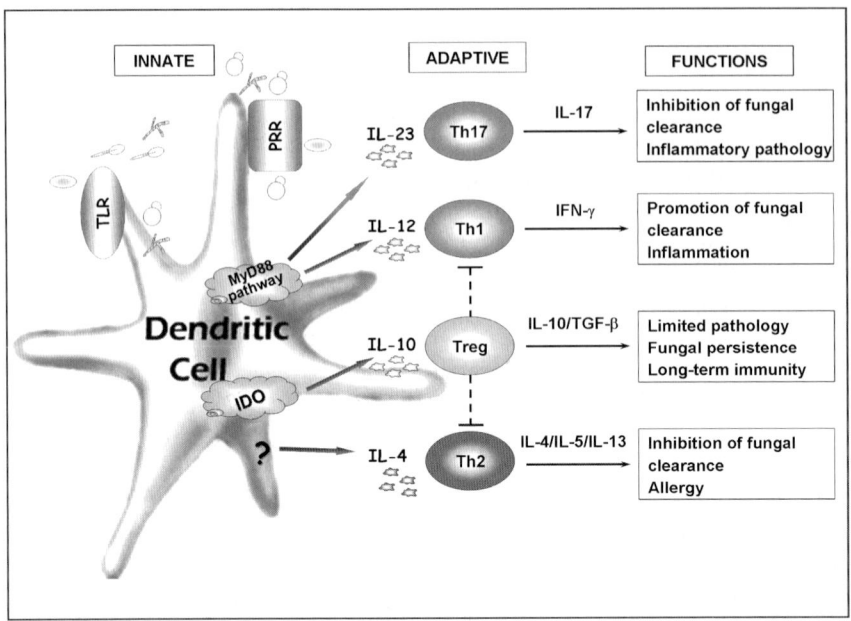

Figure 1.1 Pathways of innate and adaptive antifungal immunity: the role of dendritic cells, tryptophan catabolism, and Th subsets. The majority of fungi are detected and destroyed within hours by innate defense mechanisms. These mechanisms act immediately and are followed some hours later by an early induced response, which must be activated by infection but does not generate lasting protective immunity. These early phases help to keep infection under control. In vertebrates, however, if the infectious organism can breach these early lines of defense an adaptive immune response will ensue, with generation of antigen-specific T helper (Th) effectors and regulatory T (Treg) cells that specifically target the pathogen and induce memory cells that prevent subsequent infection with the same microorganism. Dendritic cells (DC) sample fungi at the site of colonization/infection, transport them to the draining lymph nodes and activate disparate Th/Treg cells in a morphotype- and tissue-dependent fashion. The activity of DC involves the pattern recognition receptors (PRR), including Toll-like receptors (TLR) and the enzyme indoleamine 2,3-dioxygenase (IDO)-dependent metabolic pathways leading to T cell activation and regulation. As the different Th cell subsets release a distinct panel of cytokines, capable of delivering, activating, and deactivating feedback signals to effector phagocytes, the activation of the appropriate Th subset is instrumental in the generation of a successful immune response to fungi. Counter-regulatory Treg cells may serve to dampen the excessive inflammatory reactions and to contribute to the development of memory antifungal immunity. Solid and broken lines refer to positive and negative signals, respectively

there is increasing evidence that IL-12 and IL-23 drive divergent immunological pathways (Trinchieri et al., 2003). Th cells primed for IL-17 production appear to have important roles in autoimmune diseases (Harrington et al., 2006). Moreover, although less clear, the production of high levels of IL-23/IL-17, more than IL-12/IFN-γ, better correlates with disease severity and immunopathology in diverse infections (Hunter, 2005). These studies suggest that IL-12 and IL-23 have distinct roles in promoting antimicrobial immune responses and diseases *in vivo*. Recent evidence indicated that the IL-23/IL-17 developmental pathway may act as a negative regulator of the Th1-mediated immune resistance to fungi and played an inflammatory role previously attributed to uncontrolled Th1 cell responses. Both inflammation and infection were exacerbated by a heightened Th17 response against *C. albicans* and *A. fumigatus*. Both IL-23 and IL-17 subverted the inflammatory program of neutrophils and promoted fungal virulence, which impacted severely on tissue inflammatory pathology associated with infection (author's unpublished observations). Our data support a model in which IL-23-driven inflammation promotes infection and impairs antifungal immune resistance (Figure 1.1). Thus, modulation of the inflammatory response represents a potential strategy to stimulate protective immune responses to fungi.

Dampening Inflammation and Allergy to Fungi: A Job for Treg Cells

The inflammatory response to fungi may serve to limit infection but may also contribute to pathogenicity, as documented by the occurrence of severe fungal infections in patients with immunoreconstitution disease (Cheng et al., 2000). These patients may experience intractable fungal infections despite recovery from neutropenia and the occurrence of adaptive immune responses. The above considerations imply that immunoregulation may be essential in fine-tuning inflammation and adaptive Th reactivity to fungi and fungal diseases. This imposes a new job upon the immune system. In addition to efficient control of pathogens, tight regulatory mechanisms are required in order to balance protective immunity and immunopathology. To limit the pathologic consequences of an excessive inflammatory cell-mediated reaction, the immune system resorts to a number of protective mechanisms. CD4[+] T cells making immunoregulatory cytokines such as IL-10, transforming growth factor (TGF)-β and IL-4 have long been known and discussed in terms of immune deviation or class regulation. Recently, Treg cells, capable of fine-tuning protective antimicrobial immunity in order to minimize harmful immune pathology, have become an integral component of the immune response (Montagnoli et al., 2002; Montagnoli et al., 2006; Romani & Puccetti, 2006b; Hori et al., 2002; Cavassani et al., 2006; McKinley et al., 2006). The decision of how to respond will still be primarily determined by interactions between pathogens and cells of the innate immune system, but the actions of Treg cells will feed back into this dynamic equilibrium to regulate subsequent immune responses. Usually, Treg cells serve to restrain exuberant immune reactivity, which in many chronic infections benefits the host by limiting tissue damage. However, the natural Treg cell responses may handicap the efficacy of protective immunity. Conceptually, similar to their effect on immunity against pathogens, Treg cells can also impede effective immunosurveillance of tumors. Nowadays, aberrant numbers and/or functions of Treg cells are incorporated within the view of counter-regulatory elements affecting the self versus nonself discrimination and influencing the outcome of infection, autoimmunity, transplantation, cancer, and even allergy.

A number of clinical observations suggest an inverse relationship between IFN-γ and IL-10 production in patients with fungal infections. High levels of IL-10, negatively affecting IFN-γ production, are detected in chronic candidal diseases, in the severe form of endemic mycoses and in neutropenic patients with aspergillosis (Romani, 2004b). Fungal polysaccharides are known to negatively modulate CMI through the production of IL-10, a finding suggesting that IL-10 production may be a consequence of infection (Romani & Puccetti, 2006b). However, tolerance to fungi can also be achieved through the induction of Treg cells capable of finely tuning antifungal Th reactivity. Naturally occurring Treg cells operating in the respiratory or the gastrointestinal mucosa accounted for the lack of pathology associated with fungal clearance in mice with fungal pneumonia or mucosal candidiasis (Montagnoli et al., 2002, 2006). Distinct Treg populations capable of mediating anti-inflammatory or tolerogenic effects are coordinately induced after exposure to *Aspergillus* conidia. Ultimately, the inherent resistance to *Aspergillus* diseases suggests the existence of regulatory mechanisms that provide the host with protection from infection and tolerance to allergy. It has been demonstrated that a division of labor occurs between functionally distinct Treg cells that are coordinately activated after exposure of mice to *Aspergillus* resting conidia. Early in infection, inflammation is controlled by the expansion, activation, and local recruitment of Treg cells suppressing neutrophils through the combined actions of IL-10 and cytotoxic T lymphocyte antigen-4 on the enzyme indoleamine 2,3-dioxygenase (IDO) (see below). Late in infection, and similarly in allergy, tolerogenic Treg cells which produce IL-10 and TGF-β inhibit Th2 cells and prevent allergy to the fungus.

It has long been known that the ability of *C. albicans* to establish an infection involves multiple components of the fungus, but its ability to persist in host tissue might involve primarily the immunosuppressive property of a major cell wall glycoprotein, mannan (Nelson et al., 1991). Although epitopes of mannan exist endowed with the ability to induce protective antibodies to the fungus, mannan and oligosaccharide fragments of it could be potent inhibitors of cell-mediated immunity and appear to reproduce the immune deficiency in patients with the mucocutaneous form of candidiasis (Fischer et al., 1978). CMC, although encompassing a variety of clinical entities, has also been associated with autoimmune polyendocrinopathy-candidiasis-ectodermal dystrophy (APECED) (Peterson et al., 1998). Interestingly, in APECED, the mutated gene has been proposed to be involved in the ontogeny CD25[+] Treg cells (Sakaguchi et al., 1996). In CMC, both anergy and active lymphoproliferation and variable-delayed hypersensitivity to the fungus are indeed observed (Lilic, 2002). As already discussed, this has been associated with a defective type 1 cytokine production without obvious increase in type 2 cytokine production (namely IL-4 or IL-5). However, variable, either increased or not, levels of IL-10 have also been observed, a finding that may lead to the speculation that an inherent alteration in receptor-mediated signaling in response to fungal polysaccharide, may predispose patients with CMC to a dysfunctional induction of Treg, negatively affecting the capacity of the Th1-dependent clearance of the fungus and without the activation of Th2 cells.

Collectively, these observations suggest that the capacity of Treg cells to inhibit aspects of innate and adaptive immunity may be central to their regulatory activity in fungal infections. This may result in the generation of immune responses vigorous enough to provide adequate host defense, without necessarily eliminating the pathogen (which could limit immune memory) or causing an unacceptable level of host damage. In the last two decades the immunopathogenesis of fungal infections and associated diseases was explained primarily in terms of Th1/Th2 balance. While the pathogenetic role of either subset may still hold true, the reciprocal regulation of both subsets is apparently outdated. It appears that a combination of different types of Treg cells controls the Th1, as well as the Th2 inflammatory responses (Figure 1.1).

The Central Role of the Tryptophan Metabolic Pathway in Tolerance and Immunity to Fungi

The inflammatory/anti-inflammatory state of DC is strictly controlled by the metabolic pathway involved in tryptophan catabolism and mediated by the enzyme IDO. IDO has a complex role in immunoregulation in infection, pregnancy, autoimmunity, transplantation, and neoplasia (Mellor & Munn, 2004). IDO-expressing DC are regarded as regulatory DC specialized to cause antigen-specific deletional tolerance or otherwise negatively regulating responding T cells. In experimental fungal infections IDO blockade greatly exacerbated infections, the associated inflammatory pathology and swept away resistance to reinfection, as a result of deregulated innate and adaptive immune responses caused by the impaired activation and functioning of suppressor CD4$^+$ CD25$^+$ Treg cells producing IL-10 (Bozza et al., 2005; Montagnoli et al., 2006). The results provide novel mechanistic insights into complex events that, occurring at the fungus–pathogen interface, relate to the dynamics of host adaptation by fungi. The production of IFN-γ may be squarely placed at this interface, where IDO activation likely exerts a fine control over inflammatory and adaptive antifungal responses.

   The implication for IDO in immunoregulation in candidiasis may help to accommodate several, as yet unexplained findings. As *C. albicans* is a commensal of the human gastrointestinal and genitourinary tracts and IFN-γ is an important mediator of protective immunity to the fungus, the IFN-γ/IDO axis may accommodate fungal persistence in a host environment rich in IFN-γ. In its ability to downregulate antifungal Th1 response in the gastrointestinal tract, IDO behaves in a fashion similar to that described in mice with colitis where IDO expression correlates with the occurrence of local tolerogenic responses. Alternatively, the high levels of IL-10 production, such as those seen in patients with CMC, may be a consequence of IDO activation by the fungus, impairing antifungal Th1 immunity and thus favoring persistent infection (Romani & Puccetti, 2006b). In aspergillosis, the level of inflammation and IFN-γ in the early stage set the subsequent adaptive stage by conditioning the IDO-dependent tolerogenic program of DC and the subsequent activation and expansion of tolerogenic Treg cells preventing allergy to the fungus. Therefore, regulatory mechanisms operating in the control of inflammation and allergy to the fungus are different but interdependent as the level of the inflammatory response early in infection may impact on susceptibility to allergy, in conditions of continuous exposure to the fungus. Early Treg cells, by affecting IFN-γ-production, indirectly exert a fine control over the induction of late tolerogenic Treg cells. Thus, a unifying mechanism linking natural Treg cells to tolerogenic respiratory Treg cells in response to the fungus is consistent with the revisited 'hygiene hypothesis' of allergy in infections, and may provide at the same time mechanistic explanations for the significance of the variable level of IFN-γ seen in allergic diseases and asthma and for the paradoxical worsening effect on allergy of Th1 cells. IDO has a unique and central role in this process as it may participate in the effector and inductive phases of anti-inflammatory and tolerogenic Treg cells.

Conclusions

The discovery of TLR, DC, and Treg cells have been major breakthroughs in the field of fungal immunology, which may offer new grounds for a better comprehension of the cells and immune pathways that are amenable to manipulation in patients with or at risk of fungal infections. A variety of cytokines, including chemokines and growth factors proved to be beneficial in experimental and human fungal infections (Mencacci et al., 2000; Kawakami, 2003). The Th1/Th2 balance itself can be the target of immunotherapy (Koguchi & Kawakami, 2002; Mencacci et al., 2000). It now appears that a combination of different types of Treg cells controls the Th1, as well as the Th2 inflammatory responses. Consequently, manipulation of Treg cells is thought of as a promising therapeutic approach devoid of risks associated with interference with homeostatic mechanisms of the immune system. Further understanding of the cooperation of various multiple innate immune receptors in fungal recognition potentially provides a basis for novel therapeutic strategies for immunomodulation, which will very likely contribute to successfully coping with the threat of severe fungal infections. Notwithstanding the redundancy and overlapping repertoire of antifungal effector mechanisms, the deliberate targeting of cells and pathways of antifungal CMI may represent a useful strategy in developing fungal vaccines capable of both sterilizing immunity and protecting against fungal reactivation.

Acknowledgments I thank Dr. Silvia Moretti for editorial assistance. This study was supported by the Specific Targeted Research Project 'EURAPS' (LSHM-CT-2005), contract number 005223 (FP6).

References

Akira, S. (2003). *Curr. Opin. Immunol.*, 15:5–11.
Akira, S. & Takeda, K. (2004). *Nat. Rev. Immunol.*, 4:499–511.
Alvarez, M. & Casadevall, A. (2006). *Curr. Biol.*, 16:2161–2165.
Barese, C. N., Goebel, W. S., & Dinauer, M. C. (2004). *Expert Opin. Biol. Ther.*, 4:1423–1434.

Bellocchio, S., Montagnoli, C., Bozza, S., Gaziano, R., Rossi, G., et al. (2004a). *J. Immunol.*, 172:3059–3069.

Bellocchio, S., Moretti, S., Perruccio, K., Fallarino, F., Bozza, et al. (2004b). *J. Immunol.*, 173:7406–7415.

Biondo, C., Midiri, A., Messina, L., Tomasello, F., Garufi, G., et al. (2005). *Eur. J. immunol.*, 35:870–878.

Bozza, S., Fallarino, F., Pitzurra, L., Zelante, T., Montagnoli, C., et al. (2005). *J. Immunol.*, 174:2910–2918.

Bozza, S., Gaziano, R., Lipford, G. B., Montagnoli, C., Bacci, A., et al. (2002). *Microbes Infect.*, 4:1281–1290.

Bozza, S., Montagnoli, C., Gaziano, R., Rossi, G., & Nkwanyuo, G., (2004). *Vaccine*, 22:857–864.

Brown, G. D. (2006). *Nat. Rev. Immunol.*, 6:33–43.

Buentke, E. & Scheynius, A. (2003). *APMIS*, 111:789–796.

Cassone, A., de Bernardis, F., & Torososantucci, A. (2005). *Curr. Mol. Med.*, 5:377–382.

Cavassani, K. A., Campanelli, A. P., Moreira, A. P., Vancim, J. O., Vitali, L. H., et al. (2006). *J. Immunol.*, 177:5811–5818.

Cheng, V. C., Yuen, K. Y., Chan, W. M., Wong, S. S., Ma, E. S., & Chan, R. M. (2000). *Clin. Infect. Dis.*, 30:882–892.

Claudia, M., Bacci, A., Silvia, B., Gaziano, R., Spreca, A., & Romani, L. (2002). *Curr. Mol. Med.*, 2:507–524.

Cortez, K. J., Lyman, C. A., Kottilil, S., Kim, H. S., Roilides, E., et al. (2006). *Infect. Immun.*, 74:2353–2365.

Dan, J. M. & Levitz, S. M. (2006). *Drug Resist. Updat.*, 9:105–110.

Dong, C. (2006). *Nat. Rev. Immunol.*, 6:329–333.

Edwards, L., Williams, A. E., Krieg, A. M., Rae, A. J., Snelgrove, R. J., & Hussell, T. (2004). *Eur. J. Immunol.*, 35:273–281.

Fischer, A., Ballet, J. J., & Griscelli, C. (1978). *J. Clin. Invest.*, 62:1005–1013.

Fradin, C., Mavor, A. L., Weindl, G., Schaller, M., Hanke, K., et al. (2006). *Infect. Immun.*

Gersuk, G. M., Underhill, D. M., Zhu, L., & Marr, K. A. (2006). *J. Immunol.*, 176:3717–3724.

Gross, O., Gewies, A., Finger, K., Schafer, M., & Sparwasser, T., (2006). *Nature*, 442:651–656.

Harrington, L. E., Mangan, P. R., & Weaver, C. T. (2006). *Curr. Opin. Immunol.*, 18:349–356.

Hohl, T. M., Rivera, A., & Pamer, E. G. (2006). *Curr. Opin. Immunol.*, 18:465–472.

Hori, S., Carvalho, T. L., & Demengeot, J. (2002). *Eur. J. Immunol.*, 32:1282–1291.

Huffnagle, G. B. & Deepe, G. S. (2003). *Curr. Opin. Microbiol.*, 6:344–350.

Kawakami, K. (2003). *J. Infect. Chemother.*, 9:201–209.

Kesh, S., Mensah, N. Y., Peterlongo, P., Jaffe, D., Hsu, K. M., (2005). *Ann. N. Y. Acad. Sci.*, 1062:95–1103.

Koguchi, Y. & Kawakami, K. (2002). *Int. Rev. Immunol.*, 21;423–438.

Lam, J. S., Mansour, M. K., Specht, C. A., & Levitz, S. M. (2005). *J. Immunol.*, 175:7496–7503.

Levitz, S. M. (2004). *Microbes Infect.*, 6:1351–1355.

Lilic, D. (2002). *Curr. Opin. Infect. Dis.*, 15:143–147.

Lilic, D., Gravenor, I., Robson, N., Lammas, D. A., Drysdale, P., et al. (2003). *Infect. Immun.*, 71:5690–5699.

Lin, J. S., Yang, C. W., Wang, D. W., & Wu-Hsieh, B. A. (2005). *J. Immunol.*, 174:6282–6291.

Mansour, M. K., Latz, E., & Levitz, S. M. (2006). *J. Immunol.*, 176:3053–3061.

Mansour, M. K. & Levitz, S. M. (2002). *Curr. Opin. Microbiol.*, 5:359–365.

McKinley, L., Logar, A., McAllister, F., Zheng, M., Steele, C., & Kolls, J. (2006). *J. Immunol.*, 177:6215–6226.

Mellor, A. & Munn, D. H. (2004). *Nat. Rev. Immunol.*, 4:762–774.

Mencacci, A., Cenci, E., Bacci, A., Montagnoli, C., Bistoni, F., & Romani, L. (2000). *Curr. Pharm. Biotechnol.*, 1:235–251.

Montagnoli, C., Bacci, A., Bozza, S., Gaziano, R., Mosci, P., et al. (2002). *J. Immunol.*, 169:6298–6308.

Montagnoli, C., Fallarino, F., Gaziano, R., Bozza, S., Bellocchio, S., et al. (2006). *J. Immunol.*, 176:1712–1723.

Mukhopadhyay, S., Herre, J., Brown, G., & Gordon, S. (2004). *Immunology*, 112:521–530.

Nelson, R., Shibata, N., Podzorski, R., & Herron, M. (1991). *Clin. Microbiol. Rev.*, 4:1–19.

Nemecek, J., Wuthrich, M., & Klein, B. (2006). *Science*, 312:583–588.

Netea, M., Ferwerda, G., van der Graaf, C., van der Meer, J., & Kullberg, B. (2006). *Curr. Pharm. Des.*, 12:4195–4201.

Netea, M., van der Meer, J., & Kullberg, B. (2004). *Trends Microbiol.*, 12:484–488.

Netea, M., Warris, A., van der Meer, J., Fenton, M., Verver-Janssen, T., et al. (2003). *J. Infect. Dis.*, 188:320–326.

Pachl, J., Svoboda, P., Jacobs, F., Vandewoude, K., van der Hoven, B., et al. (2006). *Clin. Infect. Dis.*, 42:1404–1413.

Peterson, P., Nagamine, K., Scott, H., Heino, M., Kudoh, J., et al. (1998). *Immunol. Today*, 19:384–386.

Puccetti, P., Romani, L., & Bistoni, F. (1995). *Trends Microbiol.*, 3:237–240.

Rivera, A., Ro, G., van Epps, H., Simpson, T., Leiner, I., Sant'angelo, D., & Pamer, E. (2006). *Immunity*, 25:665–675.

Romani, L. (1999). *Curr. Opin. Microbiol.*, 2:363–367.

Romani, L. (2001). Chapter overview of the fungal pathogens. In: S. H. *Immunology of Infectious Diseases*, Kaufmann, A. Sher and Ahmed R., Eds., ASM Press, Washington, DC, pp. 25–37.

Romani, L. (2004a). Chapter innate immunity to fungi: the art of speed and specificity. In: *Pathogenic Fungi. Host Interactions and Emerging Strategies for Control*, G. san-Blas, and R. A. Calderone Eds., Caister Academic Press, Norfolk, England, pp. 167–214.

Romani, L. (2004b). *Nat. Rev. Immunol.*, 4:1–23.

Romani, L., Bistoni, F., Perruccio, K., Montagnoli, C., Gaziano, R., et al. (2006). *Blood*, 108:2265–2274.

Romani, L., Bistoni, F., & Puccetti, P. (2002). *Trends Microbiol.*, 10:508–514.

Romani, L. & Kaufmann, S. (1998). *Res. Immunol.*, 149:277–281.

Romani, L., Mencacci, A., Cenci, E., Puccetti, P., & Bistoni, F. (1996). *Res. Immunol.*, 147:512–518.

Romani, L., Montagnoli, C., Bozza, S., Perruccio, K., Spreca, A., et al. (2004). *Int. Immunol.*, 16:149–161.

Romani, L. & Puccetti, P. (2006a). Dendritic cells in immunity and vaccination against fungi. In: *Handbook of Dendritic Cells. Biology, Diseases and Therapies*, 2 vol. Lutz, M. and Steinkasserer A. Eds., Wiley-VCH Verlag Gmbh & Co., Weinham, Germany, pp. 915–930.

Romani, L. & Puccetti, P. (2006b). *Trends Microbiol.*, 14:183–189.

Saijo, S., Fujikado, N., Furuta, T., Chung, S., Kotaki, H., et al. (2006). *Nat. Immunol.*

Sakaguchi, S., Toda, M., Asano, M., Itoh, M., Morse, S., & Sakaguchi, N. (1996). *J. Autoimmun.*, 9:211–220.

Sato, K., Yang, X., Yudate, T., Chung, J., Wu, J., et al. (2006). *J. Biol. Chem.*, 281:38854–38866.

Segal, B., Kwon-Chung, J., Walsh, T., Klein, B., Battiwalla, M., et al. (2006). *Clin. Infect. Dis.*, 42:507–515.

Segal, B., Leto, T., Gallin, J., Malech, H., & Holland, S. (2000). *Medicine (Baltimore)*, 79:170–200.

Shoham, S. & Levitz, S. M. (2005). *Br. J. Haematol.*, 129:569–582.

Steele, C., Rapaka, R., Metz, A., Pop, S., Williams, D., et al. (2005). *Plos Pathog.*, 1:e42.

Taylor, P., Tsoni, S., Willment, J., Dennehy, K., Rosas, M., et al. (2006). *Nat. Immunol.*, 6:33–43.

Trinchieri, G., Pflanz, S., & Kastelein, R. (2003). *Immunity*, 19:641–644.

Vazquez-Torres, A. & Balish, E. (1997). *Microbiol. Mol. Biol. Rev.*, 61:170–192.

Woods, J. P. (2003) *Curr. Opin. Microbiol.*, 6:327–331.

Yauch, L., Lam, J., & Levitz, S. (2006). *Plos Pathog.*, 2:e120.

Chapter 2
In Vitro Models to Analyse Fungal Infection

Susanne Gola, David M. Arana, Rebeca Alonso-Monge,
Elvira Román, and Jesús Pla

Introduction

According to the molecular Koch's postulates (Falkow, 1988), putative virulence traits can be identified in a pathogen because deletion of the gene encoding a virulence factor in an otherwise wild-type strain generates a mutant with reduced pathogenicity in a certain model of experimental infection.

Recent advances in molecular genetics have led to the generation of such altered strains in several clinically relevant fungi, including *Candida albicans, Cryptococcus neoformans, Aspergillus fumigatus*, and *Histoplasma capsulatum*. This has, in turn allowed the identification of several virulence genes involved in important physiological processes in the pathogen. These processes include, among others, the biogenesis of the cell wall, the acquisition of nutrients, the production of extracellular enzymes, and the tolerance to stress. In experimental infection models these mutants frequently display attenuated or abolished virulence. While this methodology provides global information on whether a gene is involved in virulence or not, it does not define the specific step(s) in the pathogenic process that is (are) impaired by the molecular lesion.

During the pathoenic cycle fungi interact with various types of host cells, which may lead to dissemination from the original entry site and deep-seated infection of inner organs (Figure 2.1).

This interaction is characterized by successive events at the cellular level. Fungi first attach and then enter the host cells, where they may persist or even proliferate before they leave and infect other host cells and tissues. Each of these steps may be crucial for the development of the disease, and virulence factors may contribute in each of these steps by different molecular mechanisms (Figure 2.2). *In vitro* models of infection provide a defined experimental set-up to characterize the host–pathogen interplay at a cellular level and allow us to ascertain more precisely the molecular lesion present in the mutant and the corresponding step of the pathogenic process specifically altered.

This chapter reviews the main *in vitro* models for epithelial, endothelial, and immune system cells. It outlines the available techniques to characterize and to quantify host cell–pathogen interactions following a structure as preset by the

K. Kavanagh (ed.), *New Insights in Medical Mycology.*
© Springer 2007

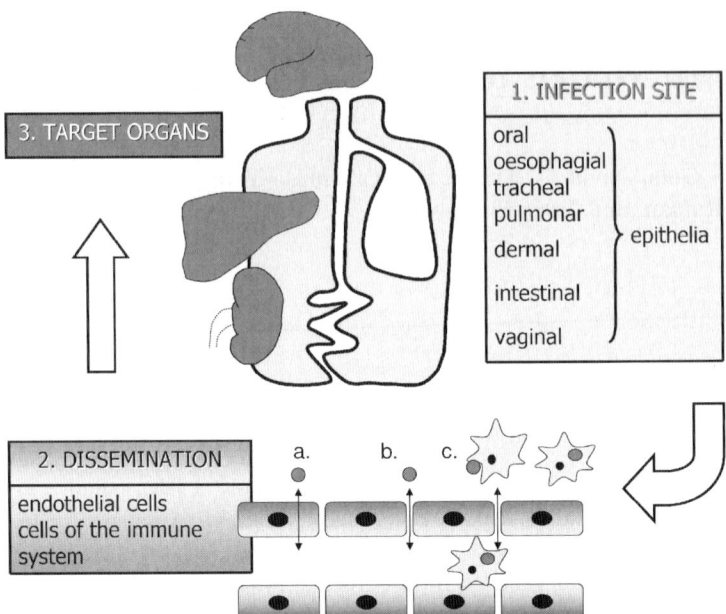

Figure 2.1 Stations of the pathogenic cycle. Fungi can enter the body interior through various epithelial sites which mark the boundary between inside and outside (1). The host's defence either limits the interaction to the epithelial surface or the fungus overcomes the barrier. The infection becomes invasive and no longer restricted to the primary infection site. A key feature initiating dissemination (2) is the interaction of the pathogen with host cells of the blood system – endothelial or immune cells. The fungus can get access to the bloodstream by transcellular (a), paracellular (b), 'Trojan horse' (c) mechanisms by which it might also leave the vascular system to attack finally vital organs as the brain, the liver, or the kidneys (3)

pathogenic course itself (Figure 2.2). Examples on how the *in vitro* methodology has contributed to clarify the effective molecular mechanism in the single steps are included for the different fungal pathogens.

Host Cells in Conjunction – Epithelial and Endothelial Models

Invading pathogens are confronted with two structural barriers which are the epithelia and endothelia located at diverse sites of the body. They are conjunctions of cells that confer, by their assembly, organ-specific physiological characteristics and functions. Thus, *in vitro* models for the investigation of fungal interactions with epithelial and endothelial cells (EC) are basically layers of varying degrees of complexity that are attached to surfaces. They aim to reflect the physiological properties

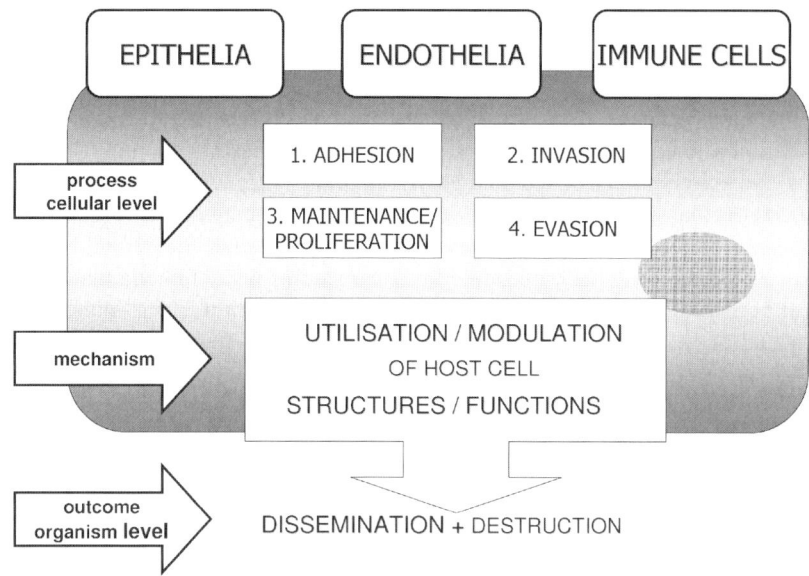

Figure 2.2 Host–pathogen interaction – players and processes. During the course of infection fungi interact with different host cells such as epithelial, endothelial, and immune system cells. This interaction is characterized by successive events at the cellular level. They include: adhesion (1), entrance (2), persistence or even proliferation inside (3), and finally exit from host cells (4). Pathogens contrive these steps by using and manipulating host cell structures and functions. Each of the processes can be crucial for the overall outcome of the interaction

of the corresponding natural organs and ongoing advances in tissue engineering have progressed towards models of an organizational level that represents an intermediate stage between single cell type culture and organ culture.

Apart from the fact that epithelial and EC form part of the host barriers they may play additional roles during the course of a fungal infection. In recent years it has been recognized that pathogens induce their own endocytosis upon interaction with these usually non-phagocytic cells. The possibility that epithelial and EC could serve as a reservoir for pathogens, which hide themselves intracellularly from the immune system is a matter of current research (for a recent review see Filler & Sheppard, 2006).

Epithelial Cell Models

Epithelia are formed by cells in close proximity to each other lying on a basal membrane. This structure is fed through the connective, highly vascularized tissue, which underlies the basal membrane. Epithelia cover every surface of the human organism, either externally or internally, as a monolayer or multilayer structure. External epithelia protect the body surfaces against mechanical damage, prevent the entry of microorganisms and control the loss of water caused by evaporation. Internal epithelia have other functions related to absorption and secretion.

Adhesion to epithelia is the first step in the pathogenesis of the microorganism. Thus, the epithelial cell models employed to study *A. fumigatus* are pulmonar or bronchoalveolar models while *C. neoformans* has been tested with olfactory mucosa or pneumocytes as these environmental fungi enter the human body normally by inhalation. There are limited *in vitro* studies with epithelia that focus on the airborne infections caused by *H. capsulatum*, although tracheal epithelium has been analysed (Eissenberg et al., 1997). As *C. albicans* is found as a commensal in the human microbiota it causes endogenously derived infections. Thus, adhesion of the fungus to epithelia of the gastrointestinal or urogenital tract, as well as to the epidermis is a critical first step in this type of infection. Next, *invasion* may occur. This step can take place at a cellular level when pathogens use a transcellular path to cross the barrier. To observe these processes in complex epithelial *in vitro* tissue models, modern microscopy assists ultrastructural and histochemical methods.

Oral and Oesophageal Epithelium Model

Different models have been developed to analyse the adhesion and invasion, especially of *C. albicans*, to oral and oesophageal epithelia *in vitro*. Human primary oral epithelial cells represent a simple *in vitro* model. Either scraping buccal mucosa (BEC) from healthy volunteers with sterile files (Jones et al., 1997; Jones et al., 1995; McCarron et al., 2004; Zhao et al., 2004) or collecting whole unstimulated saliva from healthy volunteers and enriching the population of epithelial cells by membrane retention have been reported (Steele et al., 2000).

A widely used three-dimensional model is the reconstituted human oral epithelium (RHOE). This model is constituted of human keratinocytes derived from the cutaneous carcinoma cell line TR146 and is commercialized by Skinethic Laboratory (Nice, France). The RHOE model resembles normal human oral epithelium and the pathological changes caused by *Candida* invasion are similar to the processes in human disease (Schaller et al., 1998). *C. albicans* has been found to grow intracellularly or traversing gap junctions of the epithelium, both in a similar proportion (Jayatilake et al., 2005). Using this epithelial cell model and a computer-assisted image analysis system the invasion of different *Candida* species was quantified, indicating that *C. albicans* is more invasive than others non-*albicans Candida* species (Jayatilake et al., 2006). Epithelial cells have mechanisms to sense the invasion by *Candida* spp. and respond by producing antimicrobial peptides (Lu et al., 2006). The RHOE model has been used to study this defensive response generated by oral epithelium against different *Candida* species which include the production of β-defensins 1, 2, and 3. Another host cell response to infection is the induction of cytokine expression. The pattern of cytokines and chemokines produced in presence of different *Candida* spp. correlates with their virulence potential as determined by quantitative reverse transcription-PCR (Q-RT-PCR) and enzyme-linked immunosorbent assay (ELISA) (Schaller et al., 2002).

A further improvement of the RHOE model in order to mimic the *in vivo* situation more closely is the supplement with polymorphonuclear leukocytes (PMNs), which generates simple aspects of the immune system (Schaller et al., 2004). Epithelial cells in contact with *C. albicans* induce a strong immune response that attracts PMNs to the infection site. Q-RT-PCR was employed to measure the immune response generated concluding that the extension of the system provides a useful tool for studying the immunological interaction between keratinocytes and *C. albicans*, as well as the role of PMNs in *C. albicans* pathogenesis.

Rouabhia and Deslauriers (2002) have developed an engineered human oral mucosa (EHOM). This model is a well-organized and stratified tissue in which epithelial cells are arranged to analyse pathogen-oral mucosa *in vitro*. The comparative study of infection in the *in vitro* EHOM and the RHOE model with *C. albicans* mutants of the manosyltransferase family (Pmt) allowed an assessment of the different functions for the individual Pmt isoforms in the different host niche-specific models (Rouabhia et al., 2005).

Alternatives to the RHOE and the EHOM are the EpiOral GIN-100 oral mucosa model based on primary oral keratinocytes (purchaser MaTek Corporation, Ashland, MA, USA) and a novel three-dimensional model, which uses the immortalized OKF6/TERT-2 oral epithelial cell line (Dongari-Bagtzoglou & Kashleva, 2006).

Skinethic Laboratory also provides a human reconstituted oesophageal epithelium derived from the immortalized cell line Kyse 510. These cells cultured *in vitro* develop a stratum corneum, resembling the outer cell layer of human oesophagus. This epithelial cell model was compared to the human oesophageal cell monolayer HET1-A to analyse the adherence and penetration of clinical, commensal, and reference collection *C. albicans* strains (Bernhardt et al., 2001). Clinical isolates showed an enhanced capability to adhere to HET1-A and to adhere and invade reconstituted oesophageal epithelium compared to commensal and collection strains (Bernhardt et al., 2001).

Intestinal Epithelium Model

As long-term primary culture of human intestinal epithelial cells is still a challenge, most intestinal epithelial systems rely on immortalized cell lines (Grossmann et al., 1998). In order to clarify the ability of yeast and filamentous forms to influence adherence of *C. albicans* to the intestinal epithelium, the enterocyte cell lines Caco-2 and HT29 were tested. *Candida*–enterocytes interaction was observed with high-resolution scanning electron microscopy, while yeast adherence to enterocytes was quantified by using an ELISA assay (Wiesner et al., 2002) concluding that both *C. albicans* morphologies, yeast, and hyphae, could adhere to (and perhaps invade) the apical surface of cultured enterocytes. Caco-2 cells have been also used to demonstrate that β, 1-2 and α, 1-2 oligomannosides mediate the adherence of *C. albicans* blastospores to human enterocytes (Dalle et al., 2003). Adherence was quantified by immunofluorescence and the percentage of adhesion was determined as the ratio of the number of adherent yeasts on the entire surface of the coverslip to the inoculum.

The Caco-2 line proved to give a good performance in a reconstituted intestinal epithelium model in comparison to HT29 and Lovo cells as determined by Dieterich et al. (2002). Caco-2 cells display a regularly ordered single cell layer which differentiates microvilli on the apical membrane. To construct the three-dimensional intestinal model cells were grown on a collagen matrix supplemented with primary fibroblasts and myoblasts. This model has been useful to define the defects on adhesion and invasion of the avirulent *C. albicans cph1* and *efg1* mutant, showing that double *efg1 cph1* mutants were unable to adhere or penetrate neither of the model systems tested (Dieterich et al., 2002).

Vaginal Epithelium Model

The reconstituted human vaginal epithelium (RHVE) resembles the multilayer human vaginal mucosa and is commercialized by Skinethic Laboratory (Nice, France). It is based on a cell line which was obtained by culturing transformed human keratinocytes of the cell line A431 derived from a vulval epidermoid carcinoma (Rosdy et al., 1986). The RHVE model was valid for evaluating the phenotype of mutants in the agglutinin-like sequence (ALS) family in *C. albicans*. The expression of ALSs proteins in the *in vitro* RHVE was compared with *in vivo* expression in a murine vaginitis model (Cheng et al., 2005). The RHVE model has been also used to study the epithelial cytokine response induced by *C. albicans* and the role of the Secreted Aspartyl Proteinases (Saps), important virulence factors of this fungus. Saps cause tissue damage and the different damaging potential of each Sap correlates with an epithelium-induced pro-inflammatory cytokine response, which could be crucial in the control of the vaginal infection by *C. albicans* (Schaller et al., 2005). Pro-inflammatory and other chemoattractive cytokines generated by RHVE cells were detected and quantified by quantitative reverse transcription-PCR and fluorescence-activated cell sorter (FACS) analyses.

The expression of pro-inflammatory cytokines, chemokines, and β-defensins-2 was also measured using another immortalized vaginal epithelial cell line, PK E6/E7 (Pivarcsi et al., 2005). The expression of inflammatory mediators and defensins induced in the presence of compounds from different pathogenic microorganisms, among them heat-killed *C. albicans*, was quantified by Q-RT-PCR, flow cytometry and ELISA (Pivarcsi et al., 2005).

Anti-*Candida* activity of vaginal epithelial cells was identified as a possible factor of local immunity in vulvovaginal candidiasis by using different cell models (Barousse et al., 2001; Barousse et al., 2005; Steele et al., 1999a, b). Co-inhabitants at the vaginal site of infection are bacteria of the genus *Lactobacillus* spp. which constitute the protective part of the microbiota while *Gardnerella vaginalis* is the most common bacteria associated with bacterial vaginosis. Little is known about the impact of these – and other – members of the microbial community on the virulence of *C. albicans*. A further development of the vaginal *in vitro* system could certainly be the addition of members of the microbiota to model the natural micro-environment.

Airway and Lung Epithelium Models

In order to investigate the adhesion and invasion of pathogenic fungi to airway epithelia, either human nasal or tracheal cell lines have been employed. Human nasal ciliated epithelium is obtained from healthy volunteers. Functional integrity of the cells can be measured by quantifying the beat frequency of cilia by a photometric technique (Amitani et al., 1995).

The primary culture of human nasal epithelial cells (HNEC) in air-liquid interface led to a useful model for the interaction of *A. fumigatus* with an airway epithelium (Botterel et al., 2002). HNEC formed a pseudostratified epithelium with typical airway epithelium characteristics. A similar approach was followed by Shin et al. (2006) using nasal polyp epithelial cells.

Due to the relative simplicity by which hamster tracheal epithelium (HTE) cells can be isolated and cultivated, these cells are a popular model for respiratory epithelia (Goldman & Baseman, 1980). In a biphasic chamber system they develop the epithelial-typical polarity, differentiating cilia and secreting mucin-like molecules on the apical side (Whitcutt et al., 1988). HTE cells have been used for example to analyse the persistence of the intracellular pathogen *H. capsulatum* in non-professionally phagocytic host cells (Eissenberg et al., 1997).

At the distal end of the respiratory pathway, lung epithelial cells are components of the respiratory tissue barrier. The most commonly used lung epithelial cell line is A549, a type II pneumocyte of bronchoalveolar carcinoma origin. Several studies have investigated the interaction of *A. fumigatus* with A549 cells either by scanning electron microscopy (SEM), transmission electron microscopy (TEM), or immunofluorescence-confocal microcroscopy of a GFP expressing strain (DeHart et al., 1997; Paris et al., 1997; Wasylnka & Moore, 2002, 2003). This lead to the conclusion that *A. fumigatus* conidia adhere to and are engulfed by lung epithelial cells in which they differentiate to the barrier-breaking hyphal form (Filler & Sheppard, 2006). Also *C. neoformans* adheres to and is internalized by lung epithelial cells (Merkel & Scofield, 1997). The A549 *in vitro* model enabled the identification of the capsule structure glucuronoxylomannan (GMX) and phospholipase B (*PLB1*) as important fungal factors for this process (Barbosa et al., 2006; Ganendren et al., 2006).

An improved and more physiological model consists of A549 cells grown in a two-chamber system. Here the monolayer is not immersed in medium but exposed to the air, and only separated by a thin surfactant layer produced by the polar pneumocytes (Blank et al., 2006). In three-dimensional models epithelial cells were supplemented with macrophages, monocytes, and dendritic cells (DC) at their corresponding natural apical or basal location (Radyuk et al., 2003; Rothen-Rutishauser et al., 2005). Two-layer systems that model the functional barrier of the alveolar wall include epithelial and microvascular EC (Hermanns et al., 2004). The traversal of microorganisms through the barrier could be investigated in further detail by adding monocytes to the system (Bermudez et al., 2002).

Human Epidermis

Although less common, *C. albicans* also causes cutaneous infections which in immunocompromised patients can lead to severe systemic disease (Odds, 1988). Reconstituted human epidermis for *in vitro* experiments can be purchased from Skinethic Laboratory (Nice, France). This skin model consists of stratified keratinocytes and a differentiated stratum corneum which confers the permeability barrier characteristic of the epidermis. The model was used to test the role of the secreted aspartic proteases (Saps) encoded by 10 genes in *C. albicans*. Immunoelectron microscopy allowed the localization of Saps while quantitative RT-PCR revealed the different pattern of *SAP* expression during adhesion and penetration of the epidermis (Schaller et al., 1999; Schaller et al., 2000).

A more robust model of the human epidermis was constructed by adding dermal fibroblasts embedded in a collagen matrix which facilitated differentiation of keratinocytes into a multilayered epidermis with stratum corneum (Dieterich et al., 2002). This epidermis equivalent was used to test the contribution of two transcription factors, Cph1 and Efg1, to invasion and penetration of epithelial tissues *in vitro*.

Endothelial In Vitro Models

The endothelium is a unicellular layer which constitutes the interphase between blood and tissue. As such, it forms, like epithelia, a physical barrier between 'inside' and 'outside' and fungi have to cross through it to cause a systemic disease. Located at the luminal side of blood vessels, the endothelium is a dynamic and heterogeneous organ which fulfils vital secretory, synthetic, metabolic, and immunologic functions (Cines et al., 1998; Fishman, 1982). It regulates the flow of the blood itself as well as the exchange of metabolites and various biologically active molecules between both compartments.

Endothelia exert their gate keeping role through the presence of receptors for a variety of molecules (proteins, lipid transporting particles, metabolites, and hormones) as well as through receptors and specific proteins which govern cell–cell and cell–extracellular matrix contacts (Cines et al., 1998). According to their organ-specific function endothelia are characterized by different degrees of permeability which can be relatively pronounced in discontinuous and fenestrated endothelia. In contrast, the continuous endothelium which constitutes the blood-brain barrier (BBB) is highly restrictive to the passage of molecules (Moody, 2006).

Apart from its role as physiological barrier, the endothelium is also a site of interaction with cells of the innate and the adaptive immune system. EC recruit circulating leukocytes to spots of infection where the immune cells gain access to microbes located in the tissue. Leukocytes extravasate from the vessels and initiate the inflammatory process (Ley, 1996; Vestweber, 2002).

Reflecting their various physiological functions, EC exhibit a number of cell type-specific characteristics whose recognition is a prerequisite for the production of primary EC cultures. Von Willebrand factor (vWF) is constitutively expressed in EC from veins and an important marker for cell type identity (Hoyer et al., 1973; Jaffe et al., 1974). Also, almost exclusively expressed on EC are platelet endothelial cell adhesion molecule (PECAM-1, CD31) (Albelda et al., 1990), and vascular endothelial growth factor receptors (Hewett & Murray, 1996; Thomas, 1996). Quiescent EC can be activated in response to pro-inflammatory mediators such as tumour necrosis factor α (TNFα) (Pober et al., 1986); EC respond to these stimuli by surface expression of cell adhesion molecules (e.g. ICAMs, VCAM-1, and E-selectin) (Dustin & Springer, 1988; Bevilacqua, 1993; Osborn et al., 1989). In turn, EC themselves secrete pro-inflammatory cytokines like interleukin-6 (IL-6), interleukin-8 (IL-8), and monocyte chemoattractant protein 1 (MCP-1) (Gimbrone et al., 1989; May et al., 1989; Rollins et al., 1990). The endothelial surface is decorated by a large variety of extracellular domains of membrane-bound molecules, which together constitute the glycocalyx that lines the luminal surface of the endothelium (Pries et al., 2000). Much information has been gathered with respect to specific membrane-bound molecules such as the selectins and integrins (Risau & Flamme, 1995), involved in immune reactions and inflammatory processes (Springer, 1994). Advances in electron microscopy established the concept of a thick surface layer on EC (ESL) at the luminal side, whose consequences for EC function are not yet fully understood (Pries & Kuebler, 2006). All these surface molecules are prone to be exploited by invading pathogens as interactive structures.

Macrovascular Endothelium Models

A commonly used model for the macrovascular endothelium are primary human umbilical vein endothelial cells (HUVEC) because they can be isolated relatively easily (Jaffe et al., 1973; Marin et al., 2001). In the simplest forms of single cell type assays, HUVEC are seeded for *in vitro* infection on multi-well tissue culture dishes or directly on coverslips for microscopic observations. Usually, they are grown to confluence as monolayers on a fibronectin or collagen matrix in order to reproduce the interactive nature of the endothelial cell conjunction.

Adhesion to EC is the first step in most situations for pathogens to leave the bloodstream. To quantify the adhesion step, EC monolayers are incubated with fungal cell suspensions for a defined period of time after which non-adhered cells are washed off. For small inocula, attached cells can then be visualized microscopically and quantified directly. A method described by Ibrahim et al. (1995) made use of an agar-containing medium which was allowed to solidify on the rinsed EC layer. After incubation, the number of attached organism was determined by colony counting and adhesion expressed as a percentage of the initial inoculum. For inocula with high cell concentrations, the same authors applied a radiometric assay based on incorporation of L-(^{35}S) methionine into the fungal cells (Mayer et al., 1992) to measure adherence indirectly.

Using this system for *in vitro C. neoformans* infection, the role of the capsule and its main component GMX for adherence to EC was investigated, which led to the conclusion that acapsular or poorly encapsulated forms of the fungus may be the form that crosses the epithelium to initiate invasive disease (Ibrahim et al., 1995). The contribution of the *C. albicans ALS* (agglutinin-like sequence) gene family to adhesion to EC and other host cells has been also extensively investigated (Fu et al., 1998; Hoyer, 2001; Zhao et al., 2005). Host niche-specific functions in adhesion could be assigned for single ALS members by a comparative study of different *in vitro* models of non-professional phagocytic cells (Zhao et al., 2004).

A different radiometric assay for quantification of adhesion was employed by Spreghini et al. (1999) to identify host cell structures and their corresponding *C. albicans* receptors that mediate this process. The multifunctional adhesive glycoprotein vitronectin (VN) is present in the ECM of endothelia and was known to be bound by *C. albicans* specifically (Jakab et al., 1993; Limper & Standing, 1994). ^{3}H glucose-labelled yeast cells were bound to ECM proteins or the endothelial cell line EA.hy 926 and adhesion was calculated. It could have be shown that VN is directly involved in adhesion of *C. albicans* to EC by binding to integrin-like VN receptors expressed by the fungus and, in addition, that a p105 focal adhesion kinase (Fak)-like protein plays a role in controlling yeast cell adhesion (Santoni et al., 2002).

A non-static assay is described, which adds flow as a physiological constant to the macrovascular endothelial *in vitro* system. This assay (ProteoFlow; LigoCyte Pharmaceuticals, Inc., Bozeman, MT, USA) was originally used for studying the interaction of EC with leukocytes under simulated physiological shear (Bargatze et al., 1994) and was adopted to address the question of how cell surface hydrophobicity (CSH) influences the interaction of *C. albicans* with EC under physiological conditions (Glee et al., 2001). HUVEC were grown as monolayers on the luminal surface of glass capillary tubes and yeast cell adhesion under physiological flow rates monitored by real-time video capturing. Hydrophobic cells showed more adhesion to EC than hydrophilic cells and previous activation of the EC monolayer by the cytokine IL-1β increased the binding of hydrophobic *C. albicans*, probably by induction of the integrin receptors ICAM-1 and ICAM-2 on EC. Treatment of hydrophobic cells with a monoclonal antibody against 6C5 inhibited binding events as it was shown in static assays for the binding on laminin and fibronectin before (Masuoka et al., 1999).

Invasion of EC can be addressed by several fluorescence-based assays which have been developed to quantify the uptake of fungal cells by EC. Differential fluorescence and immunostaining allow enable the identification of adhered fungal cells from those which are taken up by the endothelium monolayer directly (Levitz et al., 1987; Phan et al., 2000, 2005). To monitor unspecific particle uptake by EC, an assay with fluorescein-labeled biotinylated polystyrene beads can be included (Wasylnka & Moore, 2002). An alternative method to determine the invasion index is the nystatin-protection assay in which extracellular organisms are eliminated while intracellular fungal conidia are protected from the fungicide (Wasylnka & Moore, 2002).

In *C. neoformans*, this methodology has proved to be useful to show that acapsular, but not encapsulated strains, were internalized by HUVEC (Ibrahim et al., 1995). The phagocytotic process was shown to be dependent upon functional microfilaments as well as on the presence of a heat-labile serum factor, probably complement. This is in contrast to the situation in *C. albicans*, where internalization occurred in a serum-independent manner (Filler et al., 1995). Therefore, it was concluded that adherence and phagocytosis of both organisms are mediated by different receptors. While *C. albicans* yeast cells adhere to EC but are poorly phagocytosed (Phan et al., 2000), the hyphal form of the fungus both adheres avidly and induces its own endocytosis (Filler et al., 1995; Rotrosen et al., 1985). Recently, several lines of evidence have indicated that N-cadherin could be the host cell structure which serves as receptor that recognizes still unknown adhesins on *C. albicans* hyphae, mediating their endocytosis (Phan et al., 2005).

It is known, that EC also bind and internalize *A. fumigatus* conidia and hyphae (Paris et al., 1997). However, in contrast to *C. albicans*, it is the conidial form which is more avidly internalized *in vitro* (Lopes Bezerra & Filler, 2004). A distinct feature of the pathogenesis of this fungus is that it shows angioinvasion in the lung with subsequent thrombosis (Fraser, 1993). While quiescent endothelia express an anticoagulant (and antiadhesive) phenotype, coagulation is promoted by activated EC (Aird, 2005). Tissue factor is a component which acts at the initial steps of the coagulation cascade (Morrissey, 2001). Recently, evidence was found that internalized hyphae, but not internalized conidia, of *A. fumigatus* induce HUVEC to express tissue factor. In contrast, although *C. albicans* is known to stimulate various EC responses (Orozco et al., 2000), it did not stimulate tissue factor. Thus, the difference in the ability to stimulate prothrombotic states in EC reflects the histopathology which characterizes infections by the distinct fungi (Lopes Bezerra & Filler, 2004).

The physiological intact endothelium displays polarized functions and structures that distinguish the luminal from the abluminal side (Haller & Kubler, 1999). This polarity has to be taken into account since it may confer differences in the pathogenesis of locally invasive disease versus hematogenously disseminated infection. On the one hand, fungal pathogens enter into the bloodstream by crossing the endothelial cell line from the abluminal to the luminal surface. This is the case when *A. fumigatus* or *C. neoformans* invade the lining of the pulmonary vasculature in the lung or when *C. albicans* enters the blood system leaving its normal habitat in the intestine under host conditions that favour its switch from commensalism to pathogenicity. On the other hand, microorganisms leave the vascular compartment to colonize deeper tissues and organs in the opposite direction, i.e. they pass the endothelium from the luminal to the abluminal side. Recent EC models reflect the different modes of crossing the endothelial barrier (Kamai et al., 2006).

Endothelia from different sites of the body display different characteristics. Current research in vascular biology points to these differential phenotypes and aims to determine organ-related endothelial functions (Kallmann et al., 2002; Sana et al., 2005). Integration of this newly gained knowledge into the experimental set-up for *in vitro* interactions of fungal pathogens with EC should lead to conclusions that are relevant in the situation found *in vivo*.

Microvascular Endothelium Models – the Blood Brain Barrier

In disseminated fungal infections pathogens can reach the brain as target organ by crossing the blood brain barrier (BBB). The BBB separates the brain from the intra-vascular compartment and maintains the homeostasis of the central nervous system environment (Rubin & Staddon, 1999). Unlike EC from peripheral tissues, brain microvascular cells (BMEC) are joined by tight junctions that prevent the exchange of molecules between both compartments (Pardridge, 1999). Thus, a challenge for *in vitro* BBB models is to reflect this special physiological characteristic.

The simplest and most commonly used BBB system for fungal infections con-sists of a monolayer of primary human BMEC (HBMEC) cultured on Transwell tissue culture inserts (Chang et al., 2004; Stins et al., 2001). This system allows for separate access to the upper (blood side) and the lower (brain side) compartment to monitor the transmigration of fungal cells. Pathogens may breach the BBB by paracellular, transcellular or a 'Trojan horse' mechanism, recruiting latently infected monocytes (Figure 2.1) (Huang & Jong, 2001). For *C. neoformans* it was observed that acapsular cells efficiently adhered to and invaded HBMEC, but traversal occurred with a significantly lower frequency than for encapsulated strains (Chang et al., 2004). Since there was no damage of HBMEC and *C. neoformans* cells were never found close to junctions of the EC, these authors concluded that the fungus crosses the BBB by a transcellular mechanism. This is in contrast to a report of Chen et al. (2003) who used a similar model of the BBB. These authors found an impact of *C. neoformans* on the tight junction marker occludin (Furuse et al., 1993) inferring that the pathogen may cross the BBB by a paracellular mecha-nism (Chen et al., 2003). Also *C. albicans* adheres to and invades HBMEC, as determined by a fluorescence assay based on calcofluor white (Borg von Zepelin & Wagner, 1995). Transmigration through HBMEC occurred without affecting the monolayer integrity which suggested a transcellular mechanism of BBB crossing (Jong et al., 2001a). While adhesins may initiate attachment, hydrolytic enzymes like secreted aspartyl proteinases or phospholipases may assist the entry into host cells (Huang & Jong, 2001). However, not only these fungal-derived proteolytic activities can contribute to cellular invasion, but also specific host factors may be important. Plasminogen, an abundant factor in human plasma, is converted to its proteolytic form, plasmin. Plasmin functions *inter alia* in the degradation of ECM proteins and cellular migration and pathogens are able to recruit these host factors (Sun et al., 2006). This was also shown for *C. albicans*, which binds plasmin and plasminogen through the moonlighting protein enolase and this interaction increased crossing through the HBMEC layer (Jong et al., 2003).

Advances in brain research, especially in set-ups investigating problems of drug passage through the human BBB, have led to improved models which might be exploited also for fungal infection research in the future. For example, reconstitution of the natural micro-environment by co-culture of microvascular EC with astrocytes, astrocyte-conditioned medium or extraluminal C6-glia cells augmented the physiological properties of BBB models (Abbott, 2005; Abbott et al., 2006; Josserand et al., 2006; Megard et al., 2002; Neuhaus et al., 2006b). Also the addition of physiological flow was found to have positive effects on BBB integrity and permeability *in vitro* (Cucullo et al., 2002; Neuhaus et al., 2006b; Santaguida et al., 2006). Up to now, these technical advances barely have been implemented in investigation of fungal pathogenesis.

Techniques to Monitor Cell Viability and Tissue Integrity

To assess the functional state of cells in *in vitro* models, a variety of assays is available. They are necessary either to ascertain the quality of the deployed cell culture or to quantify the damage induced by the fungal effector cells to the host target cells as a parameter of virulence during the infection experiment. Methods to determine cell viability and function reach from simple live/dead assays over assays that focus specifically on a molecular mechanism or a cell type specific function to the most complex molecular fingerprinting to record biochemical alterations.

Commonly used techniques in *in vitro* infections are end-point live/dead assays that are based on assessment of cell membrane integrity.

The most conventional one is the trypan blue (TB) exclusion test which can be visualized by light microscopy. However, TB exclusion may lead to overestimation of cell viability (Mascotti et al., 2000). In the chromium (^{51}Cr) release assay target cells are labelled with $Na_2{}^{51}CrO_4$. The compound is taken up by the cells and retained for some time in the cytoplasm. If the target cell is damaged upon co-incubation with effector cells, gamma ray radiation generated during ^{51}Cr decay can be measured in the medium. This assay was for example adopted to determine endothelial cell damage by *C. albicans* and *C. neoformans* (Ibrahim et al., 1993; Ibrahim et al., 1995). Here, EC monolayers were incubated with $Na_2{}^{51}CrO_4$, the unincorporated tracer was removed and EC were then infected with the pathogen. By including an uninfected control, the spontaneous ^{51}Cr release by EC and the specific endothelial release of chromium due to cell injury can be precisely calculated. The system can be further refined to determine whether EC damage requires direct contact with the pathogen or is due to secreted compounds by inserting a membrane between effector and target cells.

To avoid the inconvenience which accompanies isotopic methods, two main types of non-radioactive assays, fluorescent techniques and enzymatic procedures, are available.

Viability tests by double staining with the DNA intercalating compounds acridine orange and ethidium bromide (AO/EB) or propidium iodide (AO/PI) are based on the fact that AO is taken up by both viable and nonviable cells while EB and PI only can pass through disintegrated cell membranes. Thus, viable cells appear green under darkfield fluorescence microscopy while dead cells are red (Baskic et al., 2006; Mascotti et al., 2000). In non-adherent *in vitro* models, as with certain immune cell system (see below), the fluorescence assays can be read by flow cytometry. There are commercially available products that are based on this principle of differential fluorescence as the LIVE/DEAD viability kit (Invitrogen, Molecular Probes, Carlsbad; CA, USA).

Enzymatic tests can detect the extracellular presence of intracellular proteins as indicator of cell membrane damage. Activities of the lysosomal N-acetyl-beta-D-glycosaminidase (NAG) or the lactate-dehydrogenase (LDH) are measured for this purpose (Korzeniewski & Callewaert, 1983; Niu et al., 2001; Schaller et al., 2004). The metabolic activity of cells can be quantified by ^{3}H-glucose uptake (Steele et al., 2000) or addressed by substrates that emit fluorescence after cleavage by intracellular esterases (e.g. CellTracker Green or calcein blue AM; both Invitrogen, Molecular Probes).

Widely used colorimetric tests based on enzyme function utilize the tetrazolium dyes MTT (2-(4,5-dimethyl-2-thiazolyl)-3,5-diphenyl-2H-tetrazolium bromide) and XTT (sodium 2,3-bis(2-methoxy-4-nitro-5-sulfophenyl)-5-[(phenylamino)-carbonyl]-2H-tetrazolium inner salt) (Meshulam et al., 1995; Mosmann, 1983). Both colourless salts are reduced by cellular functions to brightly coloured formazan products (for a review see Berridge et al., 2005). Thus, they also indicate metabolic activity as parameter for cell viability. While MTT forms an insoluble formazan and is therefore applicable in end-point viability tests, XTT converts into a soluble product which allows for real time assessment of viability. Both assays can be performed quantitatively in an ELISA reader format, but the crystals built by MTT must be dissolved (Meshulam et al., 1995). The exact enzymatic activities responsible for MTT and XTT reduction as well as their cellular localization are not well known, which might lead to erroneous results under certain circumstances (Berridge et al., 2005; Knight & Dancis, 2006).

A principle difference between the chromium release assay and the non-radioactive enzymatic tests has to be taken into account for the experimental design. The former permits a specific label of target cells, while the enzymatic activities measured in the latter might be present in both the target and the effector cells. Thus, appropriate controls have to be included that enable to correct for the contribution of effector cells to the assay.

Cell layer and tissue *in vitro* models require techniques that monitor their integrity specifically. A characteristic physical property of epithelia and endothelia is a high trans-epithelial/endothelial electrical resistance (TEER) established by tight junctions between the cells (Santaguida et al., 2006). The TEER has to be measured to confirm the validity of the barrier and to detect changes induced by infection with pathogens (Jong et al., 2001). The resistance values can be determined by electronic devices in real time and are expressed in ohm per square centimeter of the layer surface (Grainger et al., 2006; Neuhaus et al., 2006b). Additionally, the tightness of the layer can also be addressed by determining the permeability of indicative molecules as ³H-inulin, ³H-sucrose, Evans blue, or fluorochrome-conjugated dextrans (Jong et al., 2001; Neuhaus et al., 2006a; Raimondi et al., 2006).

Choice of the Right Model – Primary Cells or Cell Line?

Modern histology has advanced greatly in the understanding and culturing of epithelial and EC in the last decade. Primary cells can be isolated from various sites relevant in fungal infections like the skin, the lung, or the brain. Protocols have been described for epithelial cells (Steele & Fidel, Jr. 2002; Grossmann et al., 1998), macrovascular (HUVEC) (Jaffe et al., 1973; Marin et al., 2001), and microvascular EC (Bowman et al., 1983; Hewett & Murray, 1993; Lamszus et al., 1999; Richard et al., 1998; Vinters et al., 1987). However, several disadvantages are associated with the use of primary cells. Their sources are often not readily available and isolation is time consuming and generally gives low yields. Primary EC, for example, display a loss of viability and a change in typical EC marker expression during the course of their *in vitro* propagation. The source of the cells (i.e. the donor) and thereby the genetic background from the isolates might also result in differential marker expression (Stins et al., 1997).

The alternative are cell lines, either directly derived from tumours or immortalized primary cells. Methods described for immortalization include forced expression of telomerase (OKF6/TERT2 epithelial cells, (Dickson et al., 2000), fusion of primary EC with carcinoma cells (EA.hy 926, Edgell et al., 1983) or transfection with SV40-T (HPMEC-ST1.6R, Krump-Konvalinkova et al., 2001). While cell lines can reflect the behaviour of primary cells in some aspects (L'Azou et al., 2005), not all cell typical phenotypes have been found to be expressed by them *in vitro*. A comparative study of endothelial characteristics of primary EC and several cell lines revealed important differences in constitutive expression of vWF and PECAM-1 as well as in the induction of ICAM-1, VCAM-1 and E-selectin, and IL-6, IL-8 and MCP-1 upon activation by different stimuli (Unger et al., 2002).

Generally, the transfer of results obtained with cells cultured *in vitro* to the conditions found *in vivo* is hindered by limitations in the modelling of the natural cellular environment. This is especially true for endothelial cell models, since, as stated by Aird (2003), an important feature of the endothelium is that its properties vary between different sites of the vasculature and from one moment in time to the next. EC show a high degree of plasticity and this flexibility is closely linked to the extracellular environment and the available *in vitro* systems for EC generally may fail to fully capture the spatial and temporal dynamics of the endothelium (Aird, 2005). This point is stressed by a recent study where the microenvironmental modulation of EC was investigated at a molecular level. In a multidimensional protein identification assay with samples from rat lung EC and cultured rat lung EC, 41% of proteins expressed *in vivo* were not detected *in vitro* (Durr et al., 2004).

Three-dimensional models aim to reconstitute the cellular environment closely to the situation found *in vivo*. These models are useful to analyse tissue infections and allow the study of adherence, penetration, and cellular damage in detail. However, they do not yet reproduce perfectly the structure and composition of the respective human tissues. Problems may also arise with the reproducibility of experiments. When commercial models that include primary cells are used, inter-assay variation can be avoided by choosing a supplier who offers batches of tissues that come from a single donor (Netzlaff et al., 2005; Dongari-Bagtzoglou and Kashleva, 2006). Another strategy followed by different authors to validate their particular system is to compare distinct *in vitro* models or to compare it with an *in vivo* model of infection (Cheng et al., 2005; Rouabhia et al., 2005; Sanchez et al., 2004).

Models to Analyse the Interaction with Immune Cells

Both innate and acquired immunity are essential for the development of resistance to pathogenic fungi (Romani, 2004). Phagocytic cells, such as neutrophils, DCs and macrophages represent a first line of defence against microorganisms. They recognize different PAMPs (pathogen-associated molecular patterns), exposed on the surface of several pathogens by different specific PRRs receptors (Pattern Recognition Receptors), expressed on the surface of the immune cell. PAMPs–PRRs interaction is considered a prerequisite for the innate system to be able to discriminate between fungal pathogens of different nature and to develop the appropriate adaptive immune response (Underhill & Gantner, 2004).

Neutrophils are important in early immunity stages to fungal infection. The quick arrival of these cells at the site of infection depends on their ability to respond to chemoattractants released by pathogens *in situ*. Their main function involves phagocytosis, mediated by complement and Fc receptors, and killing of ingested pathogens by oxidative-mediated mechanisms (Aderem & Underhill, 1999; Underhill & Ozinsky, 2002; Witko-Sarsat et al., 2000) but, unlike macrophages and DCs, they do not function as antigen presenting cells (APCs) to T cells (Mansour & Levitz, 2002).

DCs are bone marrow-derived cells of both lymphoid and myeloid origin present in all lymphoid organs. DCs are potent APCs with a unique ability to sense and respond to fungus-associated information by means of PAMPs–PRRs patterns that result in a qualitatively different adaptive T-helper cell (T_H) immune responses (Bacci et al., 2002; d'Ostiani et al., 2000).

Macrophages are tissue specific phagocytes specialized as APCs and microbe-killing cells. These tissue-resident macrophages are found at a higher frequency than are DCs, making them more likely to encounter a pathogen upon initial infection.

Protocols for the isolation and purification of these immune cells from healthy human donors commonly making use of flow cytometry are described for neutrophils, monocytes, macrophages, and DCs (Drenth et al., 1995; Read et al., 1993). Generally, the identity of the immune cells is determined by fluorescent-activated cell sorter (FACS) analysis in which a cell type specific antibody is coupled to a second antibody that is linked to certain fluorochromes as FITC or phycoerythryne.

Isolated monocytes (CD14[+]) can differentiate into macrophages by incubation with 10% human serum or be transformed into DCs by incubation with IL-4 and GM-CSF and Langerhans cells (LC) can be obtained from cord blood CD34[+] progenitors (Sallusto & Lanzavecchia, 1994; Serrano-Gomez et al., 2004). Neutrophils are also easily isolated either from blood of healthy donors or from the blood and peritoneal cavity of mice. Murine peritoneal macrophages can be recovered from mice after intraperitoneal injection of sterile 10% thioglycolate medium (Kaposzta et al., 1999; Taylor et al., 2002). Human alveolar macrophages (HAMs) can be obtained from bronchoalveolar wash (BAL) of human donors (Dubourdeau et al., 2006) or mice (Steele et al., 2005).

In addition to the isolation of immune cells from donors or animals, several immune cell lines are available. The most commonly used macrophage cell lines are the murine peritoneal macrophages-like cell line RAW264.7 (ATCC number TIB-71), which is differentiated with IFN-γ and J774A.1 (ATCC TIB-67), a murine (BALB/c; haplotype H-2d) cell line derived from a reticulum sarcoma (Ralph et al., 1975) that display certain phenotypic characteristics similar to murine peritoneal macrophages. An alternative to the use of primary alveolar macrophages is the MH-S cell line of murine alveolar macrophages (Matsunaga et al., 2001). HL-60 is a myelomonocytic cell line which differentiates into granulocytes when the cells are grown in medium supplemented with dimethilformamide, dimethyl sulfoxide, or retinoic acid (Harris & Ralph, 1985; Mullick et al., 2004).

In addition to these standard lines, some others have been constructed to analyse specifically the role of certain receptors in the recognition of surface components of the fungal cell wall, such as the K562-DC209 (Relloso et al., 2002) and HEK293 (Gantner et al., 2005) that express the DC-SIGN and Dectin-1 receptors respectively. To determine the contribution of a given host cell receptor to the interaction with a pathogen, immune cells can be derived from mice devoid of the molecule of interest (Blander & Medzhitov, 2004).

Adhesion/Internalization

The adhesion of pathogens or isolated surface components to immune cells – as well as phagocytosis – can be quantified by flow cytometry using specific fluorochromes when the assay is performed with immune cells that do not adhere to tissue culture plates. Normally, fungal cells are labelled with a fluorescent dye (such as FITC) and able to interact with target cells under defined conditions (Newman & Holly, 2001) analysing the size and intensity of the resulting population by flow cytometry. The interaction between host cell receptors and fungal cells can be addressed by the specific inhibition of the binding using specific antibodies or compounds that are presumed or well-established ligands. For example, DC-SIGN recognizes a large array of pathogens in a mannan-dependent and Lewis oligosaccharide-dependent manner (Feinberg et al., 2001; Frison et al., 2003). Therefore, binding to the pathogen can be blocked by anti-DC-SIGN antibody or by galactomannans (Relloso et al., 2002).

Quantification of adhesion/phagocytosis is difficult given the necessity to differentiate between surface-bound particles and those cells internalized. Although cytoplasmic cells can be observed by transmission electron microscopy quantitative assays are difficult to carry out since cells and particles are not easily counted (Ezekowitz et al., 1990). Different methods rely on the use of fluorescence microscopy in which yeast cells are pre-labelled by either staining chitin with diaethanol or specific antibodies against fungal surface components or lectines (Levitz et al., 1987; Parod et al., 1986). However, internalized pathogens within living immune cells cannot be studied using this methodology. Quenching fluorescence of surface-bound pre-labelled particles with FITC, crystal violet, or TB (Hed et al., 1985; Lundborg et al., 2006) has been used. The Giemsa stain can be used after tannic acid staining to generate different colours for internalized and surface-bound particles under brightfield illumination (Giaimis et al., 1992). Alternatively, yeast particles can be also stained with TB and distinguished by differential interference contrast microscopy, thereby bypassing the requirement of two different filters. Immune complexes attached to and ingested by neutrophils have been quantified by cytofluorometry using a fluorescence quenching assay which permits differentiation between attachment and ingestion as the fluorescence intensity decreases in the phagosolysosome as a result of the low pH (Sahlin et al., 1983).

Other authors have used calcofluor staining to assess the ability of viable and nonviable *C. albicans* cells to adhere to, and to be internalized by, host mammalian cells *in vitro* (Henry-Stanley et al., 2004). Fluorescence microscopy has also been used for *C. neoformans* phagocytosis assays where specific antibodies for capsule components are used, such as 18B7 (Casadevall et al., 1998). Internalization of the *C. neoformans* cells can be verified by immunofluorescence staining with Texas Red-conjugated antibody to mouse IgG (Fan et al., 2005). The need to have appropriate antibodies, as it is the case of *A. fumigatus*, can be bypassed by a combined biotin-calcofluor-staining (BCS) technique in which biotinylated conidia are incubated with BCS macrophages. Extracellular and intracellular conidia are then visualized by calcofluor but only extracellular conidia are stained in red using Cy3-labelled streptavidin (Luther et al., 2006).

In these experiments, the phagocytosis index is commonly ascertained by determining the number of unbound cells (which are removed by washing and quantified by plate counting). Thus phagocytosis can be expressed by subtracting the colony-forming units counted in the washing fluid from those obtained from dose suspension or as percentage of fungal yeast used for macrophages infection.

Killing and Proliferation

Most of the studies of phagocyte-mediated killing are done with macrophages and neutrophils as they play a predominant role in the host defence against fungal infection. Although is has been proposed recently that DCs also play an important role, they are less potent in killing fungal pathogens than monocytes, macrophages, and neutrophils.

The classical method of determining phagocytosis and intracellular killing is the counting of the surviving fungal population (plate counting) after the incubation with host cells. Fungal cells are added to give appropriate infection ratios with host cells and the percentage of killing is estimated after lysis of the immune cells (Kullberg et al., 1993; Netea et al., 2004b; Vonk et al., 2002). As an alternative, an end-point survival assay can be used that covers a range of high multiplicities of infection. This system has proved to be useful to determine differences in the resistance to host phagocytes in mutants altered in signal transduction pathways (Marcil et al., 2002). Phagocytosis and killing assay with acapsular or poorly encapsulated strains in comparison to the capsulated *C. neoformans* strains established the role of the structure in these processes. It inhibits phagocytosis of the fungus by macrophages, DCs and neutrophils and also inhibits the killing inside the host cells (Kozel, 1977; Levitz et al., 1997).

H. capsulatum has adapted to survive within human macrophages. The fungus is able to inhibit phagosome-lysosome fusion by controlling the intraphagosomal pH and creating an intracellular environment which permits its proliferation (Eissenberg et al., 1988, 1993). In contrast, DCs seem to be efficient in killing of the pathogen. The mechanisms underlying this difference has been investigated using FITC-dextran labelled DCs (Gildea et al., 2005) while the intracellular growth of *H. capsulatum* could be quantified by measuring the incorporation of [^{3}H]-leucine (Newman & Gootee, 1992). A comprehensive compilation of methods that allows the investigation of the interaction of this fungus with immune cells can be found in Newman (2005).

Killing of target fungal cells by immune effector cells can be addressed by the methods outlined above (see section 'Techniques to monitor cell viability and tissue integrity' in this chapter). Assays based on XTT or MTT have been used in *C. albicans* to determine the damage caused by immune cells to extracellular hyphae (Netea et al., 2004a; Vonk et al., 2002, 2005) and in *A. fumigatus* to test the antifungal activity of a new compound in combination with neutrophils, monocytes, and macrophages (Choi et al., 2004).

Conclusions

In vitro models have greatly advanced our knowledge about fungal virulence by providing a means of understanding the process of pathogenicity in its single consecutive steps at a cellular level. Although any inference from these models for an *in vivo* infection has to be made with caution, the available systems reflect more and more the physiological situation found *in vivo*, thereby providing a valid matrix to model the events that occur *in vivo* under controlled experimental conditions. Improvements in reproducibility by using commercial or well-designed in-house models, and the development towards four-dimensional systems, reflecting changes in cell viability, metabolism, and model integrity in real time, make the application of the *in vitro* technology especially attractive. The advances in the development of robust models which adequately reflect reality go hand in hand with the opportunity to draw relevant conclusions from experimental infections *in vitro*. Thus, implementation of *in vitro* cell and tissue technology in fungal infection research is a multidiscipline approach which profits from the gain of knowledge in histology and physical and chemical analytical methods. However, as well as the *in vitro* models themselves as the application of analytical assays are concerned, a devastating lack of standards has to be noted which makes the transfer from single approaches to other similar experiments extremely difficult. Publications rarely address the question of improving analysis systematically, for example of images generated by microscopy by different image analysis software packages (O'Mahony et al., 2005). The application of *in vitro* models in fungal pathogenicity research would clearly profit from a prompt transfer of newly gained knowledge between the single scientific disciplines.

Acknowledgements We thank colleagues for sharing information prior to publication. S.G. is recipient of a fellowship from the German Academy of Scientists 'LEOPOLDINA'. Work in our laboratory is supported by NIH Grant RX4215-030 and BIO2006-036737.

References

Abbott, N. J. (2005). *Cell. Mol. Neurobiol.*, 25(1): 5–23.

Abbott, N. J., Ronnback, L., & Hansson, E. (2006). *Nat. Rev. Neurosci.*, 7(1): 41–53.

Aderem, A. & Underhill, D. M. (1999). *Annu. Rev. Immunol.*, 17:593–623.

Aird, W. C. (2003). *Crit. Care Med.*, 31(4 Suppl):S221–S230.

Aird, W. C. (2005). *J. Thromb. Haemost.*, 3(7)1392–1406.

Albelda, S. M., Oliver, P. D., Romer, L. H., & Buck, C. A. (1990). *J. Cell Biol.*, 110(4):1227–1237.

Amitani, R., Murayama, T., Nawada, R., Lee, W. J., Niimi, A., Suzuki, K., Tanaka, E., & Kuze, F. (1995). *Eur. Respir. J.*, 8(10):1681–1687.

Bacci, A., Montagnoli, C., Perruccio, K., Bozza, S., Gaziano, R., Pitzurra, L., Velardi, A., d'Ostiani, C. F., Cutler, J. E., & Romani, L. (2002). *J. Immunol.*, 168(6):2904–2913.

Barbosa, F. M., Fonseca, F. L., Holandino, C., Alviano, C. S., Nimrichter, L., & Rodrigues, M. L. (2006). *Microbes. Infect.*, 8(2):493–502.

Bargatze, R. F., Kurk, S., Watts, G., Kishimoto, T. K., Speer, C. A., & Jutila, M. A. (1994). *J. Immunol.*, 152(12):5814–5825.

Barousse, M. M., Espinosa, T., Dunlap, K., & Fidel, P. L., Jr. (2005). *Infect. Immun.*, 73(11):7765–7767.

Barousse, M. M., Steele, C., Dunlap, K., Espinosa, T., Boikov, D., Sobel, J. D., & Fidel, P. L., Jr. (2001). *J. Infect. Dis.*, 184(11):1489–1493.

Baskic, D., Popovic, S., Ristic, P., & Arsenijevic, N. N. (2006). *Cell Biol. Int.*, 30(11):924–932.

Bermudez, L. E., Sangari, F. J., Kolonoski, P., Petrofsky, M., & Goodman, J. (2002). *Infect. Immun.*, 70(1):140–146.

Bernhardt, J., Herman, D., Sheridan, M., & Calderone, R. (2001). *J. Infect. Dis.*, 184(9):1170–1175.

Berridge, M. V., Herst, P. M., & Tan, A. S. (2005). *Biotechnol. Annu. Rev.*, 11:127–152.

Bevilacqua, M. P. (1993). *Annu. Rev. Immunol.*, 11:767–804.

Blander, J. M. & Medzhitov, R. (2004). *Science*, 304(5673)1014–1018.

Blank, F., Rothen-Rutishauser, B. M., Schurch, S., & Gehr, P. (2006). *J. Aerosol Med.*, 19(3):392–405.

Borg von Zepelin, M. & Wagner, T. (1995). *Mycoses*, 38(9–10):339–347.

Botterel, F., Cordonnier, C., Barbier, V., Wingerstmann, L., Liance, M., Coste, A., Escudier, E., & Bretagne, S. (2002). *Histol. Histopathol.*, 17(4):1095–1101.

Bowman, P. D., Ennis, S. R., Rarey, K. E., Betz, A. L., & Goldstein, G. W. (1983). *Ann. Neurol.*, 14(4):396–402.

Casadevall, A., Cleare, W., Feldmesser, M., Glatman-Freedman, A., Goldman, D. L., Kozel, T. R., Lendvai, N., Mukherjee, J., Pirofski, L. A., Rivera, J., Rosas, A. L., Scharff, M. D., Valadon, P., Westin, K., & Zhong, Z. (1998). *Antimicrob. Agents Chemother.*, 42(6):1437–1446.

Chang, Y. C., Stins, M. F., McCaffery, M. J., Miller, G. F., Pare, D. R., Dam, T., Paul-Satyaseela, M., Kim, K. S., & Kwon-Chung, K. J. (2004). *Infect. Immun.*, 72(9):4985–4995.

Chen, S. H., Stins, M. F., Huang, S. H., Chen, Y. H., Kwon-Chung, K. J., Chang, Y., Kim, K. S., Suzuki, K., & Jong, A. Y. (2003). *J. Med. Microbiol.*, 52(Pt 11):961–970.

Cheng, G., Wozniak, K., Wallig, M. A., Fidel, P. L., Jr., Trupin, S. R., & Hoyer, L. L. (2005). *Infection and Immunity*, 73(3):1656–1663.

Choi, J. H., Brummer, E., & Stevens, D. A. (2004). *Microbes. Infect.*, 6(4):383–389.

Cines, D. B., Pollak, E. S., Buck, C. A., Loscalzo, J., Zimmerman, G. A., McEver, R. P., Pober, J. S., Wick, T. M., Konkle, B. A., Schwartz, B. S., Barnathan, E. S., McCrae, K. R., Hug, B. A., Schmidt, A. M., & Stern, D. M. (1998). *Blood*, 91(10):3527–3561.

Cucullo, L., McAllister, M. S., Kight, K., Krizanac-Bengez, L., Marroni, M., Mayberg, M. R., Stanness, K. A., & Janigro, D. (2002). *Brain Res.*, 951(2):243–254.

d'Ostiani, C. F., Del, S. G., Bacci, A., Montagnoli, C., Spreca, A., Mencacci, A., Ricciardi-Castagnoli, P., & Romani, L. (2000). *J. Exp. Med.*, 191(10):1661–1674.

Dalle, F., Jouault, T., Trinel, P. A., Esnault, J., Mallet, J. M., d'Athis, P., Poulain, D., & Bonnin, A. (2003). *Infection and Immunity*, 71(12):7061–7068.

DeHart, D. J., Agwu, D. E., Julian, N. C., & Washburn, R. G. (1997). *J. Infect. Dis.*, 175(1):146–150.

Dickson, M. A., Hahn, W. C., Ino, Y., Ronfard, V., Wu, J. Y., Weinberg, R. A., Louis, D. N., Li, F. P., & Rheinwald, J. G. (2000). *Mol. Cell Biol.*, 20(4)1436–1447.

Dieterich, C., Schandar, M., Noll, M., Johannes, F. J., Brunner, H., Graeve, T., & Rupp, S. (2002). *Microbiology*, 148(Pt 2):497–506.

Dongari-Bagtzoglou, A. & Kashleva, H. (2006). *Microb. Pathog.*, 40(6):271–278.

Drenth, J. P., Van Uum, S. H., Van, D. M., Pesman, G. J., Van, d. V., & Van der Meer, J. W. (1995). *J. Appl. Physiol*, 79(50):1497–1503.

Dubourdeau, M., Athman, R., Balloy, V., Philippe, B., Sengmanivong, L., Chignard, M., Philpott, D. J., Latge, J. P., & Ibrahim-Granet, O. (2006). *Med. Mycol.*, 44(Suppl.):213–217.

Durr, E., Yu, J., Krasinska, K. M., Carver, L. A., Yates, J. R., Testa, J. E., Oh, P., & Schnitzer, J. E. (2004). *Nat. Biotechnol.*, 22(8):985–992.

Dustin, M. L. & Springer, T. A. (1988). *J. Cell Biol.*, 107(1):321–331.

Edgell, C. J., McDonald, C. C., & Graham, J. B. (1983). *Proc. Natl. Acad. Sci. USA*, 80(12):3734–3737.

Eissenberg, L. G., Goldman, W. E., & Schlesinger, P. H. (1993). *J. Exp. Med.*, 177(6):1605–1611.

Eissenberg, L. G., Moser, S. A., & Goldman, W. E. (1997). *J. Infect. Dis.*, 175(6):1538–1544.

Eissenberg, L. G., Schlesinger, P. H., & Goldman, W. E. (1988). *J. Leukoc. Biol.*, 43(6)483–491.

Ezekowitz, R. A., Sastry, K., Bailly, P., & Warner, A. (1990). *J. Exp. Med.*, 172(6):1785–1794.

Falkow, S. (1988). *Rev. Infect. Dis.*, 10(Suppl. 2):S274–S276.

Fan, W., Kraus, P. R., Boily, M. J., & Heitman, J. (2005). *Eukaryot Cell*, 4(8):1420–1433.

Feinberg, H., Mitchell, D. A., Drickamer, K., & Weis, W. I. (2001). *Science*, 294(5549):2163–2166.

Filler, S. G. & Sheppard, D. C. (2006). *PLoS Pathog.*, 2(12):e129. doi:10.1371/journal. ppat.0020129.

Filler, S. G., Swerdloff, J. N., Hobbs, C., & Luckett, P. M. (1995). *Infect. Immun.*, 63(3):976–983.

Fishman, A. P. (1982). *Ann. N.Y. Acad. Sci.*, 401:1–8.

Fraser, R. S. (1993). *Pathol. Annu.*, 28(Pt 1):231–277.

Frison, N., Taylor, M. E., Soilleux, E., Bousser, M. T., Mayer, R., Monsigny, M., Drickamer, K., & Roche, A. C. (2003). *J. Biol. Chem.*, 278(26):23922–23929.

Fu, Y., Rieg, G., Fonzi, W. A., Belanger, P. H., Edwards, J. E. J., & Filler, S. G. (1998). *Infect. Immun.*, 66(4):1783–1786.

Furuse, M., Hirase, T., Itoh, M., Nagafuchi, A., Yonemura, S., Tsukita, S., & Tsukita, S. (1993). *J. Cell Biol.*, vol. 123(6) Pt 2:1777–1788.

Ganendren, R., Carter, E., Sorrell, T., Widmer, F., & Wright, L. (2006). *Microbes Infect.*, 8(4):1006–1015.

Gantner, B. N., Simmons, R. M., & Underhill, D. M. (2005). *EMBO J.*, 24(6):1277–1286.

Giaimis, J., Lombard, Y., Makaya-Kumba, M., Fonteneau, P., & Poindron, P. (1992). *J. Immunol. Methods*, 154(2):185–193.

Gildea, L. A., Ciraolo, G. M., Morris, R. E., & Newman, S. L. (2005). *Infect. Immun.*, 73(10): 6803–6811.

Gimbrone, M. A., Jr., Obin, M. S., Brock, A. F., Luis, E. A., Hass, P. E., Hebert, C. A., Yip, Y. K., Leung, D. W., Lowe, D. G., & Kohr, W. J. (1989). *Science*, 246(4937):1601–1603.

Glee, P. M., Cutler, J. E., Benson, E. E., Bargatze, R. F., & Hazen, K. C. (2001). *Infect. Immun.*, 69(5):2815–2820.

Goldman, W. E. & Baseman, J. B. (1980). *In Vitro*, 16(4):313–319.

Grainger, C. I., Greenwell, L. L., Lockley, D. J., Martin, G. P., & Forbes, B. (2006). *Pharm. Res.*, 23(7):1482–1490.

Grossmann, J., Maxson, J. M., Whitacre, C. M., Orosz, D. E., Berger, N. A., Fiocchi, C., & Levine, A. D. (1998). *Am. J. Pathol.*, 153(1):53–62.

Haller, C. & Kubler, W. (1999). *Z. Kardiol.*, 88(5):324–330.

Harris, P. & Ralph, P. (1985). *J. Leukoc. Biol.*, 37(4):407–422.

Hed, J., Johansson, M., & Skogh, T. (1985). *Scand. J. Immunol.*, 21(1):43–47.

Henry-Stanley, M. J., Garni, R. M., & Wells, C. L. (2004). *J. Microbiol. Methods*, 59(2):289–292.

Hermanns, M. I., Unger, R. E., Kehe, K., Peters, K., & Kirkpatrick, C. J. (2004). *Lab. Invest.*, 84(6):736–752.

Hewett, P. W. & Murray, J. C. (1993). *Microvasc. Res.*, 46(1):89–102.

Hewett, P. W. & Murray, J. C. (1996). *Biochem. Biophys. Res. Commun.*, 221(3):697–702.

Hoyer, L. L. (2001). *Trends in Microbiol.*, 9(4):176–180.

Hoyer, L. W., De los Santos, R. P., & Hoyer, J. R. (1973). *J. Clin. Invest.*, 52(11):2737–2744.

Huang, S. H. & Jong, A. Y. (2001). *Cell Microbiol.*, 3(5):277–287.

Ibrahim, A. S., Filler, S. G., Alcouloumre, M. S., Kozel, T. R., Edwards, J. E., Jr., & Ghannoum, M. A. (1995). *Infect. Immun.*, 63(11):4368–4374.

Ibrahim, A. S., Filler, S. G., Ghannoum, M. A., & Edwards, J. E., Jr. (1993). *J. Infect. Dis.*, 167(6):1467–1470.

Jaffe, E. A., Hoyer, L. W., & Nachman, R. L. (1974). *Proc. Natl. Acad. Sci. USA*, 71(5): 1906–1909.

Jaffe, E. A., Nachman, R. L., Becker, C. G., & Minick, C. R. (1973). *J. Clin. Invest.*, 52(11): 2745–2756.

Jakab, E., Paulsson, M., Ascencio, F., & Ljungh, A. (1993). *APMIS*, 101(3):187–193.

Jayatilake, J. A., Samaranayake, Y. H., Cheung, L. K., & Samaranayake, L. P. (2006). *J. Oral Pathol. Med.*, 35(8):484–491.

Jayatilake, J. A., Samaranayake, Y. H., & Samaranayake, L. P. (2005). *J. Oral Pathol. Med.*, 34(4):240–246.

Jones, D. S., McGovern, J. G., Woolfson, A. D., & Gorman, S. P. (1997). *Pharm. Res.*, 14(12):1765–1771.

Jones, D. S., Schep, L. J., & Shepherd, M. G. (1995). *Pharm. Res.*, 12(12):1896–1900.

Jong, A. Y., Chen, S. H., Stins, M. F., Kim, K. S., Tuan, T. L., & Huang, S. H. (2003). *J. Med. Microbiol.*, 52(Pt 8):615–622.

Jong, A. Y., Stins, M. F., Huang, S. H., Chen, S. H., & Kim, K. S. (2001). *Infect. Immun.*, 69(7):4536–4544.

Josserand, V., Pelerin, H., de, B. B., Jego, B., Kuhnast, B., Hinnen, F., Duconge, F., Boisgard, R., Beuvon, F., Chassoux, F., umas-Duport, C., Ezan, E., Dolle, F., Mabondzo, A., & Tavitian, B. (2006). *J. Pharmacol. Exp. Ther.*, 316(1):79–86.

Kallmann, B. A., Wagner, S., Hummel, V., Buttmann, M., Bayas, A., Tonn, J. C., & Rieckmann, P. (2002). *FASEB J.*, 16(6):589–591.

Kamai, Y., Chiang, L. Y., Lopes Bezerra, L. M., Doedt, T., Lossinsky, A. S., Sheppard, D. C., & Filler, S. G. (2006). *Med. Mycol.*, 44(Suppl.):115–117.

Kaposzta, R., Marodi, L., Hollinshead, M., Gordon, S., & da Silva, R. P. (1999). *J. Cell Sci.*, 112(Pt 19):3237–3248.

Knight, S. A. & Dancis, A. (2006). *Microbiology*, 152(Pt 8):2301–2308.

Korzeniewski, C. & Callewaert, D. M. (1983). *J. Immunol. Methods*, 64(3):313–320.

Kozel, T. R. (1977). *Infect. Immun.*, 16(1):99–106.

Krump-Konvalinkova, V., Bittinger, F., Unger, R. E., Peters, K., Lehr, H. A., & Kirkpatrick, C. J. (2001). *Lab. Invest.*, 81(12):1717–1727.

Kullberg, B. J., van 't Wout, J. W., Hoogstraten, C., & van, F. R. (1993). *J. Infect. Dis.*, 168(2):436–443.

L'Azou, B., Fernandez, P., Bareille, R., Beneteau, M., Bourget, C., Cambar, J., & Bordenave, L. (2005). *Cell Biol. Toxicol.*, 21(2):127–137.

Lamszus, K., Schmidt, N. O., Ergun, S., & Westphal, M. (1999). *J. Neurosci. Res.*, 55(3):370–381.

Levitz, S. M., DiBenedetto, D. J., & Diamond, R. D. (1987). *J. Immunol. Methods*, 101(1):37–42.

Levitz, S. M., Harrison, T. S., Tabuni, A., & Liu, X. (1997). *J. Clin. Invest.*, 100(6):1640–1646.

Ley, K. (1996). *Cardiovasc. Res.*, 32(4):733–742.

Limper, A. H. & Standing, J. E. (1994). *Immunol. Lett.*, 42(3):139–144.

Lopes Bezerra, L. M. & Filler, S. G. (2004). *Blood*, 103(6):2143–2149.

Lu, Q., Jayatilake, J. A., Samaranayake, L. P., & Jin, L. (2006). *J. Invest. Dermatol.*, 126(9):2049–2056.

Lundborg, M., Dahlen, S. E., Johard, U., Gerde, P., Jarstrand, C., Camner, P., & Lastbom, L. (2006). *Environ. Res.*, 100(2):197–204.

Luther, K., Rohde, M., Heesemann, J., & Ebel, F. (2006). *J. Microbiol. Methods*, 66(1):170–173.

Mansour, M. K. & Levitz, S. M. (2002). *Curr. Opin. Microbiol.*, 5(4):359–365.

Marcil, A., Harcus, D., Thomas, D. Y., & Whiteway, M. (2002). *Infect. Immun.*, 70(11):6319–6329.

Marin, V., Montero-Julian, F. A., Gres, S., Boulay, V., Bongrand, P., Farnarier, C., & Kaplanski, G. (2001). *J. Immunol.*, 167(6):3435–3442.

Mascotti, K., McCullough, J., & Burger, S. R. (2000). *Transfusion*, 40(6):693–696.

Masuoka, J., Wu, G., Glee, P. M., & Hazen, K. C. (1999). *FEMS Immunol. Med. Microbiol.*, 24(4):421–429.

Matsunaga, K., Klein, T. W., Friedman, H., & Yamamoto, Y. (2001). *Am. J. Respir. Cell Mol. Biol.*, 24(3):326–331.

May, L. T., Torcia, G., Cozzolino, F., Ray, A., Tatter, S. B., Santhanam, U., Sehgal, P. B., & Stern, D. (1989). *Biochem. Biophys. Res. Commun.*, 159(3):991–998.

Mayer, C. L., Filler, S. G., & Edwards, J. E., Jr. (1992). *Microvasc. Res.*, 43(2):218–226.

McCarron, P. A., Donnelly, R. F., Canning, P. E., McGovern, J. G., & Jones, D. S. (2004). *Biomaterials*, 25(12):2399–2407.

Megard, I., Garrigues, A., Orlowski, S., Jorajuria, S., Clayette, P., Ezan, E., & Mabondzo, A. (2002). *Brain Res.*, 927(2):153–167.

Merkel, G. J. & Scofield, B. A. (1997). *FEMS Immunol. Med. Microbiol.*, 19(3):203–213.

Meshulam, T., Levitz, S. M., Christin, L., & Diamond, R. D. (1995). *J. Infect. Dis.*, 172(4):1153–1156.

Moody, D. M. (2006). *Semin. Cardiothorac. Vasc. Anesth.*, 10(2):128–131.

Morrissey, J. H. (2001). *Thromb. Haemost.*, 86(1):66–74.

Mosmann, T. (1983). *J. Immunol. Methods*, 65(1–2):55–63.

Mullick, A., Elias, M., Harakidas, P., Marcil, A., Whiteway, M., Ge, B., Hudson, T. J., Caron, A. W., Bourget, L., Picard, S., Jovcevski, O., Massie, B., & Thomas, D. Y. (2004). *Infect. Immun.*, 72(1):414–429.

Netea, M. G., Gijzen, K., Coolen, N., Verschueren, I., Figdor, C., Van der Meer, J. W., Torensma, R., & Kullberg, B. J. (2004a). *Microbes. Infect.*, 6(11):985–989.

Netea, M. G., Sutmuller, R., Hermann, C., Van der Graaf, C. A., Van der Meer, J. W., van Krieken, J. H., Hartung, T., Adema, G., & Kullberg, B. J. (2004b). *J. Immunol.*, 172(6):3712–3718.

Netzlaff, F., Lehr, C. M., Wertz, P. W., & Schaefer, U. F. (2005). *Eur. J. Pharm. Biopharm.*, 60(2):167–178.

Neuhaus, W., Bogner, E., Wirth, M., Trzeciak, J., Lachmann, B., Gabor, F., & Noe, C. R. (2006a). *Pharm. Res.*, 23(7):1491–1501.

Neuhaus, W., Lauer, R., Oelzant, S., Fringeli, U. P., Ecker, G. F., & Noe, C. R. (2006b). *J. Biotechnol.*, 125(1):127–141.

Newman, S. L. (2005). *Methods Mol. Med.*, 118:181–191.

Newman, S. L. & Gootee, L. (1992). *Infect. Immun.*, 60(11):4593–4597.

Newman, S. L. & Holly, A. (2001). *Infect. Immun.*, 69(11):6813–6822.

Niu, Q., Zhao, C., & Jing, Z. (2001). *J. Immunol. Methods*, 251(1–2):11–19.

O'Mahony, R., Basset, C., Holton, J., Vaira, D., & Roitt, I. (2005). *J. Microbiol. Methods*, 61(1):105–126.

Odds, F. C. (1988). *Candida and Candidosis*, 2 edn, Baillière, Tindall, London.

Orozco, A. S., Zhou, X., & Filler, S. G. (2000). *Infect. Immun.*, 68(3):1134–1141.

Osborn, L., Hession, C., Tizard, R., Vassallo, C., Luhowskyj, S., Chi-Rosso, G., & Lobb, R. (1989). *Cell*, 59(6):1203–1211.

Pardridge, W. M. (1999). *J. Neurovirol.*, 5(6):556–569.

Paris, S., Boisvieux-Ulrich, E., Crestani, B., Houcine, O., Taramelli, D., Lombardi, L., & Latge, J. P. (1997). *Infect. Immun.*, 65(4):1510–1514.

Parod, R. J., Godleski, J. J., & Brain, J. D. (1986). *J. Immunol.*, 136(6):2048–2054.

Phan, Q. T., Belanger, P. H., & Filler, S. G. (2000). *Infect. Immun.*, 68(6):3485–3490.

Phan, Q. T., Fratti, R. A., Prasadarao, N. V., Edwards, J. E., Jr., & Filler, S. G. (2005). *J. Biol. Chem.*, 280(11):10455–10461.

Pivarcsi, A., Nagy, I., Koreck, A., Kis, K., Kenderessy-Szabo, A., Szell, M., Dobozy, A., & Kemeny, L. (2005). *Microbes. Infect.*, 7(9–10):1117–1127.

Pober, J. S., Gimbrone, M. A., Jr., Lapierre, L. A., Mendrick, D. L., Fiers, W., Rothlein, R., & Springer, T. A. (1986). *J. Immunol.*, 137(6):1893–1896.

Pries, A. R. & Kuebler, W. M. (2006). *Handb. Exp. Pharmacol.* 176(Pt 1):1–40.

Pries, A. R., Secomb, T. W., & Gaehtgens, P. (2000). *Pflugers Arch.*, 440(5):653–666.

Radyuk, S. N., Mericko, P. A., Popova, T. G., Grene, E., & Alibek, K. (2003). *Biochem. Biophys. Res. Commun.*, 305(3):624–632.

Raimondi, F., Crivaro, V., Capasso, L., Maiuri, L., Santoro, P., Tucci, M., Barone, M. V., Pappacoda, S., & Paludetto, R. (2006). *Pediatr. Res.*, 60(1):30–33.

Ralph, P., Prichard, J., & Cohn, M. (1975). *J. Immunol.*, 114(2) Pt 2:898–905.

Read, R. A., Moore, E. E., Moore, F. A., Carl, V. S., & Banerjee, A. (1993). *Surgery*, 114(2):308–313.

Relloso, M., Puig-Kroger, A., Pello, O. M., Rodriguez-Fernandez, J. L., de la, R. G., Longo, N., Navarro, J., Munoz-Fernandez, M. A., Sanchez-Mateos, P., & Corbi, A. L. (2002). *J. Immunol.*, 168(6):2634–2643.

Richard, L., Velasco, P., & Detmar, M. (1998). *Exp. Cell Res.*, 240(1):1–6.

Risau, W. & Flamme, I. (1995). *Annu. Rev. Cell Dev. Biol.*, 11:73–91.

Rollins, B. J., Yoshimura, T., Leonard, E. J., & Pober, J. S. (1990). *Am. J. Pathol.*, 136(6):1229–1233.

Romani, L. (2004). *Nat. Rev. Immunol.*, 4(1):1–23.

Rosdy, M., Bernard, B. A., Schmidt, R., & Darmon, M. (1986). *In Vitro Cell Dev. Biol.*, 22(5):295–300.

Rothen-Rutishauser, B. M., Kiama, S. G., & Gehr, P. (2005). *Am. J. Respir. Cell Mol. Biol.*, 32(4):281–289.

Rotrosen, D., Edwards, J. E., Jr., Gibson, T. R., Moore, J. C., Cohen, A. H., & Green, I. (1985). *J. Infect. Dis.*, 152(6):1264–1274.

Rouabhia, M. & Deslauriers, N. (2002). *Biochem. Cell Biol.*, 80(2)189–195.

Rouabhia, M., Schaller, M., Corbucci, C., Vecchiarelli, A., Prill, S. K., Giasson, L., & Ernst, J. F. (2005). *Infect. Immun.*, 73(8):4571–4580.

Rubin, L. L. & Staddon, J. M. (1999). *Annu. Rev. Neurosci.*, 22:11–28.

Sahlin, S., Hed, J., & Rundquist, I. (1983). *J. Immunol. Methods*, 60(1–2):115–124.

Sallusto, F. & Lanzavecchia, A. (1994). *J. Exp. Med.*, 179(4):1109–1118.

Sana, T. R., Janatpour, M. J., Sathe, M., McEvoy, L. M., & McClanahan, T. K. (2005). *Cytokine*, 29(6):256–269.

Sanchez, A. A., Johnston, D. A., Myers, C., Edwards, J. E., Jr., Mitchell, A. P., & Filler, S. G. (2004). *Infect. Immun.*, 72(1):598–601.

Santaguida, S., Janigro, D., Hossain, M., Oby, E., Rapp, E., & Cucullo, L. (2006). *Brain Res.*, 1109(1):1–13.

Santoni, G., Lucciarini, R., Amantini, C., Jacobelli, J., Spreghini, E., Ballarini, P., Piccoli, M., & Gismondi, A. (2002). *Infect. Immun.*, 70(7):3804–3815.

Schaller, M., Boeld, U., Oberbauer, S., Hamm, G., Hube, B., & Korting, H. C. (2004). *Microbiology*, 150(Pt 9):2807–2813.

Schaller, M., Korting, H. C., Borelli, C., Hamm, G., & Hube, B. (2005). *Infect. Immun.*, 73(5):2758–2765.

Schaller, M., Korting, H. C., Schafer, W., Bastert, J., Chen, W., & Hube, B. (1999). *Mol. Microbiol.*, 34(1):169–180.

Schaller, M., Mailhammer, R., Grassl, G., Sander, C. A., Hube, B., & Korting, H. C. (2002). *J. Invest. Dermatol.*, 118(4):652–657.

Schaller, M., Schackert, C., Korting, H. C., Januschke, E., & Hube, B. (2000). *J. Invest. Dermatol.*, 114(4):712–717.

Schaller, M., Schafer, W., Korting, H. C., & Hube, B. (1998). *Mol. Microbiol.*, 29(2):605–615.

Serrano-Gomez, D., Dominguez-Soto, A., Ancochea, J., Jimenez-Heffernan, J. A., Leal, J. A. & Corbi, A. L. (2004). *J. Immunol.*, 173(9):5635–5643.

Shin, S. H., Lee, Y. H., & Jeon, C. H. (2006). *Acta Otolaryngol.*, 126(12):1286–1294.

Spreghini, E., Gismondi, A., Piccoli, M., & Santoni, G. (1999). *J. Infect. Dis.*, 180(1):156–166.

Springer, T. A. (1994). *Cell*, 76(2):301–314.

Steele, C. & Fidel, P. L., Jr. (2002). *Infect. Immun.*, 70(2):577–583.

Steele, C., Leigh, J., Swoboda, R., & Fidel, P. L., Jr. (2000). *J. Infect. Dis.*, 182(5):1479–1485.

Steele, C., Ozenci, H., Luo, W., Scott, M., & Fidel, P. L., Jr. (1999a). *Med. Mycol.*, 37(4):251–259.

Steele, C., Rapaka, R. R., Metz, A., Pop, S. M., Williams, D. L., Gordon, S., Kolls, J. K., & Brown, G. D. (2005). *PLoS. Pathog.*, 1(4):e42.

Steele, C., Ratterree, M., & Fidel, P. L., Jr. (1999b). *J. Infect. Dis.*, 180(3):802–810.

Stins, M. F., Badger, J., & Sik, K. K. (2001). *Microb. Pathog.*, 30(1):19–28.

Stins, M. F., Gilles, F., & Kim, K. S. (1997). *J. Neuroimmunol.*, 76(1–2):81–90.

Sun, C. Y., Hu, Y., Wang, H. F., He, W. J., Wang, Y. D., & Wu, T. (2006). *Chin. Med. J.(Engl.)*, 119(7):589–595.

Taylor, P. R., Brown, G. D., Reid, D. M., Willment, J. A., Martinez-Pomares, L., Gordon, S., & Wong, S. Y. (2002). *J. Immunol.*, 169(7):3876–3882.

Thomas, K. A. (1996). *J. Biol. Chem.*, 271(2):603–606.

Underhill, D. M. & Gantner, B. (2004). *Microbes. Infect.*, 6(15):1368–1373.

Underhill, D. M. & Ozinsky, A. (2002). *Curr. Opin. Immunol.*, 14(1):103–110.

Unger, R. E., Krump-Konvalinkova, V., Peters, K., & Kirkpatrick, C. J. (2002). *Microvasc. Res.*, 64(3):384–397.

Vestweber, D. (2002). *Curr. Opin. Cell Biol.*, 14(5):587–593.

Vinters, H. V., Reave, S., Costello, P., Girvin, J. P., & Moore, S. A. (1987). *Cell Tissue Res.*, 249(3):657–667.

Vonk, A. G., Netea, M. G., van Krieken, J. H., Van der Meer, J. W., & Kullberg, B. J. (2002). *J. Infect. Dis.*, 186(12):1815–1822.

Vonk, A. G., Wieland, C. W., Versteegen, M., Verschueren, I. C., Netea, M. G., Joostent, L. A., Verweij, P. E., & Kullberg, B. J. (2005). *Med. Mycol.*, 43(6):551–557.

Wasylnka, J. A. & Moore, M. M. (2002). *Infect. Immun.*, 70(6):3156–3163.

Wasylnka, J. A. & Moore, M. M. (2003). *J. Cell Sci.*, 116(Pt 8):1579–1587.

Whitcutt, M. J., Adler, K. B., & Wu, R. (1988). *In Vitro Cell Dev. Biol.*, 24(5):420–428.

Wiesner, S. M., Bendel, C. M., Hess, D. J., Erlandsen, S. L., & Wells, C. L. (2002). *Crit. Care Med.*, 30(3):677–683.

Witko-Sarsat, V., Rieu, P., scamps-Latscha, B., Lesavre, P., & Halbwachs-Mecarelli, L. (2000). *Lab. Invest.*, 80(5):617–653.

Zhao, X., Oh, S. H., Cheng, G., Green, C. B., Nuessen, J. A., Yeater, K., Leng, R. P., Brown, A. J., & Hoyer, L. L. (2004). *Microbiology*, 150(Pt 7):2415–2428.

Zhao, X., Oh, S. H., Yeater, K. M., & Hoyer, L. L. (2005). *Microbiology*, 151(Pt 5):1619–1630.

Chapter 3
Insects as Models for Studying the Virulence of Fungal Pathogens of Humans

Julie Renwick and Kevin Kavanagh

Insect Models for Studying Microbial Pathogens of Mammals

Vertebrates such as birds, rabbits, guinea pigs, and rodents (Clemons & Stevens, 2005), have been employed extensively over a long period of time to study various aspects of microbial pathogenesis, innate and acquired immune responses, disease transmission, and therapy. However, these models are extremely costly to use and, in many cases, their use is no longer considered to be ethically acceptable. There is a critical requirement to develop and validate alternative models for studying microbial pathogenesis and recent studies have highlighted the use of a variety of invertebrates as models for studying microbial virulence and the efficacy of antimicrobial drugs (for reviews see: Kavanagh & Reeves, 2004; Scully & Bidochka, 2006). Due to the strong structural and functional similarities between the immune response of insects and the innate immune response of mammals, the study of microbial virulence and of the efficacy of novel antimicrobial drugs has recently begun to utilise insects in order to model the innate immune response of mammals.

Insects represent one of the most successful groups of animals, exploiting almost all niches on Earth, except the seas, and accounting for at least 1 million species and 10^{18} individuals (Vilmos & Kurucz, 1998). Insects diverged from vertebrates approximately 500 million years ago and despite this early divergence, have maintained an immune response with strong structural and functional similarities to the innate immune response of mammals (Vilmos & Kurucz, 1998; Salzet, 2001). Insects rely exclusively upon an immune system analogous to the innate immune response of mammals and consequently have become extremely valuable as models for studying vertebrate innate immune responses to many pathogenic micro-organsims.

In vertebrates the different aspects of the innate and adaptive immune responses intertwine, connect and overlap. For example, interleukin-12 (IL-12), an anti-inflammatory cytokine that acts as an immunological messenger, is produced and secreted by macrophages during an immune response. IL-12 activates T cells, which in turn activate other cells involved in the immune response. In order to investigate those responses that are solely components of the innate immune response, employing an invertebrate model such as an insect that relies exclusively on an innate immune response, is advantageous. It is also becoming increasingly

K. Kavanagh (ed.), *New Insights in Medical Mycology.*

clear that the innate immune response, the first line of defence against invading microbes, is indispensable in fighting many infections. Individuals suffering from neutropenia, caused by reduced quantities of circulating neutrophils, or chronic granulomatous disease (CGD), a deficiency in superoxide-producing neutrophils, suffer extreme microbial infections and often do not survive to adulthood (Segal, 1996). As there is no interference from an adaptive response, invertebrate models can be used to understand elements of the innate immune response to many pathogens.

There are obvious ethical concerns with using mammalian models for *in vivo* testing of microbial pathogens, many of which can be removed by employing invertebrate models. Invertebrates, such as insects, do not have a well-developed nervous system and consequently do not experience pain in the same manner as mammals. Although invertebrate models would not be the only source of *in vivo* testing, they have the potential to substantially reduce the number of mammals sacrificed. In addition using invertebrates as models to study the pathogenicity of microbes yields faster results.

Invertebrate Models for Studying Microbial Pathogenesis

Insects have an immune response structurally and functionally similar to the innate immune response of mammals and, since the innate immune response of mammals is a vital component in the overall immune response to pathogenic infections (Levy, 2001; Romani, 1999), many groups are now studying fungal virulence employing insects as model systems. Many different insect species have been employed to study microbe–host interactions and include *Drosophila melanogaster, Galleria mellonella, Bombyx mori*, and *Manduca sexta*. A wide range of micro-organisms have been studied in insects including many bacterial (Dunphy et al., 1986; Morton et al., 1987; Bergin et al., 2005) and fungal pathogens (Mylonakis et al., 2005; Reeves et al., 2004; Cotter et al., 2000). Silva et al. (2002) employed *M. sexta*, to study the pathogenicity of the nematode symbiont, *Photorhabdus luminescens*, by investigating the effects of bacterial culture supernatants on haemocyte phagocytosis.

There have also been many reports of the efficacy of insects in studying the therapeutic effects of antibiotics (Hamamoto et al., 2004; Alippi et al., 2005) and the antimicrobial effects of various drugs (Lionakis et al., 2005; Lionakis & Kontoyiannis, 2005; Tickoo & Russell, 2002). Larvae of silkworms have been used to measure the efficacy of antibiotics in killing bacteria and a positive correlation with the results obtained with mice has been demonstrated (Hamamoto et al., 2004). The fruit fly, *Drosophila melanogaster*, has also been employed to investigate antifungal drugs and *Aspergillus* virulence. Lionakis and Kontoyiannis (2005) employed *D. melanogaster* as a fast, inexpensive high-throughput screening model for anti-*Aspergillus* compounds and in investigations of the role of *Aspergillus* virulence factors in pathogenesis. Although all *Aspergillus* virulence and therapeutic investigations cannot be performed in these models they do provide primary testing systems and may reduce the number of mammals being utilised.

Invertebrates such as *Acanthamoeba castellanii, Caenorhabditis elegans, Dictyostelium discoidium, D. melanogaster*, and *G. mellonella* have been employed to study the molecular mechanisms by which *Cryptococcus neoformans* interacts with the host (London et al., 2006) and revealed that several virulence-related genes previously known to be involved in *C. neoformans* mammalian infections also played a role in virulence in these invertebrates.

The *Galleria Mellonella* Infection Model

The greater wax moth, *G. mellonella* has been widely used for screening microbial mutants and for assessing the efficacy of antimicrobial drugs (Kavanagh & Reeves, 2004) but in the wild it is a pest of beehives where larvae tunnel through the honeycombs by chewing the wood of which they are composed. Larvae can also cause severe structural damage to the hive, leaving it unusable and liable to disintegration. Although they have no direct effect on the bees in a hive they leave a silk trail, which often traps bees which eventually die.

Recent studies have shown that larvae of *G. mellonella* may be used as effective models for studying microbial virulence. We have demonstrated that larvae of the *G. mellonella* can be employed to determine the relative virulence *C. albicans* isolates and to differentiate between pathogenic and non-pathogenic yeast species (Cotter et al., 2000). A positive correlation between the virulence of *C. albicans* mutants when tested in *G. mellonella* and in BalbC mice has been also established (Brennan et al., 2002) and a number of factors affecting the virulence of *C. albicans* mutants in *Galleria* larvae have been identified (Dunphy et al., 2003). *G. mellonella* larvae were also employed to establish appropriate infection models for *Cryptococcus neoformans* (Mylonakis et al., 2005). Larvae of *G. mellonella* have been used to demonstrate a correlation between toxin production and virulence of the pulmonary pathogen *Aspergillus fumigatus* (Reeves et al., 2004) and a correlation between the stage of *A. fumigatus* spore germination and virulence has also been established (Renwick et al., 2006). Earlier work has demonstrated the use of *Galleria* larvae for investigating the immunosuppressive activities of fungal secondary metabolites (Gotz et al., 1997), by assessing anti-phagocytic properties and changes in cytoskeletal microtubule formation.

Walters & Ratcliffe (1983) studied the differential pathogenicity of *Bacillus cereus* and *Escherichia coli* using larvae of *G. mellonella* and established that *B. cereus* is more virulent than *E. coli*, as is the case in mammalian models, and that *B. cereus* trapped in nodules within the larval haemolymph were able to overcome the effects of the host's immune system. Larvae of *G. mellonella* have been used to assess the virulence of *Bacillus thuringiensis* and *B. cereus* and strong agreement has been established between the results obtained in insects and mice (Salamitou et al. 2000). A strong correlation between the virulence of *Pseudomonas aeruginosa* in *Galleria* larvae and in mice has also been demonstrated (Jander et al., 2000). In an examination of the response of *G. mellonella* and mice to a *P. aeruginosa* PA14 rpoN mutant (Hendrickson et al., 2001) it was demonstrated that similar virulence genes were required for infection in insects and mice. *G. mellonella* larvae were used to study the pathogenicity of the entomopathogenic bacterium, *Xenorhabdus nematophila*, and the effects of its metabolic secretions on the viability of *G. mellonella* larvae (Mahar et al., 2005). In this case, rather than investigating the pathogenicity of the bacteria, this study was concerned with the management of the *G. mellonella* beehive infestation. *G. mellonella* is an appropriate non-mammalian host for studying the role of the type III secretion system in *Pseudomonas aeruginosa* (Miyata et al., 2003). Bouillaut et al. (2005) employed *G. mellonella* larvae to investigate virulence factors of *B. cereus* and *B. thuringiensis*, controlled by the PlcR transcriptional regulator and determined that inactivation of *plcR* decreased but did not abolish bacterial virulence.

Inoculation of G. mellonella Larvae

While many insect models are now available for use, the larvae of *G. mellonella* have many advantages. Due to their size it is possible to inoculate individual larvae with specific doses of the pathogen in question whereas it is difficult to quantify the dose per insect when using other models such as *Drosophila*. In addition *G. mellonella* larvae can be purchased commercially and yield results in 24–48 h.

Larvae of *G. mellonella* are easy to inoculate via injection into the haemocoel through the last left proleg (Cotter et al., 2000). The base of the proleg can be opened by applying gentle pressure to the sides of the leg and this aperture will reseal after removal of the syringe needle without leaving a scar. Inoculation of larvae with test micro-organisms must be accompanied by inoculation of larvae with the buffer used to re-suspend the test micro-organisms to ensure that this has no impact on larval viability. A number of workers also suggest the 'mock-inoculation' of a number of larvae per experiment to ensure that the handling and inoculation procedures are not deleterious to the health of the larvae (Dunphy & Webster, 1984; Cotter et al., 2000). Larvae can be stored at 15°C prior to use and, once inoculated, may be maintained at temperatures up to 37°C, as long as appropriate controls are implemented to quantify the effect of temperature on survival. Larvae should be handled with care, as rough handling affects survival and also leads to the expression of stress proteins.

While larval death is often used as the end point in an experiment, other parameters may also be employed particularly when dealing with a relatively 'weak' pathogen which may not actually kill the test larvae. Fluctuations in fungal load and haemocyte density have been used as accurate indicators of fungal virulence in larvae (Bergin et al., 2003) as have changes in the expression of antimicrobial peptides (AMPs) (Bergin et al., 2006). When analysing a variety of *C. albicans* strains in larvae of *G. mellonella* isolates of high virulence could be detected by a decrease in circulating haemocyte density and an increase in fungal load whereas isolates of low virulence were detectable by a high-haemocyte density and a low fungal load (Bergin et al., 2003). Changes in the expression of selected AMPs can be used as an indicator of the immune response to infection and assist in differentiating pathogenic from non-pathogenic infections. (Bergin et al., 2006). Irrespective of which endpoint is used (larval death, fungal load, haemocyte density, AMP expression) results can be obtained within 48 h (Cotter et al., 2000; Brennan et al., 2002; Bergin et al., 2003, 2006).

Insect Immune System and Parallels with the Innate Immune Response of Mammals

Insects have maintained a parallel immune response to the innate immune response of vertebrates, despite their divergence approximately 500 million years ago. The preservation of this immune response highlights its success and efficacy in combating microbial infection (Kavanagh & Reeves, 2004; Salzet, 2001; Leclerc & Reichhart, 2004; Kanost et al., 2004). It was initially thought that the innate immune response was inferior to the adaptive immune response, as the adaptive immune system has the ability to 'remember' pathogenic encounters. However, it is now generally agreed that the innate immune response is a very important arm of the immune system, being the first line of defence encountered by invading microbes. The importance of the innate immune response in fighting infection has been highlighted by disorders in which aspects of this immune system are not functioning, such as neutropenia and CGD (Segal, 1996; Meyer & Atkinson, 1983; Gerson & Talbot, 1984; Latge, 1999).

Insects rely exclusively on an innate immune response, and because of this, can be employed to investigate the innate immune response of vertebrates to many pathogenic micro-organisms without the difficulty of having to separate out the adaptive immune responses as would occur in mammals (Figure 3.1). Many parallels exist between the immune response of insects and the innate immune response of vertebrates, including similar coagulation processes (Theopold et al., 2002), phagocytosis (Costa et al., 2005; Tojo et al., 2000), superoxide production (Imamura et al., 2002; Adema et al., 1994) and AMP production pathways (De Gregorio et al., 2002). Insects also have a blood equivalent, a fat body that is an analogous tissue to vertebrate liver, immune cell equivalents (Tojo et al., 2000), immunological receptors (Cherry et al., 2006; Pandey et al., 2006; Yoshida et al., 1986; Duvic & Soderhall, 1992) and AMPs comparable to vertebrate immuno-related proteins (Cociancich et al., 1993; Johns et al., 2001; Thompson et al., 2003; Yu et al., 2002).

The insect immune system consists of cellular and humoral responses. The cellular responses are mediated by haemocytes, which are found in the insect haemolymph and operate by engulfing or binding to foreign cells that entry through the cuticle. Haemocytes are structurally and functionally analogous to human neutrophils. The Humoral elements include the ability to produce antimicrobial peptides to combat a pathogen and to produce melanin (melanisation) which acts to immobilise and/or kill the pathogen.

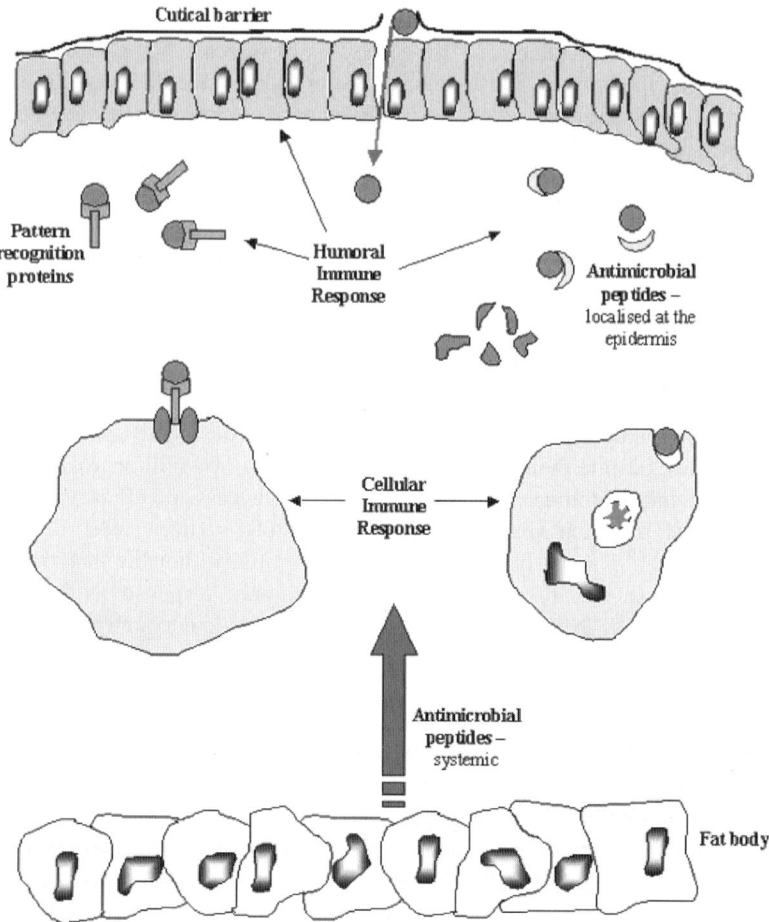

Figure 3.1 An overview of the insect immune response. Once a microbe has managed to breach the thick exoskeleton of the insect and has crossed the epidermal tissue, the insect immune response is activated. The humoral response provides recognition molecules to alert the cellular response, and antimicrobial peptides (AMPs), are produced to enzymatically degrade the invaders. Specialised phagocytes engulf and kill microbes and, along with the fat body, produce more AMPs. The immune cells of the insect also perform other functions such as encapsulation

The Insect Cuticle: The First Line of Defence

The cuticle forms a mechanical barrier, made of chitin fibrils embedded in a protein matrix, on the exterior and the interior midgut, trachea, and genital tract regions of the insect (Wigglesworth, 1972). The outer layer of the cuticle (the epicuticle) is covered in a waxy layer containing lipids, fatty acids, and sterols. This chitinous exoskeleton is a matrix of carbohydrates and protein secreted from a monolayer of epidermal cells covering the entire surface area of the insect (Wigglesworth, 1972).

An intact cuticle effectively prevents entry of micro-organisms and injury of the cuticle activates the humoral responses such as melanisation around injured area and the production of AMPs to fight microbes at the site of entry (Kavanagh & Reeves, 2004).

Cellular Immune Responses

Haemocytes and their Role in the Immune Response

Haemocytes are the immune cells of the insect and are found attached to internal organs of the insect, such as the fat body, trachea, or the digestive system, and circulating freely within the haemolymph. The density of haemocytes circulating in the haemolymph is indicative of infection, with low haemocyte densities being associated with infection and high haemocyte densities associated with healthy insects. This has been explicitly demonstrated by Bergin et al. (2003), where infection of *G. mellonella* larvae with the yeast *Candida albicans* resulted in greatly decreased numbers of circulating haemocytes, in contrast to larvae inoculated with the non-pathogenic yeast *Saccharomyces cerevisiae*, whose haemocyte decreased only slightly in number.

There are at least six types of haemocytes identified in lepidopteran insects (Figure 3.2) (e.g. *G. mellonella*). Here, the haemocyte classification of Price and Ratcliff (1974) is described. Haemocytes can be divided into the following types; prohaemocytes (6–13 µM), plasmatocytes (40–50 µM), granulocytes (45 µM), coagulocytes, sperulocytes (25 µM), and oenocytoids. Prohaemocytes are small round cells with large nuclei, which divide and are thought to differentiate into other cell types. Both plasmatocytes and granulocytes are involved in phagocytosis, nodule formation, and encapsulation (Tojo et al., 2000). Plasmatocytes are the most abundant haemocyte, have a leaf-like shape and contain lysosomal enzymes. Granulocytes have a small nucleus and a granule-rich cytoplasm. Spherules are not uniform in shape and have many small spherical inclusions. Oenocytoids are large, binucleate, non-phagtocytic cells, which may contain prophenoloxidase. Coagulocytes, as their name suggests, are involved in the clotting process. Haemocytes are also involved in the production of AMPs during the humoral response (Meister et al., 1994; Samakovlis et al., 1990).

Haemocytes with functions and characteristics similar to other vertebrate lymphocytes have been identified. Chernysh et al. (2004) reported haemocytes in *Calliphora vicina* with activities similar to those of mammalian cytotoxic lymphocytes. These haemocytes recognised the human myelogenous leukaemia K562 cells as non-self, attached to their surface and induced target destruction by inducing apoptosis of the K562 leukemia cells. Exposure of the K562 cells to these cytotoxic haemocytes caused a series of cytoplasmic-bulging movements termed zeiosis or membrane blebbing of the K562 cells. These effects were similar to effects observed of natural killer cells on K562 cells.

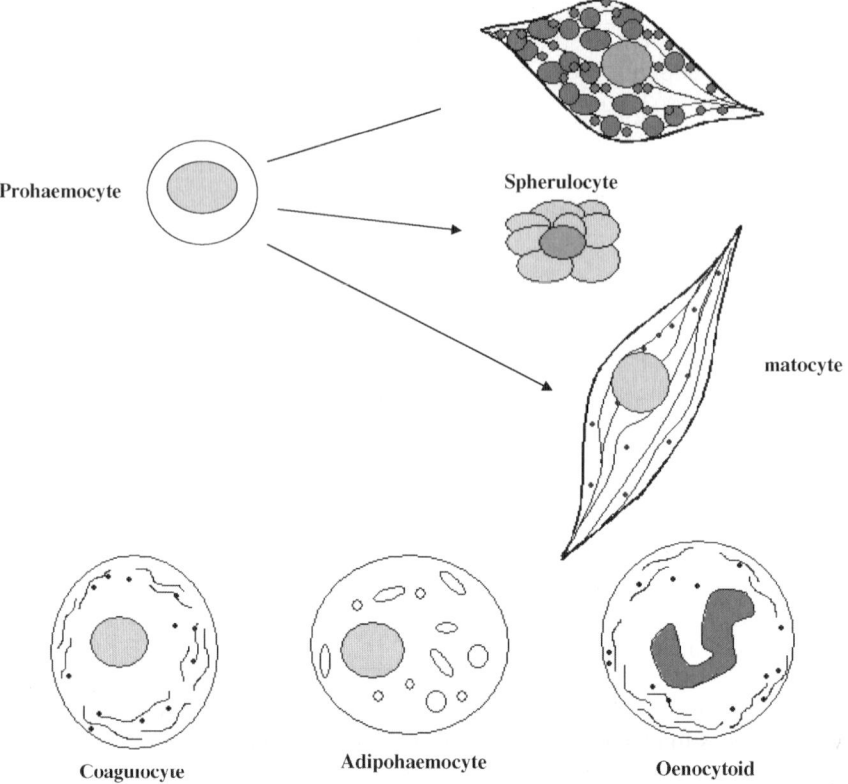

Figure 3.2 Morphology of haemocytes involved in the cellular immune response. Possible differentiation of prohaemocytes into granulocyte, plasmatocytes and spherulocytes (marked by arrows). Not to scale

Phagocytosis

In vertebrates phagocytosis occurs upon recognition and subsequent internalisation of invading microbes, by migrating phagocytes derived from the myeloid cell line which are represented by the neutrophils and the macrophages. Before phagocytosis can occur, phagocytic cells are recruited to the site of infection by sequential signal transduction events, where phagocytes must recognise the microbe as foreign (Baggiolini & Wymann, 1990). Microbes are then ingested into a phagosome where they encounter many toxic substances and degrading enzymes.

The process of phagocytosis in insects and mammals is very similar. In both cases phagocytosis is preceded by binding of opsonic ligands to the surface of the microbe, followed by recognition by specific receptors. Recognition results in engulfment of the foreign body. Activation of the ProPO cascade is required for granulocytes to bind to non-self matter and internalise it, while calcium is required for adherence of plasmatocytes (Kavanagh & Reeves, 2004). Recognition of non-self and phagocytosis is lectin-mediated in vertebrate phagocytes and in insect phagocytes (Wilson et al., 1999; Chen et al., 1999). Insects produce different lectins for recognising a range of different bacterial cell surfaces (Wilson & Ratcliff, 2000), giving the insect immune response the diversity required to recognise various microbes.

In vertebrate phagocytes, reactive oxygen species (ROS) are produced during phagocytosis and these ROSs and proteolytic enzymes are released into the phagosome from cytoplasmic granules to destroy engulfed microbes. This process has not yet been fully investigated in invertebrate systems; however, there is a lot of evidence to support the existence of similar processes in insect haemocytes. The production of ROS by insect haemocytes has been studied by nitroblue tetrazolium reduction (Glupov et al., 2001) and was observed in haemocytes of *G. mellonella, Aporia crataegi, Dendrolimus sibericus*, and *Gryllus bimaculatus*. In addition, Nappi et al. (1998) and Slepneva et al. (1999) reported evidence of both superoxide and its dismutase product hydrogen peroxide in *D. melanogaster* and *G. mellonella* haemocytes, respectively. Moreover, all the essential components of the nicotinamide adenine dinucleotide phosphate (NADPH) oxidase responsible for producing superoxide have been identified in *G. mellonella* haemocyte (Bergin et al., 2005). The presence of these NADPH oxidase components and ROSs in insect haemocytes implies that a parallel process occurs following phagocytosis in haemocytes as in vertebrate phagocytes.

Vertebrate neutrophils also undergo a degranulation process when challenged with microbes too large to ingest, such as parasites and hyphae. Granules contain proteolytic enzymes, such as elastin, cathepsin G, lysozyme, defensin, transferrin, and myeloperoxidase, which are released into either the phagosome where they actively destroy the ingested microbe or are exocytosed into the extracellular matrix, in a process referred to as degranulation. Once these enzymes are in close contact with the microbe they begin to effectively digest the invading microbes. A similar degranulation process occurs in insect haemocytes. Smith and Soderhall (1983) reported that the freshwater crayfish, *Astacus astacus* haemocytes undergo a similar process of degranulation in the presence of various bacteria. They also suggested that proteins of the phenoloxidase cascade are strong non-self signals for the haemocytes and may initiate this degranulation process. Many of these proteolytic enzymes have been identified in insect haemocytes. Lysozyme has been found in many insects, including *Hylphora cecropia* (Hultmark et al., 1980), *Manduca sexta* (Spies et al., 1986), *D. melanogaster* (Regel et al., 1998), *Musca domestica* (Ito et al., 1995), *H. virescens* (Lockey et al., 1996), and *Aedes aegypti* (Rossignol et al., 1986), *G. mellonella, Bombyx mori*, and *Agrius convolvuli* (Yu et al., 2002). Lysozyme has been found within insect haemocytes and the intrahaemolymph lysozyme levels have been shown to increase upon infection indicating the possible exocytosis of intracellular lysozyme upon contact of the haemocyte with a microbe (Wilson & Ratcliff, 2000). Other enzymes associated with neutrophils granules have been identified in insects, such as defensin from

G. mellonella (Lee et al., 2004) and *Dermacentor variabilis* (Johns et al., 2001), transferrin from the termite *Mastotermes darwiniensis* (Thompson et al., 2003) and from *D. melanogaster* (Yoshida et al., 2000). The myeloperoxidase homologue peroxynectin has also been isolated from many insects (Lin et al., 2006). The presence of a homologue of MPO in insect haemocytes is of great significance as MPO released from the granules into the phagosome converts hydrogen peroxide to one of the most toxic products found within the phagolysosome, hypochlorous acid.

Encapsulation

Large foreign bodies such as protozoa, nematodes, eggs, and larvae of parasitic insects are too large for haemocytes to phagocytose. Instead, a process called encapsulation occurs in which a capsule of overlapping layers of cells is formed around the object within 30 min of entry of the pathogen (Bowman & Hultmark, 1987). The encapsulation process initially involves the attachment of granular cells to the target. These granulocytes release plasmatocyte-spreading peptides (PSP) (Vilmos & Kurucz, 1998), which recruit plasmatocytes to the site of capsule formation. Plasmatocytes then bind to the layered granulocytes of the capsule. Choi et al. (2002) identified a *G. mellonella* protein with amino acid sequence homology to human and *Drosophila* calreticulin, which was enriched in the early stages of encapsulation. Interestingly, calreticulin has recently been reported to partially localise on the surface of neutrophils, and antimicrobial bound calreticulin was observed to transmit a signal to cells via G-protein to activate neutrophils to generate superoxide anion (Cho et al., 1999). The enrichment of calreticulin seen in *G. mellonella* capsules may have similar implications to the localisation of human calreticulin on the surface of neutrophils, in that enrichment of calreticulin in haemocytes could potentially activate haemocytes to generate ROS within the capsule.

Nodulisation

Upon invasion of insects by different micro-organisms, nodules are formed. Nodules are amalgamations of viable and degraded haemocytes, non-self material and melanised debris. Nodules are found attached to tissue or surrounded by haemocytes. The exact processes and events involved in the formation of nodules is not yet fully characterised, however, it is known that it is a lectin-mediated process (Vilmos & Kurucz, 1998). A lectin called scolexin is thought to be involved in nodule formation in *Manduca sexta* (see Kavanagh & Reeves, 2004). There is still a lot to be learnt of the process of nodulisation and it is likely that this process could represent a significant aspect of insect immunity.

Pattern Recognition Receptors

The task of recognition of different micro-organisms or essentially recognising non-self and the activation of the appropriate pathways is allocated to pattern recognition receptors (PRRs) (Janeway, 1989). PRRs recognise general microbial components such as LPS and β-1,3 glucans, and these complexes stimulate immune responses. Many PRRs exist in insects such as LPS-binding proteins (Kim et al., 2000; Koizumi et al., 1999), many different lectins (Yu et al., 2002; Koizumi et al., 1999), hemolin (Yu et al., 2002; Rasmuson et al., 1979; Lee et al., 2002), β-1,3 glucan binding protein (Ochiai & Ashida, 2000; Ma et al., 2000; Jiang et al., 2003), Gram-negative bacterial recognition protein (Kim et al., 2000), peptidoglycan recognition protein (Zhu et al., 2003; Werner et al., 2000; Yu et al., 2002), and hemocytin (Kotani et al., 1995). The Toll and the immune deficiency (IMD) pathways integrate the signals from the PRRs and initiate the production of AMPs (Lau et al., 2003).

The Toll Pathway

The production of AMPs by insect haemocytes and the fat body is an essential part of the insect immune response (Khush et al., 2001; Silverman & Maniatis, 2001). The production of these AMP has largely been attributed to the activation of a NF-κ-like transcription factor (Engstrom, 1999). The Toll (Figure 3.3) and IMD pathways in insects activate two distinct NF-κ-like transcription factors. The Toll pathway was first discovered in *Drosophila* (Lemaitre et al., 1996; Horng & Medzhitov, 2001; Tauszig-Delamasure et al., 2002) and later a number of TLRs were identified in vertebrates. The Toll pathway is similar to the TLR/interleukin-1 receptor (IL-1R) pathway involved in the innate immune response of vertebrates (Khush et al., 2001; Silverman & Maniatis, 2001). Toll is a transmembrane protein receptor with an extracellular leucin-rich repeat (LRR) domain and an intracellular domain similar to that of the IL-1R in vertebrates. This domain is referred to as the Toll/IL-1R (TIR) domain. The Toll receptor is not considered a PRR, as it does not directly recognise antimicrobial surfaces, rather a series of proteolytic reactions initiated by the presence of invading microbes results in the generation of the active form of the Spatzle protein, which is the ligand for the Toll receptor (Lemaitre et al., 1996; Ferrandon et al., 2004). Toll activation is triggered by Gram-positive or fungal infections and is initiated by the dimerisation of the Toll molecules (Weber et al., 2003). This dimerisation is thought to induce the recruitment of hetero-trimeric complexes composed of Myd88, Pelle, and Tube. The Myd88 protein and Pelle are homologues of the human Myd88 and IL-1-receptor-associated kinases (IRAK), respectively. The Myd88 protein contains a death domain (DD) and a TIR domain, the later of which interacts with Toll through its TIR domain. Tube and Pelle also contain DDs required for interaction of Tube with Myd88 and Pelle, but Myd88 and Pelle to not directly interact (Sun et al., 2002). Formation of

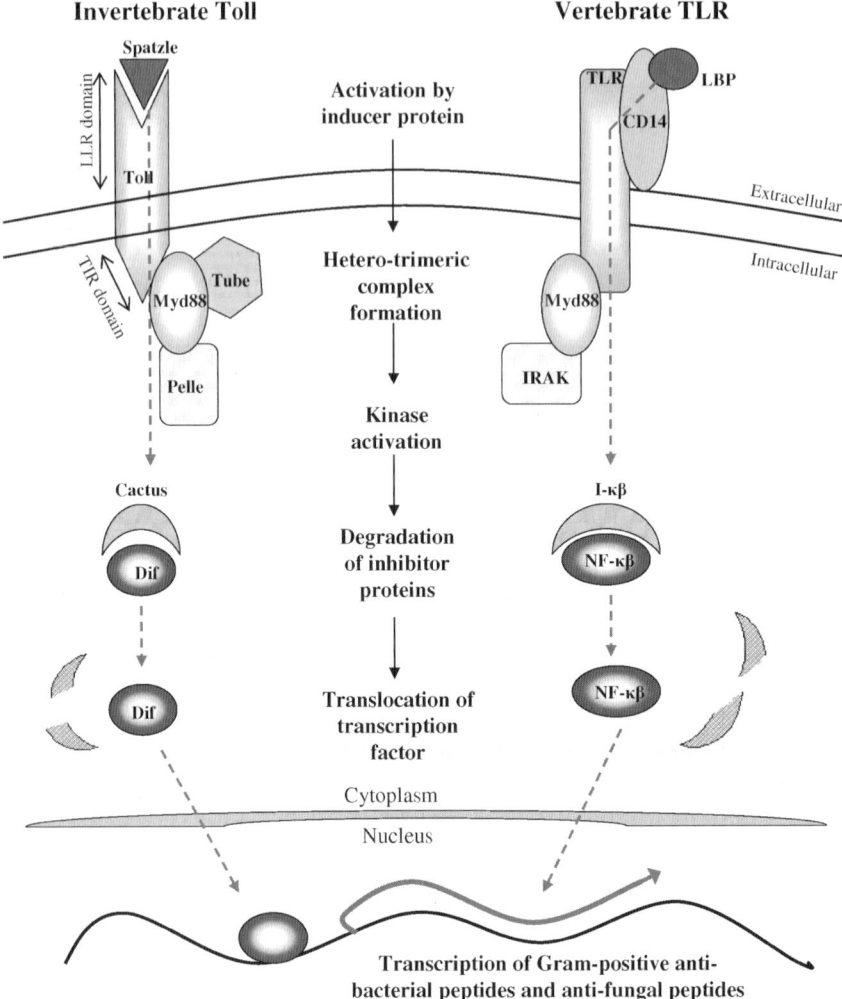

Figure 3.3 Diagrammatic representation of the similarities between the invertebrate and verte-brate Toll cascades. The activation of transcription factors NF-κβ in vertebrates and the NF-κβ-like factor, Dif, in *Drosophila* is initiated by the activation of homologous Toll receptors. This initiates a cascade of reactions ultimately resulting in the breakdown of parallel inhibitory proteins I-κβ in vertebrates and Cactus in *Drosophila*. TLR pathway in vertebrates activates the production of pro-inflammatory cytokines such as IL-1, IL-6, and IL-8

the hetero-trimeric adaptor complex results in the activation of the Pelle kinase and the subsequent hydrolysis of the cactus inhibitor. Cactus is homologous to the NF-k inhibitor of vertebrates, namely I-κβ (Geisler et al., 1992), and functions in a simi-lar manner by inhibiting the activity of the dorsal-related immune factor (DIF), a homologue of vertebrate NF-k (Ip et al., 1993). Hydrolysis of cactus results in the release of DIF, which enters the nucleus and activates the expression of many

AMPs including the antifungal defensin drosomycin (Nicolas et al., 1998; Tauszig-Delamasure et al., 2002; Belvin et al., 1996). The kinase activity of Pelle is required for cactus degradation (Towb et al., 2001) and cactus must be phosphorylated in order to be degraded (Fernandez et al., 2001), however, Pelle has not been shown to directly phosphorylate cactus and another unknown kinase may be involved. Considering published data, the Drosophila Toll family consists of eight members; among them are the 18-wheeler receptor, the Toll receptor, and Toll-3 to Toll-8 (Imler & Hoffmann, 2000).

The Toll pathways activation is also involved in other aspects of the insect immune response, other than AMP production. The Toll pathway activates NF-k which is required for the elimination of the Serpin27A serine protease inhibitor required to keep ProPO in its inactive form (Petros et al., 2002). This indicates crosstalk between the Toll pathway and the phenoloxidase cascade in insects.

The IMD Pathway

The immune deficiency (IMD) pathway is reminiscent of the vertebrate tumour necrosis factor (TNF)-α pathway (Khush et al., 2001; Silverman & Maniatis, 2001) and derives its name from the first mutation with impaired response to Gram-negative bacteria (Lemaitre et al., 1995). It is thought that the peptidoglycan recognition protein (PGRP) is a co-receptor in the IMD pathway (Gottar et al., 2002). Also *Drosophila* homologues of the TNF-α receptor, Wengen, and its ligand, Eiger, have been isolated but there is no current knowledge of their involvement in the IMD pathway (Kauppila et al., 2003; Kanda et al., 2002; Igaki et al., 2002). The IMD pathway activates the NF-k-like transcription factor Relish (Hedengren et al., 1999). In order for Relish to translocate to the nucleus, its C-terminal inhibitory portion has to be removed (Stoven et al., 2000). In contrast to vertebrate counterparts and to cactus, Relish activation does not require the proteolytic breakdown of its inhibitor, just its dissociation. In mammals, a TNF-α stimulus induces the assembly of the receptor, a part of a large receptor–adaptor complex, which includes the receptor integrating protein (RIP), Fas-associated DD (FADD) and caspase 8. The homologues of these proteins, IMD, dFADD (Hu et al., 2000; Leulier et al., 2002; Naitza et al., 2002) and the death-related ced-3/NEDD2-like protein (DREDD) (Leulier et al., 2000) are required for Relish activation after Gram-negative infection. The favoured hypothesis of IMD activation is that the IMD-dFADD-DREDD complex is formed following PGRP activation (Leclerc & Reichhart, 2004). The activation of this adaptor complex induces the activation of *Drosophila* transforming growth factor-activated protein kinase 1 (dTAK1) (Vidal et al., 2001). This dTAK1 is a homologue of TAK1, a mitogen-activated protein kinase kinase kinase (MAPKKK) involved in the signal transduction pathways of vertebrates, including the c-Jun N-terminal kinase (JNK) and the NF-α pathways. The IMD pathway in insects is activated on the most part by Gram-negative bacteria and initiates the transcription of mainly antibacterial peptides such as diptericin

(Lee et al., 2001). This immune-inducing pathway is still poorly understood and there are probably multiple factors involved in the pathway still unidentified.

Humoral Immune Responses

Coagulation of Insect Haemolymph

The insect body cavity (haemocoel) contains haemolymph, which is analogous to vertebrate blood. Haemolymph transports nutrients, waste products, AMPs and signal molecules, and immune cells around the body. The haemolymph is also capable of coagulating or clotting upon contact with an invading microbe or upon physical injury. There are two clotting mechanisms, which occur in insects. The first method involves the clottable proteins, lipophorins, and the vitellogenein-like proteins (Theopold et al., 2002). The second mechanism involves a haemocyte-derived clotting cascade where clottable proteins are released from the cytoplasmic L-granules of the haemocytes into the haemolymph in response to activation by cell wall components of invading microbes. Gram-negative bacteria activate factor C while fungi activate factor G, resulting in conformational changes in these two proteins (Sritunyalucksana & Soderhall, 2000) and subsequent cleavage of factor B by a serine protease. This leads to the cleavage of pro-clotting enzyme (PCE). This enzyme catalyses the cleavage of a soluble protein, coagulogen into an insoluble aggregate, coagulin forming a viscous clot that traps invading microbes (Gorman & Paskewitz, 2001).

Melanisation and the Phenoloxidase Cascade

Melanisation occurs when the cuticle is injured or upon contact of the haemolymph with an invading microbe. This element of insect immunity is vital as it conveys protection against a wide range of micro-organisms. Melanisation is initiated by the binding of foreign surface proteins to soluble PRRs in the haemolymph. This initiates the serine protease cascade, which catalyses the conversion of prophenoloxidase (ProPO) to the oxidoreductase phenoloxidase (PO) and the cross-linking and melanisation of proteins (Leclerc & Reichhart, 2004) (Figure 3.2). This process is referred to as the prophenoloxidase activating system (ProPO-AS) and is proposed to be a non-self recognition system. The inactive form ProPO is found in insect haemocytes, where it is released by exocytosis (Soderhall et al., 1996) and is either actively transported to the cuticle or deposited around wounds or encapsulated parasites. Once activated by the serine protease cascade, PO catalyses the *o*-hydroxylation of monophenols and the subsequent oxidation of phenols to quinines (Figure 3.4). These quinines then polymerise non-enzymatically to form melanin (Soderhall & Cerenius, 1998), which is deposited on the surface of the microbe

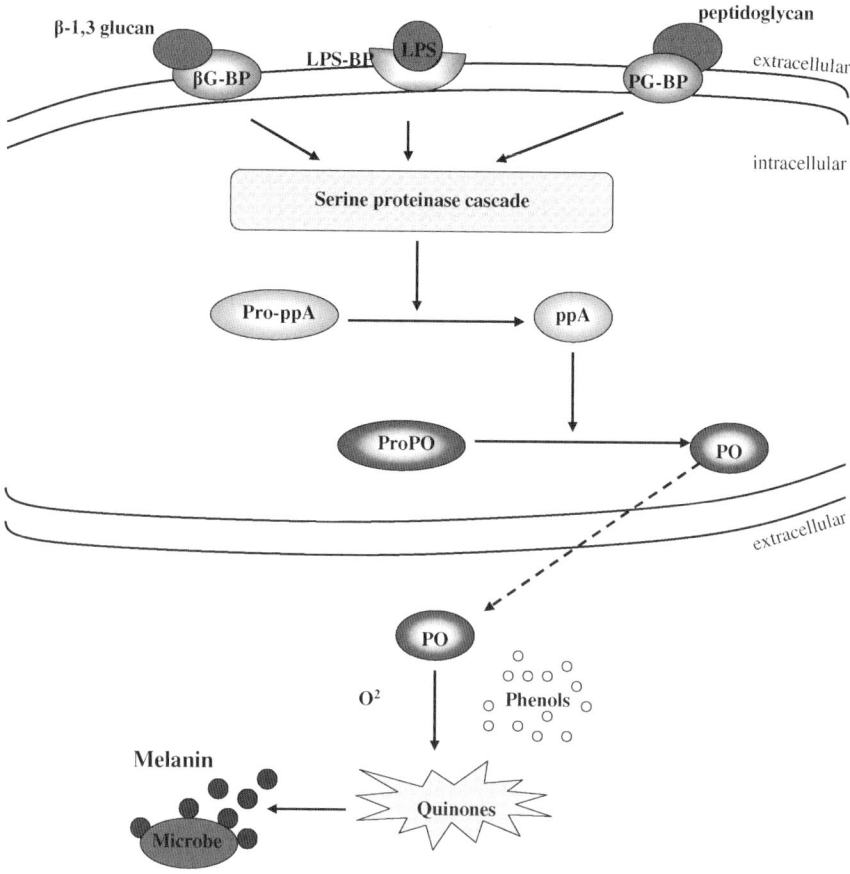

Figure 3.4 Overview of the phenoloxidase activating system. The system is activated by pattern recognition receptors (β-1,3 glucan-binding protein [βG-BP], lipopolysaccharide-binding protein [LPS-BP], peptidoglycan-binding protein [PG-BP]), or other factors produced upon tissue damage. A cascade of serine proteinases results in the cleavage of the pro-form of the prophenoloxidase-activating enzyme (pro-ppA) into the active form, ppA. Alternatively some serine proteinases can directly convert prophenoloxidase (proPO) to phenoloxidase (PO). Then in the presence of phenols and O_2, quinones are produced and subsequently melanin is produced

(Ratcliff, 1985). The formation of melanin is catalysed by phenoloxidase-monophenyl-L-dopa: oxygen oxidoreductase (Soderhall & Cerenius, 1998). As melanin is highly toxic to microbes and the insect host, its overactivity could be potentially damaging to the host. Therefore, the serine protease cascade that initiates the ProPO-AS system is highly controlled by protease inhibitors to confine activity to the site of infection (Soderhall & Cerenius, 1998). There is evidence to suggest interaction between the insect haemolymph-clotting mechanism and the phenoloxidase system, with many of the same proteins being involved in both pathways (Li et al., 2002). This study highlights the complexity and high level of interconnection between the different responses of the insect immune response.

PRRs activate the compliment system in vertebrates and the proPO-AS system in insects. Many PRRs have been identified in insects; the peptidoglycan recognition protein (Yoshida et al., 1986; Mellroth et al., 2005), C-type lectins (Watanabe et al., 2006; Tanji et al., 2006; Ourth et al., 2005) and the β-1,3 glucan-binding protein known to be involved in activation of the ProPO-AS pathway (Wang & Jiang, 2006; Lee et al., 2003). Mammalian TLRs are considered PRRs, as they recognise LPS and other microbial products. Many Toll-related genes exist in the *Drosophila* genome (Toll3–Toll8, 18-wheeler) (Leclerc & Reichhart, 2004). The involvement of Toll in the activation of the ProPO-AS would imply the dual production of AMPs and production of melanin. Metros et al. (2002) linked Toll activation and melanisation by revealing that Toll activation was dependent upon the removal of the *Drosophila* serine protease inhibitor Serpin27A. They also reported that microbial challenge induced removal of the serine protease inhibitor Serpin27A via Toll-dependent transcription of an immune reaction component and therefore activating the phenoloxidase cascade and melanisation. Insect ProPO has sequence similarity to the thiol-ester region of vertebrate complement proteins C3 and C4 (Nair et al., 2005), indicating a possible relationship between these proteins and perhaps the ProPO-AS in insects and the complement system of vertebrates.

Antimicrobial Peptide Production

A vital feature of the insect immune response is the induction of AMP synthesis by the fat body and circulating haemocytes and secretion of these AMPs into the haemolymph to destroy invading microbes (Ratcliff, 1985). This aspect of the innate immune response is also important in vertebrates; however, its presence in insects is crucial as insects lack any adaptive response. The activation of the Toll and IMD pathways initiates the transcription of many AMPs, which can act by directly damaging microbial cell surfaces. The insect immune response can discriminate between pathogens and is able to mount the appropriate responses and produce the most suitable AMPs when it encounters different micro-organisms. This is partly performed by the initiation of specific pathways by different microbes; Gram-positive bacteria and fungi activate the Toll pathway and the appropriate AMPs, and Gram-negative bacteria activate the IMD pathway and also the AMPs associated with activation of this pathway, which are appropriate for destroying these specific micro-organisms. Lemaitre et al. (1996) and Rutschmann et al. (2002) found that *Drosophila* deficient in the Toll pathway are susceptible to fungal and Gram-positive infections, while Lemaitre et al. (1995) found that *Drosophila* deficient in the IMD pathway were susceptible to Gram-negative infections.

There have been many classes of AMPs identified in insects with similarities to vertebrate AMPs (Vilmos et al., 1998; Salzet, 2001). In *Drosophila* there are seven classes of AMPs, which are divided into three groups depending on their microbial targets. Drosomycins and metchnikowins are active against fungi, defensins against Gram-positive bacteria, and attacins, cecropins, diptericins, and drosocins against Gram-negative bacteria. All of the AMPs produced by insects are ampiphatic molecules that act in a detergent-like manner, forming pores in microbial membranes. Attacins have a very small spectrum of antibacterial and antifungal activity, and it is believed that the primary function of attacins is to facilitate the action of lysozyme and cecropins (Engstrom et al., 1984). Proline-rich peptides, glycin-rich peptides, and diptericin are small AMPs (Hoffmann, 1995), which peptides appear to function by increasing membrane permeability and cause lysis of Gram-negative bacteria. Lysozyme is a cationic protein with a broad spectrum of activity against Gram-positive bacteria. Lysozyme acts by hydrolysing cell wall components and it found in the granules of neutrophils, monocytes, macrophages, blood plasma, tears, saliva, and airway secretions. Lysozyme effectively hydrolyses β-(1,4) glycosidic bonds in peptidoglucan of bacterial cell walls (Suzuki and Rode, 1969; Thorne et al., 1976). LPS-binding proteins facilitate bacterial clearance in insects by promoting nodule formation. LPS-binding proteins are known to activate the vertebrate TLR pathway and may act on similar pathways in insects (see Kavanagh & Reeves, 2004). Transferrin has an iron-binding domain and may function by sequestering iron from pathogens thus inhibiting their growth (Lowenberger, 2001). Human lactoferrin is a member of the transferring family and displays antibacterial activities by limiting the availability of environmental iron (Bullen, 1981). Defensins are cysteine-rich cationic peptides with antimicrobial activities in insects and are active against a variety of bacteria, fungi and some viruses (Lehrer et al., 1989, 1993; Ganz et al., 1985). Defensins represent an early defence mechanism, with production occurring within 3 h of infection. These AMPs lyse bacterial cells by forming voltage-dependent ion channels in the cytoplasmic membrane of the bacteria, leading to leakage of potassium and other ions (Hoffmann, 1995). Cociancich et al. (1993) reported that an insect defensin disrupted the permeability barrier of the cytoplasmic membrane of *Micrococcus luteus*, resulting in a loss of cytoplamsic potassium, a partial depolarisation of the inner membrane, a decrease in cytoplasmic ATP and an inhibition of respiration. Additionally, Johns et al. (2001) isolated a small inducible cationic peptide with 83% similarity to scorpion defensin, from the American dog tick *Dermacentor variabilis*. Interestingly, the active peptide constituted 0.1% of the total protein in the haemolymph plasma and was present as early as 1 h post-*Bacillus subtilis* and *B. burgdorferi* infection. This gives an idea of the promptness of the appearance of AMPs in an infection and suggests their importance in initial defence. The fact that defensins are produced by a wide variety of organisms (Broekaert et al., 1995; Hoffmann & Reichhart, 1997), suggests their early evolutionary origin (Ganz & Lehrer, 1995). In vertebrates, AMPs are stored in neutrophil granules where they are released upon cell activation into the extracellular matrix. There is evidence of a similar phenomenon occurring in insects, where lysozyme is stored within intracellular vesicles and released in a degranulation process upon haemocyte activation (Munoz et al., 2002).

In addition to AMP, there have been cytokine-like molecules identified in various invertebrate haemocytes such as the echinoderm *Asterias forbesi* (Beck et al., 1986), tunicates (Beck et al., 1989), molluscan haemocytes (Ottaviani et al., 1993), and haemocytes of *Calliphora vomitoria* (Franchini et al., 1996). The only cytokines identified in invertebrates so far are IL-1, IL-2, IL-6, and TNF. These are all pro-inflammatory cytokines and have remained highly conserved throughout evolution. Vertebrate IL-1 is produced by circulating monocytes and macrophages and has many functions. IL-1 is involved in cell growth regulation and differentiation in bone marrow and the thymus. IL-1 is also involved in the stimulation of cytokine secretion, secretion of components of the complement system, acute phase proteins and prostaglandin 2 (PGE_2). TNF is also mainly produced by circulating monocytes and macrophages, and is also a multifunctional cytokine. TNF, besides its tumour-suppressing capabilities, regulates inflammatory reactions by interacting with IL-1 and IL-6. Other activities of TNF are its ability to increase the phagocytosis and cytotoxicity activities of monocytes, macrophages, and granulocytes. TNF also causes the expression of adhesion molecules on endothelial cells and granulocytes, the growth and differentiation of B cells, the production and secretion of cytokines and the expression of MHC-molecules in fibroblasts and endothelial cells. Both of these molecules, IL-1, and TNF, have been identified in *G. mellonella* plasmatocytes and granulocytes and were observed to be released upon bacterial stimulation (Wittwer et al., 1999). As IL-1 is produced by the TLR/IL-1R pathway in vertebrate lymphocytes and TNF is produced by activation of the TNF activation pathway, and considering the fact that homologues of both these pathways exist in invertebrates, namely the Toll pathway and IMD pathway (Lamaitre et al., 1995; De Gregario et al., 2002; Medzhitov & Janeway, 1998), it is tempting to suggest that the IL-1-like molecule and TNF-like molecule are produced through activation of the invertebrate Toll pathway and the IMD pathway, respectively. The invertebrate cytokines have functional similarity to vertebrate cytokines, in that they are able to modulate natural-killer cell activity of molluscan haemocytes (Franceschi et al., 1991) and to induce the release of biogenic amines and to affect cell migration, phagocytosis, and induction of nitric oxide synthase (Ottaviani et al., 1997).

Conclusion

The insect immune response bears many striking structural and functional similarities to the innate immune response of mammals. This fact can, and has been, exploited by employing insects as models for studying microbial pathogens of mammals and in the evaluation of the antimicrobial potential of novel drugs. Results obtained using mammals can be extrapolated to those which might be obtained using mammals thus reducing the need to use as many mammals in such testing. While the efficacy of using insects in screening microbial mutants and testing drugs is now well established, another aspect of insect immunology deserves increased attention. The insect immune system is capable of dealing rapidly and effectively with microbial loads that would prove fatal to mammals which, in addition to having the innate immune response also have the of advantage of possessing an adaptive response. Consequently study of cellular and humoral elements of the insect immune response may provide us with insights into the functioning of our own innate immune system and offer increased understanding of how it functions without the need to take account of the adaptive immune system. This has the potential to inform us of new means to increase the efficacy of the innate immune system and help engineer it to better deal with microbial infections.

References

Adema, C. M., Van, Deutekom-Mulder, E. C., van der Knaap, W. P., & Sminia, T. (1994, November). *Parasitology*, 109(Pt 4):479–485.

Alippi, A. M., Albo, G. N., Reynaldi, F. J., & De Giusti, M. R. (2005, Aug 10) *Vet. Microbiol.*, 109(1–2):47–55.

Baggiolini, M. & Wymann, M. P. (1990, Feb) *Trends Biochem. Sci.*, 15(2):69–72.

Beck, G, Vasta, G. R., Marchalonis, J. J., & Habicht, G. S. (1989). *Comp. Biochem. Physiol. B.*, 92(1):93–98.

Beck, G. & Habicht, G. S. (1986, Oct). *Proc. Natl. Acad. Sci. USA*, 83(19):7429–7433.

Belvin, M. P. & Anderson, K. V. (1996). *Annu. Rev. Cell Dev. Biol.*, 12:393–416.

Bergin, D, Brennan, M, & Kavanagh, K. (2003, Dec). *Microbes Infect.*, 5(15):1389–1395.

Bergin, D, Murphy, L, Keenan, J, Clynes, M., & Kavanagh, K. (2006, Jul). *Microbes Infect.*, 8(8):2105–2112.

Bergin, D, Reeves, E. P., Renwick, J, Wientjes, F. B., & Kavanagh, K. (2005). *Infect. Immun.*, 73:4161–4170.

Bouillaut, L, Ramarao, N, Buisson, C, Gilois, N, Gohar, M, Lereclus, D, & Nielsen-Leroux, C. (2005, Dec). *Appl. Environ. Microbiol.* 71(12):8903–8910.

Bowman, H. G. & Hultmark, D. (1987). *Ann. Rev. Microbiol.*, 41:103–126.

Brennan, M, Thomas, D. Y., Whiteway, M, & Kavanagh, K. (2002, Oct 11). *FEMS Immunol Med Microbiol.*, 34(2):153–157.

Broekaert, W. F. (1995). *Plant Physiol.*, 108:1353–1358.

Chen, C, Rowley, A. F., Newton, R. P., & Ratcliffe, N. A. (1999). *Comp. Biochem. Physiol. B. Biochem. Mol. Biol.*, 122(3):309–319.

Chernysh, S. I., Filatova, N. A., Chernysh, N. S., & Nesin, A. P. (2004, Sep). *J Insect Physiol.*, 50(9):777–781.

Cherry, S. & Silverman, N. (2006). *Nat. Immunol.*, 7:911–917.

Cho, J. H., Homma, K. I., Kanegasaki, S., & Natori, S. (1999). *Eur. J. Biochem.*, 266:878–885.

Choi, J. Y., Whitten, M. M., Cho, M. Y., Lee, K. Y., Kim, M. S., Ratcliffe, N. A., Lee, B. L. (2002, May). *Dev Comp Immunol.*, 26(4):335–343.

Clemons, K. V. & Stevens, D. A. (2005, May). *Med Mycol.*, 43(Suppl. 1):S101–110. Review.

Cociancich, S., Ghazi, A., Hetru, C., Hoffmann, J. A., & Letellier, L. (1993, Sep 15). *J. Biol. Chem.*, 268(26):19239–19245.

Costa, S. C., Ribeiro, C, Girard, P. A., Zumbihl, R, & Brehelin, M. (2005, Jan). *J Insect Physiol.*, 51(1):39–46.

Cotter, G, Doyle, S., & Kavanagh, K. (2000, Feb). *FEMS Immunol. Med. Microbiol.*, 27(2): 163–169.

De Gregorio, E., Spellman, P. T., Tzou, P., Rubin, G. M., & Lemaitre, B. (2002, Jun 3). *EMBO J.*, 21(11):2568–2579.

Dunphy, G. & Webster, J. (1984). *J. Insect. Physiol.*, 30:883–889

Dunphy, G., Morton, D., Kropinski, A., & Chadwick, J. (1986). *J. Invert. Pathol.*, 47:48–55.

Dunphy, G. B., Oberholzer, U., Whiteway, M., Zakarian, R. J., & Boomer, I. (2003). *Can J. Microbiol.*, 49:514–524.

Duvic, B. & Soderhall, K. (1992, Jul 1). *Eur. J. Biochem.*, 207(1):223–228.

Engstrom, P., Carlsson, A., Engstrom, A., Tao, Z., & Bennich, H. (1984). *EMBO J.*, 3:3347–3351.

Engstrom, Y. (1999, Jun–Jul). *Dev. Comp. Immunol.*, 23(4–5):345–358. Review.

Fernandez, N. Q., Grosshans, J., Goltz, J. S., & Stein, D. (2001, Aug). *Development*, 128(15):2963–2974.

Ferrandon, D., Imler, J. L., & Hoffmann, J. A. (2004, Feb). *Semin. Immunol.*, 16(1):43–53. Review.

Franchini, A., Miyan, J. A., & Ottaviani, E. (1996, Oct). *Tissue Cell*, 28(5):587–592.

Ganz, T., Selsted, M. E., Szklarek, D., Harwig, S. S., Daher, K., Bainton, D. P., & Lehrer, R. I. (1985). *J. Clin. Invest.*, 76:1427–1435.

Ganz, T. & Lehrer, R. I. (1995). *Pharmacol. Ther.*, 66:191–205.

Geisler, R., Bergmann, A., Hiromi, Y., Nusslein-Volhard, C. (1992, Nov). *Cell*, 13; 71(4):613–621.

Gerson, S. L., Talbot, G. H., Hurwitz, S, Strom, B. L., Lusk, E. J., & Cassileth, P. A. (1984, Mar). *Ann. Intern. Med.*, 100(3):345–351.

Glupov, V. V., Khvoshevskaya, M. F., Lozinskaya, Y. L., Dubovski, I. M., Martemyanov, V. V., & Sokolova, J. Y. (2001). *Cytobios*, 106(Suppl. 2):165–178.

Gorman, M. J. & Paskewitz, S. M. (2001, Mar 1). *Insect Biochem. Mol. Biol.*, 31(3):257–262.

Gottar, M., Gobert, V., Michel, T., Belvin, M., Duyk, G., Hoffmann, J. A., Ferrandon, D., & Royet, J. (2002, Apr 11). *Nature*, 4.1.6(6881):640–644.

Gotz, P., Matha, V., Vilcinskas, A. (1997, Nov). *J Insect Physiol.*, 43(12):1149–1159.

Hamamoto, H., Kurokawa, K., Kaito, C., Kamura, K., Razanajatovo, I. M., Kusuhara, H., Santa, T., & Sekimizu, K. (2004). *Antimicrob. Agents Chemother.*, 48:774–779.

Hendrickson, E., Plotnikova, J., Mahajan-Miklos, S., Rahme, L., & Ausbel, F. (2001). *J. Bacteriol.* 183:7126–7134.

Hedengren, M., Asling, B., Dushay, M. S., Ando, I., Ekengren, S., Wihlborg, M., & Hultmark, D. (1999, Nov). *Mol. Cell.*, 4(5):827–837.

Hoffmann, J. (1995). *Curr. Opin. Immunol.*, 7:4–10.

Hoffmann, J. A. & Reichhart, J. M. (1997). *Trends Cell Biol.*, 7:309–316.

Horng, T. & Medzhitov, R. (2001, Oct 23). *Proc. Natl. Acad. Sci. USA*, 98(22):12654–12658.

Hu, W. H., Johnson, H., & Shu, H. B. (2000, Apr 14). *J. Biol. Chem.*, 275(15):10838–10844.

Hultmark, D., Steiner, H., Rasmuson, T., & Boman, H. G. (1980, May). *Eur. J. Biochem.*, 106(1):7–16.

Igaki, T., Kanda, H., Yamamoto-Goto, Y., Kanuka, H., Kuranaga, E., Aigaki, T., & Miura, M. (2002, Jun 17). *EMBO J.*, 21(12):3009–3018.

Imamura, M., Yang, J., & Yamakawa, M. (2002, Jun). *Insect Mol. Biol.*, 11(3):257–265

Imler, J. L. & Hoffmann, J. A. (2000). *Rev Immunogenet.*, 2(3):294–304. Review.

Ito, Y., Nakamura, M., Hotani, T., & Imoto, T. (1995, Sep). *J. Biochem. (Tokyo)*, 118(3):546–551.

Jander, G., Rahme, L. G., & Ausubel, F. M. (2000, Jul). *J. Bacteriol.*, 182(13):3843–3845.

Johns, R., Sonenshine, D. E., & Hynes, W. L. (2001, Jul 26). *Insect Biochem. Mol. Biol.*, 31(9):857–865.

Kanda, H., Igaki, T., Kanuka, H., Yagi, T., & Miura, M. (2002, Aug 9). *J. Biol. Chem.*, 277(32):28372–28375.

Kanost, M. R., Jiang, J., & Yu, X. O. (2004, Apr). *Immunol. Rev.*, 198:97–105. Review.

Kauppila, S., Maaty, W. S., Chen, P., Tomar, R. S., Eby, M. T., Chapo, J., Chew, S., Rathore, N., Zachariah, S., Sinha, S. K., Abrams, J. M., Chaudhary, P. M. (2003, Jul 31). *Oncogene*, 22(31):4860–4867.

Kavanagh, K. & Reeves, E. P. (2004, Feb). *FEMS Microbiol. Rev.*, 28(1):101–112. Review.

Khush, R. S., Leulier, F., & Lemaitre, B. (2001, May). *Trends Immunol.*, 22(5):260–264.

Kim, S. Y., Ryu, J. H., Han, S. J., Choi, K. H., Nam, K. B., Jang, I. H., Lemaitre, B., Brey, P. T., & Lee, W. J. (2000, Oct 20). *J. Biol. Chem.*, 275(42):32721–32727.

Koizumi, N., Imamura, M., Kadotani, T., Yaoi, K., Iwahana, H., & Sato, R. (1999, Jan 25). *FEBS Lett.*, 443(2):139–143.

Latge, J. P. (1999). *Clin. Micro. Rev.*, 12:310–350.

Lau, G. W., Goumnerov, B. C., Walendziewicz, C. L., Hewitson, J., Xiao, W., Mahajan-Miklos, S., Tompkins, R. G., Perkins, L. A., & Rahme, L. G. (2003, Jul). *Infect Immun.*, 71(7):4059–4066

Leclerc, V. & Reichhart, J. M. (2004, Apr). *Immunol Rev.*, 198:59–71.

Lee, J. H., Cho, K. S., Lee, J., Yoo, J., Lee, J., & Chung, J. (2001, Jun, 27). *Gene*, 271(2): 233–238.

Lee, K. Y., Horodyski, F. M., Valaitis, A. P., & Denlinger, D. L. (2002, Nov). *Insect Biochem. Mol. Biol.*, 32(11):1457–1467.

Lee, M. H., Osaki, T., Lee, J. Y., Baek, M. J., Zhang, R., Park, J. W., Kawabata, S., Soderhall, K., & Lee, B. L. (2004, Jan 30). *J. Biol. Chem.*, 279(5):3218–3227.

Lehrer, R. I., Lichtenstein, A. K., & Ganz, T. (1993). *Annu. Rev. Immunol.*, 11:105–128.

Lemaitre, B., Kromer-Metzger, E., Michaut, L., Nicolas, E., Meister, M., Georgel, P., Reichhart, J. M., & Hoffmann, J. A. (1995, Oct 10). *Proc Natl Acad Sci USA*, 92(21):9465–9469.

Lemaitre, B., Nicolas, E., Michaut, L., Reichhart, J. M., & Hoffmann, J. A. (1996, Sep 20). *Cell*, 86(6):973–983.

Leulier, F., Rodriguez, A., Khush, R. S., Abrams, J. M., & Lemaitre, B. (2000, Oct). *EMBO Rep.*, 1(4):353–358.

Leulier, F., Vidal, S., Saigo, K., Ueda, R., & Lemaitre, B. (2002, Jun 25). *Curr. Biol.*, 12(12):996–1000.

Levy, J. A. (2001). *Trends Immunol.*, 22:312–316.

Li, D., Scherfer, C., Korayem, A. M., Zhao, Z., Schmidt, O., & Theopold, U. (2002, Aug). *Insect Biochem. Mol. Biol.*, 32(8):919–928.

Lin, X., Cerenius, L., Lee, B. L., Soderhall, K. (2006, Jul 25). *Biochem. Biophys. Acta.* [Epub ahead of print.].

Lionakis, M. S., Lewis, R. E., May, G. S., Wiederhold, N. P., Albert, N. D., Halder, G., & Kontoyiannis, D. P. (2005, Apr 1). *J. Infect. Dis.*, 191(7):1188–1195.

Lionakis, M. S. & Kontoyiannis, D. P. (2005, May). *Med. Mycol.*, 43(Suppl. 1):S111–114. Review.

Lockey, T. D. & Ourth, D. D. (1996, Mar 27). *Biochem. Biophys. Res. Commun.*, 220(3):502–508.

London, R., Orozco, B. S., & Mylonakis, E. (2006, Jun). *FEMS Yeast Res.*, 6(4):567–573.

Lowenberger, C. (2001). *Insect Biochem. Mol. Biol.*, 31:219–229.

Ma, C. & Kanost, M. R. (2000, March 17). *J. Biol. Chem.*, 275(11):7505–7514.

Mahar, A. N., Munir, M., Elawad, S., Gowen, S. R., Hague, N. G. (2005, Jun). *J. Zhejiang. Univ. Sci. B.*, 6(6):457–463.

Medzhitov, R. & Janeway, C. A. Jr. (1998, Oct). *Semin. Immunol.*, 10(5):351–353. Review.

Meister, M., Braun, A., Kappler, C., Reichhart, J. M., Hoffmann, J. A. (1994, Dec 15). *EMBO J.* 13(24):5958–5966.

Mellroth, P., Karlsson, J., Hakansson, J., Schultz, N., Goldman, W. E., Steiner, H. (2005, May 3). *Proc. Natl. Acad. Sci. USA*, 102(18):6455–6460.

Meyer, J. D. & Atkinson, K. (1983, Oct). *Clin. Haematol.*, 12(3):791–811.

Miyata, S., Casey, M., Frank, D. W., Ausubel, F. M., & Drenkard, E. (2003, May). *Infect. Immun.*, 71(5):2404–2413.

Morton, D. B., Dunphy, G. B., & Chadwick, J. S. (1987, Winter). *Dev. Comp. Immunol.*, 11(1):47–55.

Munoz, M., Vandenbulcke, F., Saulnier, D., & Bachere, E. (2002, Jun). *Eur. J. Biochem.*, 269(11):2678–2689.

Mylonakis, E., Moreno, R., El Khoury, J. B., Idnurm, A., Heitman, J., Calderwood, S. B., Ausubel, F. M., & Diener, A. (2005, Jul). *Infect. Immun.*, 73(7):3842–3850.

Naitza, S., Rosse, C., Kappler, C., Georgel, P., Belvin, M., Gubb, D., Camonis, J., Hoffmann, J. A., & Reichhart, J. M. (2002, Nov). *Immunity*, 17(5):575–581.

Nappi, A. J. & Vass E. (1998, Dec). *J. Parasitol.*, 84(6):1150–1157.

Nicolas, E., Reichhart, J. M., Hoffmann, J. A., & Lemaitre, B. (1998, Apr 24). *J. Biol. Chem.*, 273(17):10463–10469.

Ochiai, M. & Ashida, M. (2000, Feb 18). *J. Biol. Chem.*, 275(7):4995–5002.

Ottaviani, E., Franchini, A., & Franceschi, C. (1993, Sep 15). *Biochem Biophys Res Commun.*, 195(2):984–988.

Ottaviani, E., Franchini, A., & Franceschi, C. (1997). *Int. Rev. Cytol.*, 170:79–141. Review.
Ourth, D. D., Narra, M. B., & Chung, K. T. (2005, Oct 7). *Biochem. Biophys. Res. Commun.*, 335(4):1085–1089.
Pandey, S. & Agrawal, D. K. (2006, Aug). *Immunol. Cell. Biol.*, 84(4):333–341.
Price, C. D. & Ratcliff, N. A. (1974). *Z. Zellforsch. Mikrosk. Anat.*, 147:537–549.
Ratcliff, N. (1985). *Immunol. Lett.*, 10:253–270.
Reeves, E. P., Messina, C. G., Doyle, S., & Kavanagh, K. (2004, Jul). *Mycopathologia*, 158(1):73–79.
Regel, R., Matioli, S. R., & Terra, W. R. (1998, May–Jun). *Insect Biochem. Mol. Biol.*, 28(5–6): 309–319.
Renwick, J., Reeves, E. P., Frans B. W., & Kavanagh, K. (2007). *Dev. Comp. Immunol.*, 31: 347–359.
Romani, L. (1999). *Curr. Opin. Microbiol.*, 2:363–367.
Rossignol, P. A. & Lueders, A. M. (1986). *Comp. Biochem. Physiol. B.*, 83(4):819–822.
Rutschmann, S., Kilinc, A., & Ferrandon, D. (2002, Feb 15). *J. Immunol.*, 168(4):1542–1546.
Salamitou, S., Ramisse, F., Brehelin, M., Bourguet, D., Gilois, N., Gominet, M., Hernandez, E., & Lereclus, D. (2000 Nov). *Microbiology*, 146(Pt 11):2825–2832
Salzet, M. (2001, Jun). *Trends Immunol.*, 22(6):285–288. Review.
Samakovlis, C., Kimbrell, D. A., Kylsten, P., Engstrom, A., & Hultmark, D. (1990, Sep). *EMBO J.*, 9(9):2969–2976
Scully, L. R. & Bidochka, M. J. (2006). *FEMS Microbiol Letts.*, 263:1–9.
Segal, A. W. (1996, Mar). *Mol. Med. Today*, 2(3):129–135. Review.
Silva, C. P., Waterfield, N. R., Daborn, P. J., Dean, P., Chilver, T., Au, C. P., Sharma, S., Potter, U., Reynolds, S. E., & French-Constant, R. H. (2002, Jun). *Cell Microbiol.*, 4(6):329–339.
Silverman, N. & Maniatis, T. (2001, Sep 15). *Genes Dev.*, 15(18):2321–2342. Review
Slepneva, I. A., Glupov, V. V., Sergeeva, S. V., & Khramtsov, V. V. (1999, Oct 14). *Biochem Biophys Res Commun.*, 264(1):212–215.
Smith, V. J., & Soderhall, K. (1983). *Cell Tissue Res.*, 233(2):295–303.
Soderhall, K., & Cerenius, L. (1998, Feb). *Curr. Opin. Immunol.*, 10(1):23–28.
Spies, A. G., Karlinsey, J. E., & Spence, K. D. (1986). *Comp. Biochem. Physiol. B.*, 83(1):125–133.
Sritunyalucksana, K. & Soderhall, K. (2000). *Aquaculture*, 191:53–59.
Stoven, S., Ando, I., Kadalayil, L., Engstrom, Y., & Hultmark, D. (2000, Oct). *EMBO Rep.*, 1(4): 347–352.
Sun, H., Benjamin, N., Bristow, Qu. G., & Wasserman. S. A. (2002, Oct 1). *PNAS*, 99: 12871–12876.
Suzuki, Y. & Rode, L. J. (1969). *J. Bacteriol.*, 98:238–245.
Tanjim T., Ohashi-Kobayashi, A., & Natori, S. (2006, May 15). *Biochem. J.*, 396(1):127–138.
Tauszig-Delamasure, S., Bilak, H., Capovilla, M., Hoffmann, J. A., & Imler, J. L. (2002, Jan). *Nat. Immunol.*, 3(1):91–97.
Theopold, U., Li, D., Fabbri, M., Scherfer, C., & Schmidt, O. (2002, Feb). *Cell. Mol. Life Sci.*, 59(2):363–372. Review.
Thompson G. J., Crozier Y. C., & Crozier R. H. (2003, Feb). *Insect Mol. Biol.*, 12(1):1–7.
Thorne K. J., Oliver R. C., & Barrett A. J. (1976). *Infect. Immun.*, 14:555–563.
Tickoo S., & Russell S. (2002, Oct) *Curr. Opin. Pharmacol.*, 2(5):555–560.
Tojo S., Naganuma F., Arakawa K., & Yokoo S. (2000). *J. Insect Physiol.*, 46, 1129–1135.
Towb, P., Bergmann, A., & Wasserman, S. A. (2001, Dec). *Development*, 128(23):4729–4736.
Vidal, S., Khush, R. S., Leulier, F., Tzou, P., Nakamura, M., & Lemaitre, B. (2001, Aug 1). *Genes Dev.*, 15(15):1900–1912.
Vilmos, P. & Kurucz, E. (1998, Jun). *Immunol. Lett.*, 62(2):59–66.
Walters, J. B. & Ratcliffe, N. A. (1983). *J. Insect Physiol.*, 29:417–424.
Wang, Y. & Jiang, H. (2006, Apr 7). *J Biol Chem.*, 281(14):9271–9278. Epub 2006 Feb 6.
Watanabe, A., Miyazawa, S., Kitami, M., Tabunoki, H., Ueda, K., & Sato, R. (2006, Oct 1). *J. Immunol.*, 177(7):4594–4604.

Weber, A. N., Tauszig-Delamasure, S., Hoffmann, J. A., Lelievre, E., Gascan, H., Ray, K. P., Morse, M. A., Imler, J. L., & Gay, N. J. (2003, Aug). *Nat. Immunol.*, 4(8):794–800. Epub 2003 Jul 20.

Werner, T., Liu, G., Kang, D., Ekengren, S., Steiner, H., & Hultmark, D. (2000, Dec 5). *Proc. Natl. Acad. Sci. USA*, 97(25):13772–13777.

Wigglesworth, V. B. (1972). *Principles of Insect Physiology*, London, Chapman & Hall.

Wilson, R., Chen, C., & Ratcliffe, N. A. (1999, Feb 1). *J. Immunol.*, 162(3):1590–1596.

Wilson, R. & Ratcliffe, N. A. (2000, May 1). *J. Insect. Physiol.*; 46(5):663–670.

Wittwer, D., Franchini, A., Ottaviani, E., & Wiesner, A. (1999, Sep) *Cytokine*, 11(9):637–642.

Yoshida, H., Ochiai, M., & Ashida, M. (1986, Dec 30). *Biochem. Biophys. Res. Commun.*, 141(3):1177–1184.

Yoshida, L. S., Abe, S., & Tsunawaki, S. (2000, Feb 24). *Biochem. Biophys. Res. Commun.*, 268(3):716–723.

Yu, K. H., Kim, K. N., Lee, J. H., Lee, H. S., Kim, S. H., Cho, K. Y., Nam, M. H., & Lee, I. H. (2002, Oct). *Dev. Comp. Immunol.*, 26(8):707–713.

Chapter 4
Novel Antifungal Therapies

Khaled H. Abu-Elteen and Mawieh M. Hamad

Introduction

Advances in cancer medicine, transplantation biology, management of AIDS patients, and diabetics are responsible for the alarming expansion rates in the number of immunocompromised patients susceptible to life-threatening fungal infections. In response to these concerns, researchers have successfully labored at modifying some of the existing antifungals (polyenes and azoles) and introducing novel therapies (peptides, oligonucleotides, and monoclonal antibodies (MAbs)) that have collectively expanded the spectrum of activity and minimized associated side effects. Nonetheless, the classical approach of dealing with fungal infections by targeting the pathogen remains prone to failure over time owing to the extensive genetic flexibility of fungi to evade or resist antifungal therapeutics. Conditioning or modulating the immune system of the host may help circumvent the resistance problem. Vaccines, cytokines, and adoptive T cell transfer are the backbone of this approach.

Antifungal Therapies Targeting the Host

The degree to which a specific immune component participates in responding to fungal pathogens varies depending on the species of the pathogen, its morphotype (yeast, pseudohyphae or hyphae), the anatomical site, and form of the infection (Shoham & Levitz, 2005; Matthews & Burnie, 2004) (Figure 4.1). Phagocytic cells present within the target organ and those recruited by chemotactic mediators participate in reducing fungal burden by killing significant percentages of fungal cells. Intracellular and extracellular killing of fungi is mediated by reactive oxygen species and antimicrobial peptides. The importance of neutrophils is evidenced by increased susceptibility of neutropenic patients and experimental animals to candidiasis. Along with their capacity to express T cell co-stimulatory molecules, dendritic cells (DCs) capture and process fungal antigens making them potent fungal antigen presenting cells. Hence, DCs can initiate a cascade of innate and acquired

K. Kavanagh (ed.), *New Insights in Medical Mycology.*
© Springer 2007

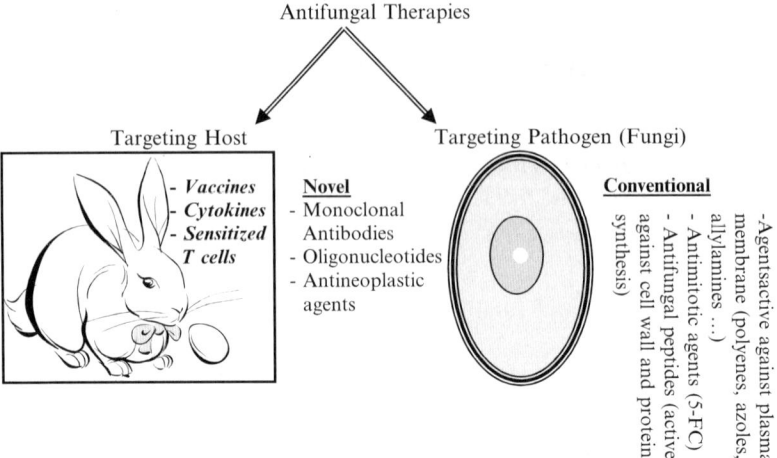

Figure 4.1 General summary of antifungal therapies

immune responses against fungi. Defects in cell-mediated immunity (CMI) are strongly associated with chronic mucocutaneous candidiasis. Differentiation of Th cells into either Th1 or Th2 type often determines susceptibility or resistance to several fungal infections. Differentiation of Th cells into Th1 is a prerequisite for protection against vaginal candidiasis. The production of protective monoclonal antibodies against *Candida albicans* and *Cryptococcus neoformans* infections is well established. Additionally, systemic candidiasis often results in the production of protective polyclonal antisera.

Fungal vaccines

Vaccination strategies against fungal infections involving live or attenuated fungi, cell wall, or cytoplasmic extracts and the transfer of passive or adoptive immunity had all been tried since the 1930s with little success. The coccidioidomycosis vaccine clinical trial conducted in the early 1980s by the Valley Fever Vaccine Study Group, did not succeed in producing distinct differences between immunized and placebo subjects (Pappagianis, 1993). The discovery of protective MAbs against *C. albicans* (Han & Cutler, 1995; Matthews et al., 1995) and *C. neoformans* (Dromer et al., 1987; Sanford et al., 1990; Mukherjee et al., 1992) has intensified the focus on antifungal humoral immunity. Although the list of potential fungal immunogens is expanding (Deepe, 1997), thus far, not a single fungal vaccine has been approved for clinical use.

MAbs to the capsular polysaccharide of *C. neoformans* were shown to prolong survival and decrease fungal burden in mice (Mukherjee et al., 1994, 1995). Peptides derived from the ribotoxin of *A. fumigatus* modulated immunity in mice consistent with a Th1-type response (Svirshchevskaya et al., 2000). Active immunization with a tetanus toxoid conjugate of glucuronoxylomannan resulted in the production of protective Abs in animal models of cryptococcosis (Murphy, 1992). Dipheria toxoid CRM197 conjugated with laminarin (Lam), a poorly immunogenic β-glucan from *Laminaria digitata*, was shown to be protective against various forms of systemic and mucosal *C. albicans* infections in mice (Torosantucci et al., 2005). Recovery from disseminated *C. albicans* infection in mice has been correlated with the production of protective Abs against the molecular chaperone heat-shock protein 90 (hsp90). Abs against pathogen-specific hsp90 epitopes and those against epitopes common to both pathogen and host hsp90 are frequently reported in patients recovering from systemic candidiasis (Matthews, 1992, 1994). Vaccination with the recombinant N-terminal domain of Als1p (rAls1p-N) was protective against disseminated candidiasis in Balb/c mice and in several other outbred mouse strains (Ibrahim et al., 2006).

Various DNA-based vaccination approaches have been introduced as alternatives to conventional vaccination practice. Administration of a recombinant strain of *B. dermatiditis* lacking the WI-1 adhesin gene protected against pathogenic strains of the fungus via the induction of Th1-type responses (Wüthrich et al., 2000). A hsp90-expressing DNA vaccine tested on a Balb/c mouse model of systemic candidiasis was shown to be protective. The vaccine elicited a protective anti-candida-hsp90 IgG type Ab response. DCs pulsed with fungi, fungal extracts or fungal RNA could be used to develop fungal vaccines (Bozza et al., 2003). Fungal RNA has the capacity to activate DCs via nucleotide receptor signaling cascades evidenced by the upregulation of several Toll-like receptors (TLR) on DCs (Bozza et al., 2004). Th1-mediated protection has been reported in mice challenged with a mixture of DCs pulsed with *A. fumigatus* extract and CpG-ODN adjuvant (Bozza et al., 2002). *A. fumigatus* RNA-transfected DCs induced resistance to subsequent challenges with the pathogen via Th1-derived IFN-γ (Bozza et al., 2004). Effective immunization is understandably the preferred means of protection as it sensitizes the immune system and enhances sterilization of infected organs thus preventing dormant infections granted that the subject is immunocompetent. Groups at risk of developing serious fungal infections like AIDS patients, diabetics, cancer patients on chemotherapy, and transplantees on immunosuppressants are immunocompromised. Therefore, the logic of developing fungal vaccines would seem dubious if the intended target population can not benefit from it.

T cells and Cytokines

The role of CD4⁺ T cells as major players against fungal infections is well established (Lindell et al., 2005; Harmsen & Stankiewicz, 1990; Phair et al., 1990). Depletion of CD4⁺ T cells results in increased susceptibility to *P. pneumonia* and other mycotic infections both in mice (Shellito et al., 1990) and rats (Thullen et al., 2003). Differentiation of naive Th into Th1 or Th2 type represents a major determinant of resistance or susceptibility of the host to fungal infections (Shoham & Levitz, 2005). Tipping the balance towards Th1-type responses by the orchestrated effort of IFN-γ, IL-6, and IL-12 in the relative absence of IL-4 and IL-10, confers significant protection against different forms of mycoses (Romani, 2002; Hebart et al., 2002; Cenci et al., 1998). A positive lymphoproliferative response characterized by overproduction of IFN-γ in cultured cells was noted in healthy individuals receiving cellular extracts of *A. fumigatus*, or its 88 kDa dipeptidase or the 90 kDa catalase antigens (Hebart et al., 2002). Adoptive transfer of CD4[+] splenocytes from mice sensitized against *A. fumigatus* into naïve mice resulted in prolonged survival following intravenous challenge with viable *A. fumigatus* conidia (Cenci et al., 2000). The combination of anti-CD40 and IL-2 significantly prolonged the survival of mice infected with *C. neoformans* (Zhou et al., 2006). This protection was correlated with increased serum concentration of IFN-γ and tumor necrosis factor alpha (TNF-α) and decreased fungal burden in kidney and brain tissues, which was found to be dependent on IFN-γ as evidenced by lack of protection in IFN-γ knockout mice.

The role of CD8[+] T cells in protection against infections (fungal and otherwise) is generally dependent upon CD4[+] T cell help. CD4 deficiency can impair CD8[+] activation or function in nonfungal infections (Ribedly et al., 2000; Wang & Livingstone, 2003). However, recent evidence suggest that generation of fungi-specific effector CD8[+] T cells capable of producing protective levels of IFN-γ against pulmonary *C. neoformans* infections is achievable in CD4[-] mice (Lindell et al., 2005). Neutralization of IFN-γ in CD4[-]CD8[+] mice increased macrophage infection by *C. neoformans*. Depletion of CD4[+] T cells in mice did not affect the increased serum concentration of IFN-γ induced by anti-CD40 antibody agonist/IL-2 combination treatment (Zhou et al., 2006).

Administration of recombinant human hematopoietic cytokines like granulcocyte colony stimulating factor (G-CSF), Monocyte-CSF (M-CSF), GM-CSF or IFN-γ were reported to shorten the duration of neutropenia (G-CSF and GM-CSF) and enhance the phagocytic and killing activity of neutrophils, monocytes and macrophages (G-CSF, M-CSF and GM-CSF) (Georgopapadakou and Walsh, 1996; Chiou et al., 2000). In cancer patients on corticosteroids, neutropenia predisposes to invasive fungal infections (Ribaud et al., 1999). However, the efficacy of cytokine adjunctive therapy varies depending upon the level/nature of immunosuppression and the antifungal agent used. In *A. fumigatus*-infected outbred ICR mice pretreated with hydrocortisone, administration of recombinant human G-CSF prior to infection antagonized the action of the azole SCH56592. In these mice, large lung abscesses with polymorphonuclear cells and significant lung burden were evident. Absence of G-CSF in similarly treated mice allowed for effective limitation of the infection characterized by reduced lung fungal burden and longer survival rate (Graybill et al., 1998a). Mice made neutropenic by 5-FC, G-CSF augmented the activity of the azole. *In vitro* testing of the killing and fungistatic activities of voriconazole versus fluoconazole against *C. albicans* in the presence of GM-CSF has shown that GM-CSF significantly enhanced the killing by PMNs and the collaboration between PMNs and either of the two azoles to kill *C. albicans* cells (Vora et al., 1998).

However, given that voriconazole is more potent (>tenfold) anti-*Candida* azole than fluoconazole, the synergistic effect of GM-CSF was more evident when combined with voriconazole. G-CSF was shown to reverse neutrophil dysfunction against *Aspergillus* hyphae in HIV-infected nonneutropenic patients (Roilides et al., 1993).

The proinflammatory activity of polyenes (AMB and NY) has been related to their capacity to induce cytokine secretion. AMB is recognized as a pathogen-associated molecular pattern (PAMP) by Toll-like receptor 2 (TLR-2), a pattern recognition receptor (PRR) (Razonable et al., 2005a; Sau et al., 2003). Binding of AMB to TLR-2 on phagocytes causes the release of cytokines only from TLR-2 expressing cells (Sau et al., 2003). Nystatin (NY) mediate its proinflammatory activity through a TLR-1- or TLR-2-induced release of IL-1β, IL-8, and TNF-α (Razonable et al., 2005b). These findings provide new avenues for recruiting cytokine-based modulation of antifungal immunity using drugs that can influence cytokine release.

Antifungal Therapies Targeting the Pathogen

Novel Therapies: Antifungal Monoclonal Antibodies

Gordon and Lapa (1964) established that the administration of serum Abs enhances the outcome of antifungal chemotherapy in cryptococcosis. Several fungal infections are marked by significant Ab responses, however, the extent to which they confer protection varies depending on Ab isotype (Casadevall, 1995) and the MHC background of the host (Rivera & Casadevall, 2005). Variations in Ab V_H gene usage were reported to significantly influence the specificity and efficacy of Abs, hence the resistance or susceptibility to cryptococcosis (Maitta et al., 2004). Protective Abs may not be made in quantities sufficient to alter the course of the infection; their action may also be countered by blocking or competing nonprotective Abs (Nussbaum et al., 1996).

Unlike vaccines, therapeutic MAbs provide immediate protection irrespective of the immunostatus of the host. Their inherent passive activity to counter fungal infections permits their use in treating immunocompromised hosts. Derivation of effective MAbs as alternative antifungals relies on the identification of immunodominant fungal antigens against which MAbs can be raised. The fungal cell wall represents an interface of host–pathogen interactions and hence known to contain a number of structural moieties of significance in fungal pathogenesis and virulence. MAb A9, an IgG1 raised against cell wall extracts of *A. fumigatus*, was shown to bind to a surface peptide on the hyphal and yeast forms of the fungus, inhibit hyphal development and reduce spore germination time. A9 protected against IA in mice by reducing fungal burden and enhancing survival rates (Chaturvedi et al., 2005). MAbs raised against the glucosylceramide (GlcCer) moiety *N*-2′-hydroxyhexadecanoyl-1-β-D-glucopyranosyl-9-methyl-4,8-sphingadenine of *Fonsecaea pedrosoi* reduced fungal growth and enhanced phagocytosis and killing of fungal cells by murine macrophages (Nimrichter et al., 2004). Growth of *Fusarium* was inhibited by the presence of a fusion protein consisting of recombinant chicken-derived single chain Abs specific to surface antigens and antifungal peptides (Peschen et al., 2004). Expression of the fusion protein in transgenic *Arabidopsis thaliana* plants was protective against *F. oxysporum* sp. G15, a human IgM MAb raised in xenomice transgenic for human IgM, IgG2, and light chain κ challenged with the *C. neoformans* serotype D strain 24067 could significantly prolong the survival of mice challenged with a lethal dose of the fungus (Maitta et al., 2004). The Ab recognizes an epitope on the capsular polysaccharide glucuronoxylomannan (GXM) of *C. neoformans*.

MAbs raised against a number of intracellular or expressed fungal cell components have also been developed and tested in animal models. Anti-gp70, a MAb raised against the 70 kDa intracellular/secreted glycoprotein component of *P. brasiliensis*, abolished lung granuloma formation in infected mice (De Mattos Grosso et al., 2003). *C. albicans* strains susceptible to the yeast killer toxin (KT) activity take cover by Abs generated in the host against KT (competing Abs) or its receptor (KTR) on *C. albican* surface (blocking Abs). Anti-idiotypic MAbs that neutralize anti-KT or anti-KTR Ab activity thus allowing KT activity against the fungus to ensue can circumvent the evasiveness of pathogenic *C. albicans*. An IgM anti-idiotypic MAb generated in mice immunized with the anti-KT MAb (KT4) was reported to have a potent killing activity against KT-sensitive *C. albicans* strains (Polonelli et al., 1997). A decapeptide containing the first three amino acids of the light chain CDR1 of a KT anti-idiotype optimized by single residue replacement by alanine exerted a strong anticandidacidal activity *in vitro*, hence the peptide was termed KP for killer peptide (Polonelli et al., 2003). In a rat model of vaginal candidiasis, post-challenge local administration of KP resulted in rapid clearance of infections caused by fluconazole-resistant or susceptible strains of *C. albicans*. Administration of KP into Balb/c or SCID mice infected with lethal intravenous doses of *C. albicans* significantly prolonged survival beyond 60 days compared with 3–5 days in control mice (Polonelli et al., 1997).

The molecular chaperon protein hsp90, essential for yeast cell viability, represents an immunodominant antigen that elicits significant antibody responses in animals and in humans. A human recombinant anti-hsp90 MAb (Mycograb) has been derived from the anti-hsp90 antibody cDNA of patients recently recovering from invasive candidiasis (IC) (Matthews & Burnie, 2001). Mycograb was protective against *C. tropicalis* but not *C. albicans*, *C. krusei*, *C. glabrata*, or *C. parapsilosis* infections in mice (Matthews et al., 2003). Combination therapy consisting of Mycograb and AMB resulted in complete resolution of *C. albicans*, *C. krusei*, and *C. glabrata* infections. In mice infected with *C. parapsilosis*, however, Mycograb–AMB combination therapy cleared the liver and spleen but not the kidneys (Matthews et al., 2003). MAb G5, an anti-*C. albicans* IgA, was shown to have potent candidacidal activity *in vitro* and potent prophylactic activity *in vivo* (Kacishwar & Shulka, 2006).

The safety and pharmacokinetics of MAb 18B7, a monoclonal against the *C. neoformans* capsular polysaccharide, was tested as an adjunctive therapy in HIV patients successfully treated for cryptococcal meningitis (Larsen et al., 2005). Despite some minor side effects in subjects receiving a single infusion of 1–2 mg/ kg body weight of the drug, serum antigen titers declined by a median of twofold at week 1 and threefold at week 2 post-infusion. The halflife of the drug in serum was 53 h and the CSF of recipients was free of 18B7. A multinational, double-blind, placebo-controlled clinical trial on the efficacy of Mycograb as an adjunctive therapy in patients receiving LAMB for treating culture-proven IC is now underway (Matthews & Burnie, 2004).

Antifungal Oligonucleotides

Targeting RNA with oligonucleotides is emerging as an important therapeutic strategy to treat certain forms of cancer and some infectious states (Disney et al., 2001; Testa et al., 1999). The antisense oligonucleotide Vitravene (Galderisi et al., 1999), has proven to be effective in treating cytomegalovirus retinitis in AIDS patients intolerant, unresponsive or have contraindications to other treatment(s) of the infection. Gentasense, anti-bcl-2 anti-sense messenger RNA construct, and other oligonucleotide preparations (Childs et al., 2002) are proven potent against acute myeloid leukemia. Features that influence oligonucleotide affinity to bind target RNA and its nuclease stability can be accommodated into the sequence design (Freier & Altman, 1997). Recent evidence suggest that *C. albicans* takes up significant quantities of oligonucleotides in an energy-dependent manner, taken up oligonucleotides remain stable for >12 h (Disney et al., 2003). A 19-mer with a 2′-*O-methyl* backbone (19-mer2′-OMe) hairpin oligonucleotide can inhibit the growth of *C. albicans* in culture at pH < 4.0 which is easily tolerated in anatomic sites subject to candidiasis (Disney et al., 2003). Take up of the oligonucleotide by COS-7 mammalian cells in culture was reported to be tenfold lower than that of *C. albicans*; oligonucleotide stability inside COS-7 cells was minimal. The oligonucleotide failed to inhibit the growth of COS-7 cells suggesting some sort of selectivity in oligonucleotide uptake and stability in favor of fungal cells over mammalian cells (Disney et al., 2003). *In vivo* dimethylsulfate modifications of rRNA and the decreased rate of protein synthesis suggest that the hairpin oligonucleotides alter ribosomal activity independent of base pairing with target RNA.

RNA targets like rRNA, RNase, P RNA, group I and group II introns and mRNAs with untranslated regulatory sequences require secondary or tertiary folding to function properly (Testa et al., 1999; Disney et al., 2001). Generation of potential oligonucleotides is now achievable by oligonucleotide directed misfolding of RNA (ODMiR), which uses short oligonucleotides to stabilize the inactive form of target RNA (Childs et al., 2002). The oligonucleotides ^{L}(TACCTTTC) and T^{L}CT^{L}AC^{L}GA^{L}CG^{L}GC^{L}C that target group I introns of *C. albicans* were generated by ODMiR and tested on *C. albicans* in culture. Both oligonucleotides induced misfolds in group I introns and inhibited 50% of its splicing in transcription mixtures 150 and 30 nM, respectively (Childs et al., 2002). Antisense repression of key genes in pathogenic fungi has been shown to alter the growth and reproduction of fungi (Gorlach et al., 2002). Serotype D yeast transformed with a plasmid containing the calcineurin A (*CNA1*) cDNA in an antisense orientation under the control of the inducible *GAL7* promoter demonstrated a temperature-sensitive phenotype only when grown on galactose, which was shown to be associated with decreased native *CNA1* transcript levels. Furthermore, it was possible to modestly impair the growth of *C. neoformans* at 37°C by a 30 bp antisense oligonucleotide targeting *CNA1* (Gorlach et al., 2002).

Unconventional Agents with Antifungal Activity

Many antineoplastic agents, natural plant products, and other drugs have been shown to exhibit wide spectrum antimycotic activity (Cardenas et al., 1999). Camptothecin, a topoisomerase I inhibitor, and some of its derivatives (Del Poeta et al., 1999), and eflornithine, an irreversible suicide inhibitor of ornithine decarboxylase (McCann and Pegg, 1992) exhibit significant activity against *C. neoformans*. Antifungal activity has been associated with immunosuppressants rapamycin, cyclosporin A, and FK506, the phosphatidylinositol 3-kinase inhibitor wortmannin, the angiogenesis inhibitors fumagillin and ovalicin, the hsp90 inhibitor geldanamycin and drugs that inhibit sphingolipid metabolism. The sodium channel blocker amiodarone, which treats arrhythmia was reported to have significant anti-*S. cerevisiae* activity (Courchesne, 2002; Courchesne & Oztruk, 2003). The drug was most potent against yeast mutants lacking key calcium transporter proteins (Cupta et al., 2003). The drug seems to mediate its cytotoxic activity by inducing gradual and sustained release of calcium from internal stores (Cupta et al., 2003). Although the antifungal activity was noticed only at toxic doses, combining a low dose of the drug with miconazole reduced the survival rate of *S. cerevisiae* to <10%. A low dose of the drug combined with fluoconazole resulted in close to 100% killing of *C. albicans* and *C. neoformans in vitro*.

Conventional Antifungal Therapies

Antifungal Peptides

Natural antifungal peptides are derived from bacteria (iturin, bacillomycin, syringiotoxins, cepacidines), fungi (echinocandins), insects (cecropins A and B, drosomycin, dermaseptin), plants (zeamatin, the cyclopeptide alkaloids amphibine H, franguflo-line, nummularine), and mammals (α- and β-defensins, HNP series, gallinacin) (De Lucca and Walsh, 2000). Antifungal peptides are either linear sequences or folded structures with hydrophobic or amphipathic properties. The recombinant defensin Tfgd1 synthesized from a cDNA cloned from *Trigonella foenum-graecum* exhibits broad spectrum antifungal activity (Olli and Kirti, 2006). The synthetic peptide D4E1 is active against *Aspergillus* with a 50% lethal dose of 2.1–16.8 μg/mL (De Lucca and Walsh, 2000). MICs of the synthetic peptide halocidin analog di-K19Hc against clinical isolates of *C. albicans* and *Aspergillus* sp. are below 4 and 16 8 μg/mL, respectively (Jang et al., 2006). It rapidly (<30 s) kills *C. albicans* by binding to cell wall β-(1,3)-glucan and forming membrane ion channels. The synthetic lactoferrin-derived peptide (Lfpep) and the kaliocin-1 peptide, which includes the sequences 18–40 and 153–183 of the human lactoferin protein, respectively, are fungicidal against fluconazole- and AMB-resistant *C. albicans* strains (Viejo-Diaz et al., 2005). Lfpep kills *C. albicans* by permealizing the plasma membrane evidenced by the ability of propidium iodide to permeate Lfpep-treated cells (Viejo-Diaz et al., 2005). Generally speaking, antifungal peptides kill or inhibit the growth of fungi by lysis, binding, and disruption of plasma membranes, pore formation and leakage, or by interacting with cytoplasmic or nuclear targets (De Lucca and Walsh, 2000).

Echinocandins

Echinocandins are fatty acid derivatives of cyclic hexapeptides and consist of a diverse family of lipopeptides including caspofungin, micafungin, and anidulafungin. Echinocandins are noncompetitive inhibitors of the $(1,3)$-β-D-glucan synthase (Morrison, 2006; Turner et al., 2006; Randhawa & Sharma, 2004). Fungistatic effects result from blockage of cell wall synthesis that reduces cell growth and fungicidal effects result from loss of cell wall integrity, loss of mechanical strength and inability to resist intracellular osmotic pressure. *Caspofungin acetate* (MK-991, L-743,872) is a semisynthetic water soluble lipopeptide produced from the fermentation products of the fungus *Glarea lozoyensis* as a derivative of pneumocandin B_0 (Figure 4.2). It has been approved to treat IA patients unresponsive or intolerant to conventional antifungal therapy in the USA (2001) and Europe (2002) (Letscher-Bru & Herbrecht, 2003). The fungicidal and fungistatic activities of caspofungin have been demonstrated in *Candida* spp. and *Aspergillus* spp., respectively (Espinel-Ingroff, 2003) (Table 4.1). MIC values range is 0.015–4 mg/L depending on the species (Randhawa & Sharma, 2004; Perea et al., 2002b). The drug selectively acts on the extremities of hyphae, which are the sites of cell wall synthesis essential for fungus apical growth (Kurtz & Douglas, 1997; Hossain et al., 2003).

The drug is active against cystic *P. carinii* in animals (Morrison, 2006). Caspofungin is active against *Paecilomyces variotii* (MIC ≥ 0.5 mg/L) and *Scedosporium apiospermum* but not against *Paecilomyces lilacinus* (MIC 3–100 mg/L) or *S. prolificans*, (Pfaller et al., 1998b). Furthermore, it is active against rare molds such as *Acremonium, Curvularia, Bipolaris, Trichoderma* and *Alternaria* (Kahn et al., 2006). The proposed susceptibility breakpoint for the drug against *Candida* spp. is at MIC ≤1 μg/mL. The drug is metabolized by peptide hydrolysis and *N*-acetylation to inactive metabolites in the liver without the involvement of the

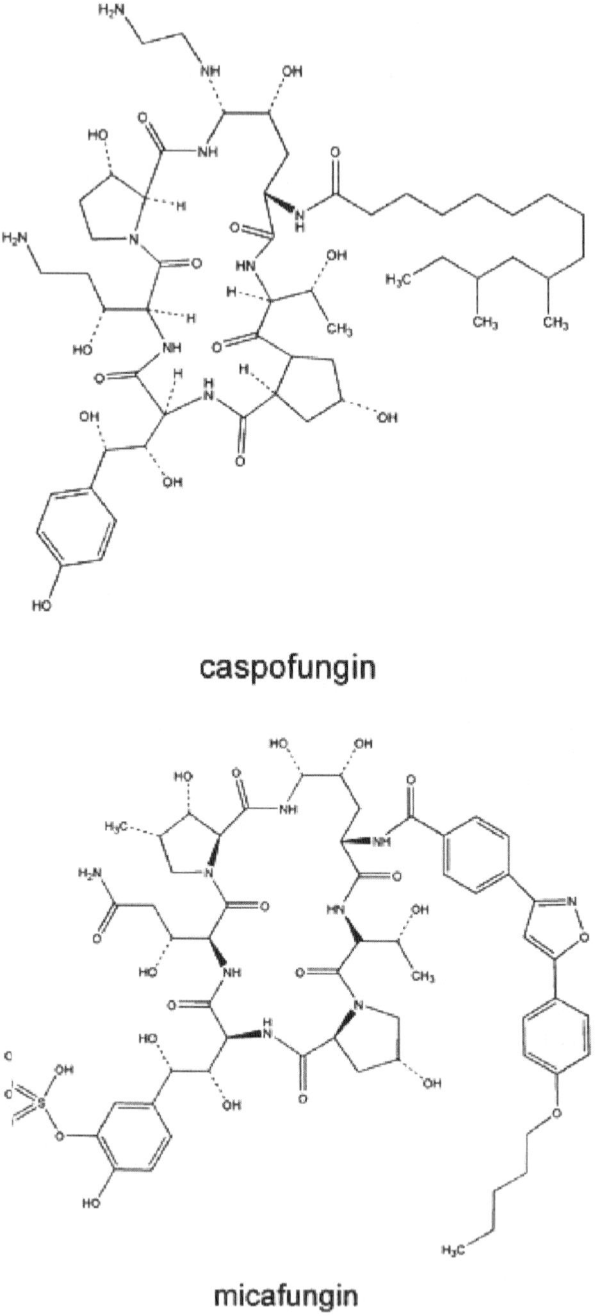

caspofungin

micafungin

Figure 4.2 Echinocandins. (A) Caspofungin, (B) Micafungin

Table 4.1 Antifungal activity of the echinocandin

Highly active	Very active	Some activity	Inactive
C. albicans	*C. parapsilosis*	*Coccidioides immitis*	Zygomycetes
C. glabrata	*C. gulliermondii*	*Blastomyces dermatididis*	*Cryptococcus neoformans*
C. tropicalis	*C. lusitaniae*	*Scedosporium* spp.w	*Fusarium* spp.
C. krusei	*A. fumigatus*	*Paecilomyces variotii*	*Trichosporon* spp.
C. kefyr	*A. flavus*	*Histoplasma capsulatum*	
Pneumocystis carinii*	*A. terreus*		

* Only active against cyst form, and probably only useful for prophylaxis.

CYP450 system (Sandhu et al., 2004; Keating & Jarvis, 2001). The efficacy of caspofungin is similar to that of AMB with successful outcome in 73.4% of patients treated with caspofungin versus 61.7% in those treated with AMB. Caspofungin was as effective as AMB in patients who had candidemia with favorable responses in 71.7% and 62.8% of patients, respectively (Mora-Duarte et al., 2002). The efficacy and safety of caspofungin (50 mg) versus fluconazole (200 mg IV) as assessed in adults with esophageal (EC) (Villanueva et al., 2001a, b) was 81% favorable response in patients treated with caspofungin versus 85% in patients treated with fluconazole. Combining voriconazole with caspofungin can extend the survival rate in patients with aspergillosis unresponsive to initial AMB therapy by $\simeq 3$ months (Marr et al., 2004).

Micafungin sodium (FK463) $C_{56}H_{70}N_9NaO_{23}S$ (Figure 4.2), is isolated from cultured *Coleophoma empedri*. It is fungicidal against C. *albicans, C. glabrata, C. krusei*, and C. *tropicalis* where most isolates exhibit MICs in the range of 0.03–0.06 µg/mL (Ostrosky-Zeichner et al., 2003). Higher MIC values have been reported for C. *lusitaniae, C. guilliermondii*, and C. *parapsilosis* (Takakura et al., 2004; Espinel-Ingroff, 2003). Micafungin is fungistatic against *Aspergillus* spp. (Turner et al., 2006; Matsumoto et al., 2000; Mikamo et al., 2000; Uchida et al., 2000). The relationship of AUC to micafungin is linear over a daily dose range of 50–150 mg, 3–8 mg, or 12.5–200 mg/kg body weight (Chandrasekar and Sobel, 2006; Hiemenz et al., 2005; Carver, 2004). Apparent clearance is 0.15–0.53 mL/min/kg body weight and apparent volume of distribution is 0.33–0.79 L/kg body weight (Hiemenz et al., 2005). It is metabolized to M-1 (catechol) by arylsulfatase with further metabolism to M-2 (methoxy) by catechol-*O*-methyltransferase; M-5 that forms by hydroxylation at the side chain ω-1 position is catalyzed by CYPP450 isozymes. Whereas the average ratio of metabolite to parent exposure AUC at a dose of 150 mg/day is 6% M-1, 1% M-2 and 6% M-5, it is 11% M-1, 2% M-2, and 12% M-5 in patients with EC (Chandrasekar and Sobel, 2006).

Micafungin administered once daily in HIV patients with EC has yielded a side effects-free clinical response rate of 33% at 12.5 mg dose, 54% at 25 mg dose, $\simeq 85\%$ at 50–75 mg dose and 95% at 100 mg dose (Pettengell et al., 2004). Treatment of HIV patients with EC using micafungin at 50, 100 and 150 mg/day compared to fluconazole at 200 mg/day IV-infused once daily for 14–21 days resulted in a response rate of 68.6% at 50 mg, 77.4% at 100 mg and 89.8% at 150 mg versus 86.7% for fluconazole (De Wet et al., 2004, 2005). The overall success rate in a study which has compared micafungin at 50 mg IV daily dose and fluconazole at 400 mg IV daily dose in SCT patients receiving either syngeneic or allogeneic cells was 80.7% and 73.7%, respectively (Van Burik et al., 2004). Linear pharmacokinetics and increased clearance occurred in neutropenic pediatric patients receiving a dose of 0.5–4.0 mg/kg/day (Seibel et al., 2005). IV infusion of 12.5–150 mg/day micafungin for up to 56 day in patients with *Aspergillus-* or *Candida* spp.-induced deep-seated mycosis resulted in clinical response rates of 60% in IPA patients, 67% in patients with chronic necrotizing PA, 55% in those with PA, 100% in candidemics and 71% in those with EC (Kohno et al., 2004).

Anidulafungin (LY303366, V-echiniocandin), approved for clinical use by the FDA in early 2006, consists of the echinocandin nucleus with a terphenyl head and a C5 tail (Figure 4.2). Anidulafungin MIC values for *C. parapsilosis, C. famata,* and *C. guillermondii* are relatively higher ($\leq$0.03–4 μg/mL) than those for other *Candida* spp. (Pfaller et al., 2005; Pfaller, 2004; Uzun et al., 1997; Zhanel et al., 1997; Cuenca-Estrella et al., 2000). Combinations of fluconazole with caspofungin or anidulafungin produced different but not antagonistic effects against isolates of *C. albicans, C. glabrata, C. tropicalis, C. krusei,* and *C. neoformans* (Roling et al., 2002). Anidulafungin has a minimum effective concentration of 0.02 mg/mL for 90% of clinical isolates and an MIC_{90} of 10.24 mg/mL against *Aspergillus* spp. (Zhanel et al., 1997). Excellent activity has been observed against *Aspergillus* spp., *Penicillium* spp., and *Curvularia* spp. (Messer et al., 2004; Espinel-Ingroff, 2003). Although, traces of the drug or its degradation products can be detected in human and animal urine; most degradation products pass into feces via bile (Vazquez and Sobel, 2006). In humans, anidulafungin exhibits linear pharmacokinetics after a single oral dose of 100–1,000 mg. The drug is well tolerated at doses of up to 700 mg with adverse gastrointestinal effects defining the maximal tolerated dose (Pfaller, 2004; Chiou et al., 2000; Vazquez & Sobel, 2006; Vazquez, 2005). In a randomized study that compared the efficacy and safety of IV anidulafungin to that of oral fluconazole in AIDS patients with EC, success rate in anidulafungin-treated patients was 97.2% with 9.3% of patients showing side effects (Krause et al., 2004a). Patients with IC or candidemia who received 3 IV regimens of 50, 75, or 100 mg anidulafungin once daily for 2 weeks beyond resolution or signs of improvement, showed EOT success rates of 84%, 90%, and 89%, respectively (Krause et al., 2004b).

Aminocandin (HMR 3270, IP-960) is active against *C. albicans,* non-*albicans Candida* spp., and *Aspergillus* spp. Aminocandin at the dose of $\geq$1.0 mg/kg/day is as effective as AMB in improving survival and reducing organ fungal burden in immunocompromised mice with disseminated *C. tropicalis* (Warn et al., 2005). In mice with *Candida* spp. infections (*C. glabrata, C. guillermondii,* or *C. tropicalis*), it can reduce kidney fungal burden at doses 20–50 mg/kg making it superior to fluconazole (Turner et al., 2006). The degree of protein binding is >99% in both mice and humans (Andes et al., 2003a).

Chitin Synthase Inhibitors

Chitin is synthesized on the cytoplasmic surface of the plasma membrane as a linear homopolymer of β-(1,4)-linked-*N*-acetyl-D-glucosamine (GlcNAc) residues. It extrudes perpendicularly from the cell surface as microfibrils to crystallize outside the cell through extensive hydrogen bonding as á-chitin. Polymerization of GlcNAc is catalyzed by the membrane-bound enzyme chitin synthase (Wills et al., 2000; Hector, 1993). Absence of chitin from mammalian cells makes it a potential target for therapeutic use. The two structurally related groups of secondary fermentation metabolites that act as specific inhibitors of chitin synthase, nikkomycins, and polyoxins, bear strong resemblance to uridine diphosphate (UDP)-GlcNAc, a chitin precursor substrate (Figure 4.3). The antifungal activity of different chitin synthase inhibitors varies depending on their differential capacity to permeate the cell wall.

Nikkomycin ($C_{20}H_{25}N_5O_{10}$) is produced by *Streptomyces tendae* TU901; currently, 14 naturally derived nikkomycin (Bx, Bz, Cx, Cz, D, E, I, J, M, N, X, Z, Pseudo-J, and Pseudo-Z) variants are known (Wills et al., 2000). Nikkomycin has fungicidal activity against the dimorphic fungi *C. immitis* and *B. dermatitidis*; *C. albicans*, and *C. neoformans* are Nikkomycin resistant (Hector et al., 1990). MICs for Nikkomycins X and Z against *C. immitis* are 0.77 and 0.1 µg/mL while that against *B. dermatitidis* are 8 and 30 µg/mL, respectively (Graybill et al., 1998b; Hector et al., 1990). Additive *in vitro* effects are attainable against filamentous fungi (*Aspergillus*) by combining nikkomycin Z with caspofungin (Ganesan et al., 2004; Stevens, 2000). Pairwise combinations of nikkomycin Z with caspofungin or micafungin result in synergistic activity against *A. fumigatus* (Ganesan et al., 2004).

Aureobasidins is an 18 member family with a common structure of 8 lipophilic amino acid residues and an α-hydroxy acid synthesized by *Aureobasidium pullulans*. Auroebasidins lyse target cells by altering chitin assembly and sphingolipid synthesis (Endo et al., 1997; Nagiec et al., 1997). Aureobasidins A, B, C, E, S_{2b}, S_3, and S_4 are active against *Candida* spp. and *C. neoformans* with MICs of 0.05–3.12 µg/mL; the MICs for *H. capsulatum* and *B. dermatitidis* are <0.63 µg/mL. Aureobasidin A is active against dematiceous fungi at ≤2.5 µg/mL but has little activity against *Aspergillus* spp. (Kurmoe et al., 1996). Synthetic aureobasidin A is highly fungicidal to *Candida* and *C. neoformans* with an MIC of 0.01–1.6 µg/mL (Kurmoe et al., 1996).

Antifungal compounds targeting protein synthesis: Both fungal and mammalian cells require two elongation factors (EF), EF1α and EF2, for polypeptide chain elongation. However, with the discovery of EF3, a new essential factor for protein synthesis unique to yeast cells, fungal translation has evolved as a desirable target. This 120–125 kDa protein is present in most fungi, including *C. albicans* and *P. carinii* to provide an ATPase activity specifically required by the 40S ribosomal subunit of yeast cells to translocate growing polypeptides (Kovalchuke & Chakraburtty, 1994). Sordarins, a new class of antifungals prepared from the fermentation broth of *Sordaria araneosa*, inhibit protein synthesis in pathogenic fungi. The primary targets for sordarins are EF2 and the large ribosomal subunit

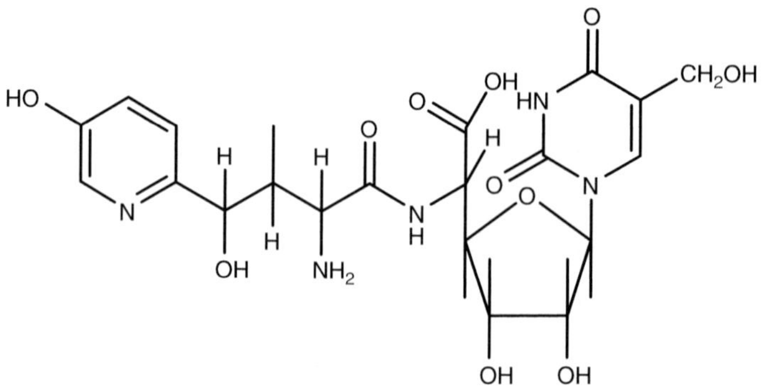

Nikkomycin Z

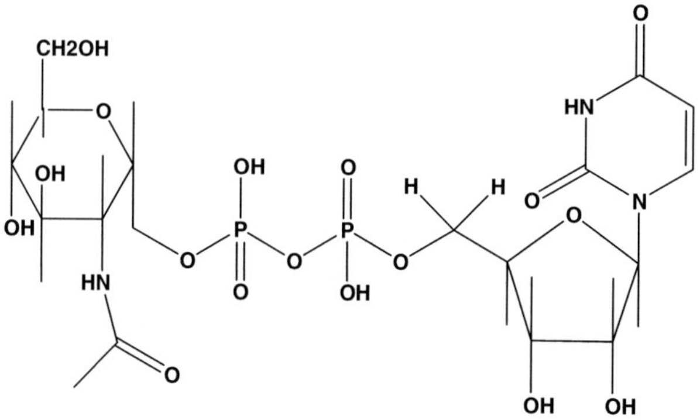

Polyoxin B

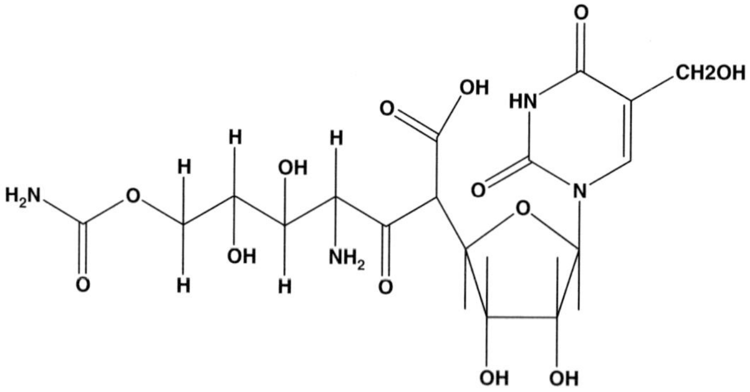

UDP-N-acetylglucosamine

Figure 4.3 Chemical structures of nikkomycin Z, polyoxin B, and uridine diphosphate (UDP)-
N- acetylglucosamine

stalk rpPO (Santos et al., 2004; Odds et al., 2003; Odds, 2001). EF2 promotes the displacement of tRNA from the A site to the P site and the movement of ribosome along the mRNA thread (Odds, 2001). A number of new sordarins, GM193663, GM-211676, GM-222712, GM-237354, and GR-135402 (Figure 4.4), are in different phases of preclinical or clinical evaluation. Sordarins are to *C. albicans, C. glabrata*, and *C. tropicalis* but inactive against *C. krusei* and *C. parapsilosis* (Herreros et al., 1998). While the activity of sordarins against *C. neoformans, P. carinii*, and *H. capsulatum* is well demonstrated, limited activity against *A. fumigatus* has been suggested (Gargallo-Viola, 1999). Significant *in vivo* activity of sordarins was reported in immunocompetent mouse models of disseminated and mucosal candidiasis (Martinez et al., 2000), histoplasmosis (Graybill et al., 1999), and coccidioidomycosis (Clemons & Stevens, 2000). Sordarins GM-193663, GM-211676, and GM-237354 were reported to be equivalent or superior to fluconazole in treating murine experimental systemic coccidioidomycosis (Graybill et al., 1999; Clemons & Stevens, 2000). Sordarins synergize with AMB, itraconazole, and voriconazole to yield better activity against *Aspergillus* spp. and *S. apiospermum* (Andriole, 1999).

Azasordarin exhibits superior activity against mucosal *C. albicans* and other *Candida* spp. infections (Herreros et al., 2001). GW-471558 (Figure 4.4) is active against *C. albicans, C. glabrata*, and *C. tropicalis* including some isolates insensitive to fluconazole but inactive agaisnt *C. parapsilosis, C. krusei, C. guilliermondii*, and *C. lusitaniae* (Cuenca-Estrella et al., 2001). GW-471552 and GW-471558 have significant therapeutic efficacy against vulvovaginal *C. albicans* infections (Martinez et al., 2001). R-35853, a sordarin derivative with a 1,4-oxazepane ring moiety (Figure 4.4), has good *in vitro* activity against fluconazole-susceptible *C. albicans* strains, *C. glabrata, C. tropicalis*, and *C. neoformans* (Kamai et al., 2005). MICs of R-135853 against dose-dependent fluconazole-susceptible and fluconazole-resistant strains of *C. albicans* are 0.03–0.06 μg/mL. R-135853 exhibits dose-dependent efficacy against *C. albicans*-induced murine experimental hematogenous candidiasis when administered subcutaneously or orally.

Antifungal Agents Targeting Plasma Membrane Synthesis or Metabolism

Polyenes

Amphotericin B (AMB) and NY remain the most effective and widely used compounds in the treatment of presystemic and systemic fungal infections. Polyenes form complexes with ergosterol and disrupt the fungal plasma membrane resulting in increased membrane permeability and cell death. Auto-oxidation of ergosterol, results in the formation of free radicals and increased membrane permeability. Low affinity to cholesterol minimizes toxicity to mammalian cells due to polyene's (Abu-Elteen and Hamad, 2005; Groll and Kolve, 2004; Odds et al., 2003; Groll and Walsh, 2002). AMB lipid-based delivery systems enhance the therapeutic index by

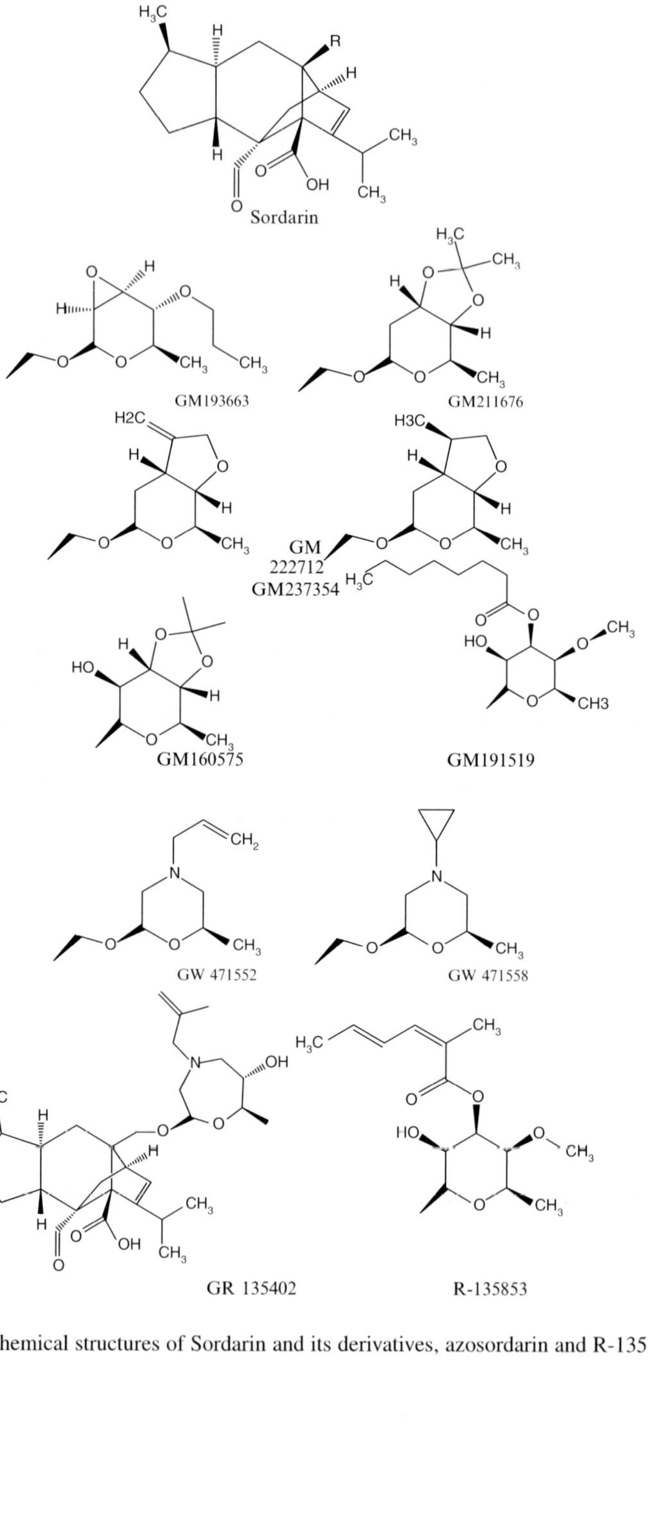

Figure 4.4　Chemical structures of Sordarin and its derivatives, azosordarin and R-135853

decreasing toxicity and increasing target specificity (Abuhammour and Habte-Gaber, 2004; Ng et al., 2003). Incorporating AMB into phospholipid vesicles (Liposomes) or cholesterol esters permit delivery of larger amounts of the drug with minimal nephrotoxicity. AMB lipid complex (AMBLC) was the first lipid-formulated AMB product to be approved by the FDA for clinical. The lipid-formulated AMB colloidal dispersion (AMBCD), which consists of disc-like structures of cholesteryl sulfate complexed with AMB in a 1:1 molar ratio received FDA approval in 1996. Addition of cholesterol to the phospholipid bilayer enhances liposome stability, decrease rate of clearance, and prolong halflife. The clinical response of patients with invasive aspergillosis (IA) was slightly better when treated by a lipid formulation compared with conventional AMBD (Bowden et al., 2002; Reichenberger et al., 2002; Wingard et al., 2000). Lipid formulations used empirically in neutropenic cancer patients resulted in decreased nephrotoxicity while maitaining clinical efficacy comparable to that of AMBD. Use of AMBLC in patients with aspergillosis, disseminated candidiasis, zygomycosis, and fusariosis, who were refractory/ intolerant to conventional antifungal therapy produced reasonable efficacy (Bowden et al., 1996, 2002).

NY binds ergosterol and causes alterations in membrane permeability that lead to cell death. NY is active against *Candida, Aspergillus, Histoplasma* spp., and *Coccidioides immitis*. Significant antifungal activity and reduced toxicity was reported following intravenous (IV) administration of NY (Arikan et al., 2002; Groll et al., 1999a, b). L-NY increases survival, reduces tissue injury, clears the infection and produces tolerable side effects in neutropenic rabbits with pulmonary aspergillosis (PA) (Groll et al., 1999a, b). At 1–7 mg/kg/day dose in humans, L-NY exhibits linear plasma pharmacokinetics with peak plasma levels above the MIC of most relevant fungi and a terminal halflife of 5–7 h (Groll et al., 1999a). Currently, clinical trials are targeting non-neutropenic patients with candidemia, neutropenia, persistent fever, and patients with invasive fungal infection refractory to standard therapy (Groll and Kolve, 2004; Ng et al., 2003).

SPA-S-753 (*N*-dimethylaminoacetyl-partricin A2-dimethyl-aminoethylamide diascorbate), a new water soluble partricin-A derivative semisynthetic polyene antifungal agent is active against strains of *Candida, Cryptococcus*, and *Saccharomyces* spp. SPA-S-753 is effective against murine candidiasis, aspergillosis and cryptococcosis (Rimaroli et al., 2002). Although its spectrum of activity is similar to that of AMB, it yields higher survival rates, longer survival times and comprehensive sterilization of kidneys, liver, and spleen (Rimaroli et al., 2002; Rimaroli & Bruzzese, 2000;1998).

Azoles: Fluconazole derivatives, voriconazole (UK-109–496) and ravuconazole (BMS-207147, ER-30346), and the hydroxylated analog of itraconazole, posaconazole (SCH56592), are significant additions to the azole family (Figure 4.5). Voriconazole, active both orally and parenterally, has one triazole moiety replaced by a fluropyrimidine group and a methyl group added to the propanol backbone (Figure 4.5). Like other azoles, Voriconazole inhibits ergosterol synthesis by inhibiting the P450-dependent 14 α–demethylase demethylation step *C. albicans* and *A. fumigatus* lysates at 1.6 and 160 folds greater than the parent molecule. Voriconazole

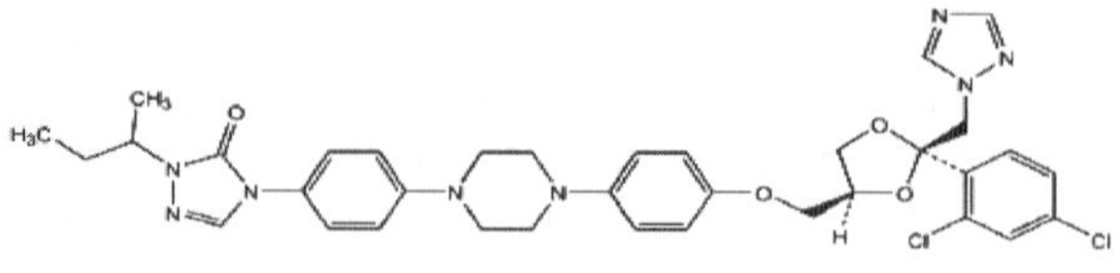

itraconazole

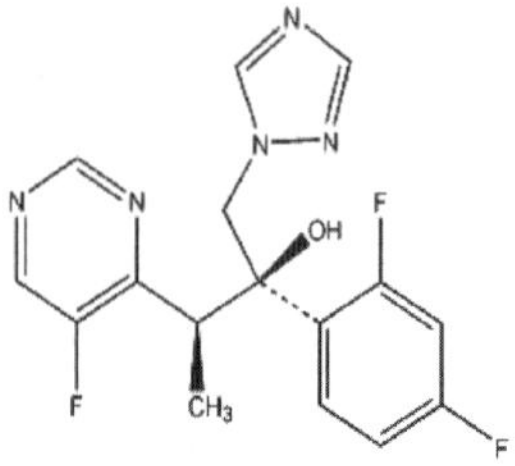

voriconazole

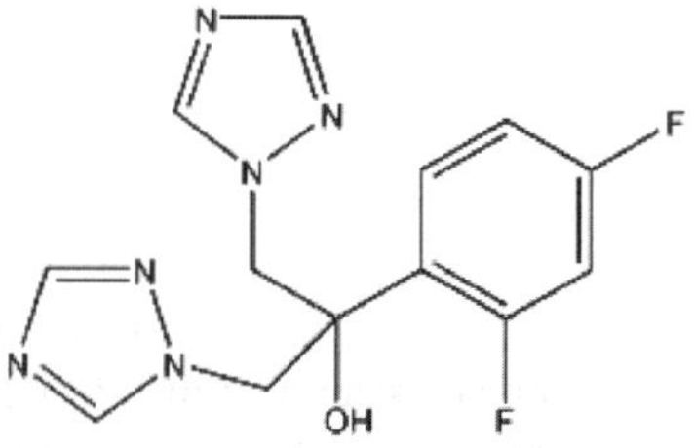

fluconazole

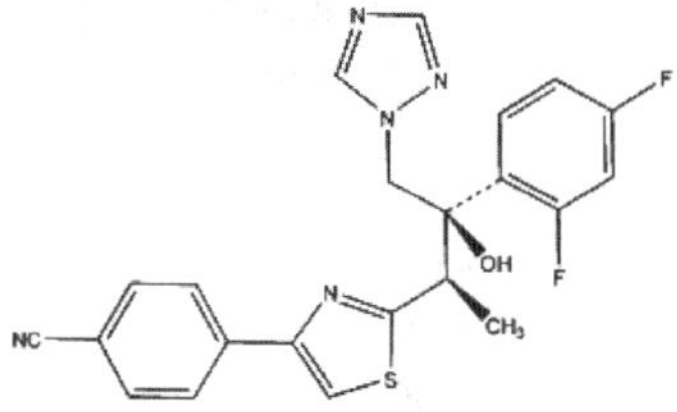

ravuconazole

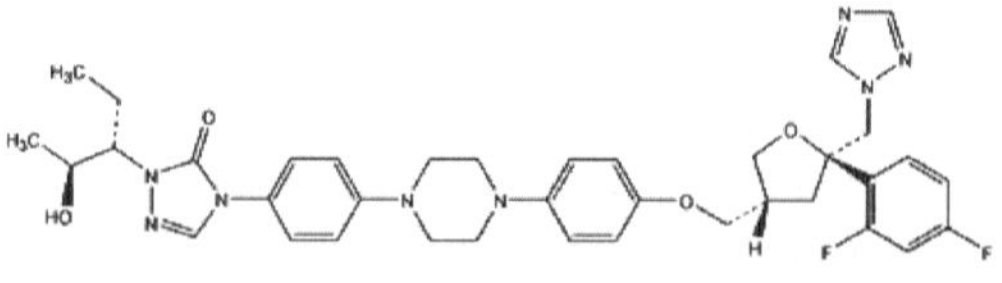

posaconazole

Figure 4.5 Chemical structures of new triazole antifungal agents

is active against *Candida* spp. (Marco et al., 1998; Pfaller et al., 1998a), *Aspergillus* spp. (Clancy & Nguyen, 1998; Murphy et al., 1997), *C. neoformans* (both flucona-zole-susceptible and -resistant strains) (Nguyen & Yu, 1998), and dimorphic fungi (Groll et al., 2003). Voriconazole is more active than AMB and flucytosine (5-FC) against all *Candida* spp. except for *C. glabrata* (Marco et al., 1998). It is more active than itraconazole against strains of *A. fumigatus* and *A. niger* but less active against strains of *A. flavus* (Murphy et al., 1997). Data from phases II & III clinical trails indicate that voriconazole is active against oropharyngeal candidiasis (OPC) and EC (Groll & Walsh, 2002; Ally et al., 2001), IC and IA (Denning et al., 2002; Walsh et al., 2002a).

The drug is superior to AMBD or LAMB evidence by improved survival, fewer breakthrough infections and decreased infusion-related toxicity and nephrotoxicity in patients receiving voriconazole (Groll & Walsh, 2002). Excellent CNS-penetration properties make voriconazole a good choice for treating cerebral mold infections (Denning et al., 2002; Walsh et al., 2002b). At 58% plasma protein binding rate, the concentration of voriconazole in saliva reaches 65% that of plasma (Groll & Klove, 2004; Chiou et al., 2000). The drug is extensively metabolized; while 78–88% of a single dose appears in urine, <5% appears as unchanged suggesting that elimination occurs by oxidative hepatic metabolism primarily by CYP2C19 (Pearson et al., 2003; Denning et al., 2002).

Posaconazole (SCH 56592) is a hydroxylated analog of itraconazole with a 1,3-dioxolone skeleton (Figure 4.5). The oral formula is currently used to prevent and treat invasive mycoses (Cuenca-Estrella et al., 2006; Vazquez et al., 2006; Courtney et al., 2004a). Posaconazole is active against *Candida* spp., *C. neofor-mans, Aspergillus* spp., filamentous fungi including zygomycetes, *Fusarium* spp., and some dematiaceous molds (Carrillo-Munoz et al., 2006; Cuenca-Estrella et al., 2006; Van Burik et al., 2006; Raad et al., 2006a; Torres et al., 2005; Barchiesi et al., 2004; Herbrecht, 2004; Pfaller et al., 2004). In a study that investigated 2000 bloodstream *Candida* isolates, most isolates gave low posaconazole MICs (0.03–0.13 µg/mL) with higher MICs noted for *C. glabrata* and *C. krusei* (Ostrosky-Zeichner et al., 2003). For *C. parapsilosis* and *C. krusei*, MICs were in the range of 0.015–1 and 0.12–2 µg/mL, respectively (Pfaller et al., 2004). Posaconazole can inhibit 100% of *C. neoformans* isolates including fluconazole-resistant ones at 1 µg/mL dose (Cuenca-Estrella et al., 2006; Pfaller et al., 2004; Barchiesi et al., 2004). Posaconazole displays linear, dose-proportional pharmacokinetics at a single dose of 800 mg/kg/day (Table 4.2) (Courtney et al., 2003; 2004a, b; Krieter et al., 2004). It is orally bioavailable with maximum concentration reached about 6–10 h post-dosing (Krieter et al., 2004; Courtney et al., 2003). Absorption is linear at 400 mg/12 h (Ezzet et al., 2005) and can be enhanced 2.6–4 folds if coadministered with food (Courtney et al., 2004b). The drug binds albumin at high proportions (97–99%) without affect-ing its penetration and distribution into CSF (Groll & Walsh, 2006; Al-Abdely et al., 2005). Posaconazole degrades in the liver by glucuronidation to produce inactive metabolites for excretion in feces and urine (Herbrecht, 2004; Krieter et al., 2004). Therefore, it could potentially represent an appropriate alternative to AMB in patients with impaired renal function (Torres et al., 2005).

Table 4.2 Comparison of pharmacokinetic characteristics of Voriconazole, Posaconazole, and Ravuconazole

Characteristic	Voriconazole	Posaconazole	Ravuconazole
Derivative	Fluconazole	Itraconazole	Fluconazole
Formulation	Oral and intravenous	Oral and topical	Oral and intravenous
Bioavailability (%)	Bioavailability decreases to about 80% with fatty meals	Bioavailability increases with food or nutrition supplements	Bioavailability increases with food
Metabolism	Hepatic, primarily via N-oxidation. Metabolized via several hepatic CYP isoenzymes, including CYP2C19, CYP2C9, and CYP3A4	Hepatic. Inhibits hepatic CYP3A4 but no other isoenzymes	ND
Plasma C_{max} (mg/L)	2–4	More than 1	ND*
Plasma protein binding (%)	58	97–99	95.8
Terminal elimination halflife (h)	6–9	15–35	3.9–4.8
Current status	Approved	Approved	Approved

* ND = not determined.

Posaconazole can successfully treat refractory fungal infections including fusariosis, zygomycosis, pseudallescheriasis, endemic mycosis, chromoblastomycosis, and mycetoma (Raad et al., 2006b). In a randomized trial (Vazques et al., 2006) conducted to evaluate the capacity of posaconazole to treat OPC in HIV subjects, clinical success occurred in 91.7% of posaconazole recipients versus 92.5% of fluconazole recipients; sustained clinical success following treatment cessation was noted. In patients with invasive fusariosis treated with oral posaconazole (800 mg/ day), successful outcome was >45% (Herbrecht, 2004). Leukemic patients who received posaconazole for <3 days, an overall success rate of 50% for those who recovered from myelo-suppression and 20% for those with persistent neutropenia was noted (Raad et al., 2006a). A 70% clinical response rate was reported in posaconazole-treated patients with zygomycosis intolerant or refractory to standard therapy (Greenberg et al., 2006). The drug may be useful in treating mycetoma coccidioidomycosis, chromoblastomycosis, hyalohyphomycosis, and phaeohyphomycosis (Keating, 2006; Torres et al., 2005).

Ravuconazole is an oral derivative of fluconazole (Figure 4.5) with expanded spectrum of *in vitro* activity. Its inhibitory potency and binding affinity to yeast P-450 dependent 14α-demethylase is similar to that of itraconazole. Ravuconazole is active against *Candida* spp., *A. fumigatus, C. neoformans*, most hyaline hyphomycetes (except *Fusarium* spp. and *P. boydii*), dermatophytes, and dematiaceous fungi (Cuenca-Estrella et al., 2006; Espinel-Ingroff, 2003; Yamazumi et al., 2000; Pfaller et al., 1998a). After 8 h of oral administration of 10 mg/kg body weight, maximum plasma concentration reaches 1.68 µg/mL (Mikamo et al., 2002; Andes et al., 2003b). In neutropenic mice with candidiasis, area under the concentration-time curve (AUCs)/MIC ratio for ravuconazole was similar to that of fluconazole (Andes et al., 2003b). Serum elimination halflife is 3.9–4.8 h, protein binding is 95.8%. Except for minor headaches, the drug is well tolerated at 800 mg single dose or 400 mg/day dose for up to 14 days (Ernst, 2001).

Other azoles: Systematic modifications of the piperazine moiety resulted in the discovery of several novel triazoles like SYN-2869 (Figure 4.6). SY-2869 is orally active against isolates of *Candida* spp., *Aspergillus* spp., *C. neoformans*, and several dematiaceous molds (Johnson et al., 1999). SYN-2869 derivatives, SYN-2836 (has a *P*-trifluoromethyl moiety at the benzyl group) and 3′-fluoro-substituted analogs of SYN-2836 (SYN-2903) and SYN-2869 (SYN-2921) have activities comparable to those of other azoles (Figure 4.6). SYN azoles (SYN-2836, SYN-2869, SYN-2903, and SYN-2921) are rapidly absorbed into the circulation reaching maximum concentration in mice following a 50 mg/kg body weight oral dose with >45% bioavailability and 4.5–6 h halflife of (Khan et al., 2000). Higher lung concentration of SYN-2869 enhances its capacity to manage respiratory tract infections (Khan et al., 2000). R-126638 (Figure 4.6) is activite against dermatophytes, *Candida* spp. and *Malassezia* spp. (Odds et al., 2004). R-126638 is superior to itraconazole in treating dermatophytic infections (ED_{50} being three- to eightfold lower) and cutaneous model of *Trichophyton mentagrophytes* dermatophytosis (ED_{50} bieng fivefold lower) (Odds et al., 2004). The *in vitro* activity of BAL4815 (the active component of BAL8557) was compared with that of itraconazole, voriconazole, caspofungin, and AMB against isolates of *Aspergillus* spp. *fumigatus, terreus, flavus*, and *niger* (Warn et al., 2006).

Allylamines and thiocarbamates: The allylamines terbinafine and naftifine and the thiocarbamate tolnaftate are synthetic fungicidals that act as reversible, noncompetitive inhibitors of squalene epoxidase, the enzyme responsible for the cyclization of squalene to lanosterol (Figure 4.7) (Andriole, 1999; 1998). Terbinafine is effective against dermatophytes, *Aspegillus* spp., *Fusarium* spp. *Penicillium marneffei* and other filamentous fungi (Garcia- Effron et al., 2004). Mean MIC of terbinafine against *C. albicans* is 1.2 µg/mL (Perea et al., 2002a; McGinnis et al., 2000). Terbinafine shows strong *in vitro* activity against *Penicillium* spp., *Paecilomyces* spp., *Trichoderma* spp. *Acremonoim* spp., and *Arthrographis* spp. with a mean MIC of <1 mg/L. *Scedosporium* spp., *Fusarium* spp., *Scopulariopsis brevicaulis*, and most of *Mucorales* exhibit high MICs of the allylamine with a mean MIC of ≥4 mg/L (Garcia-Effron et al., 2004). Although terbinafine is relatively active against murine aspergillosis especially when combined with AMB, pharmacokinetic properties confine the clinical efficacy of allylamines and thiocarbamates to dermatophytes (Gupta et al., 2006;2005; 2003).

Antimitotic antifungal agents: 5-fluorocytosine (5-FC) is active against *Candida* and *Cryptococcus* spp., dematiaceous fungi causing chromomycosis like *Phialophora* and *Cladosporium* spp. and *Aspergillus* spp. (Vermes et al., 2000). MICs of 5-FC is 0.1–25 mg/L for these fungal species; MIC values for 5-FC ≥25 mg/L signifies resistance. 5-FC inhibits pyrimidine metabolism by interfering with RNA and

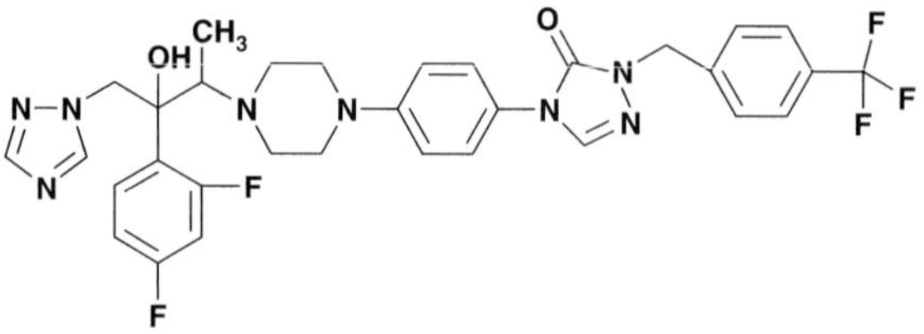

Figure 4.6 Structures of novel triazole antifungal agents SYN-2836, SYN-2869, SYN-2903, SYN-2921, and R-126638

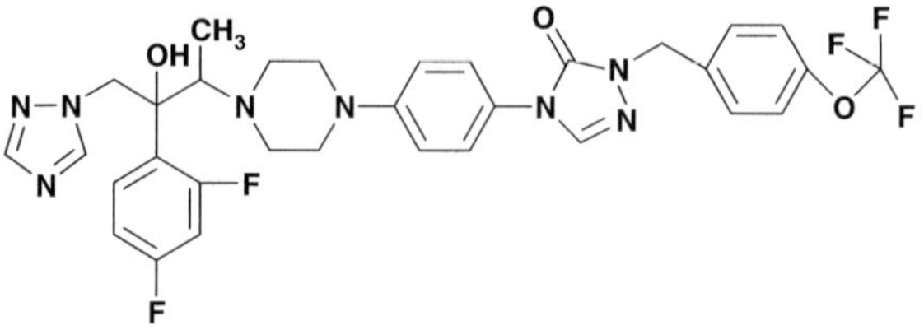

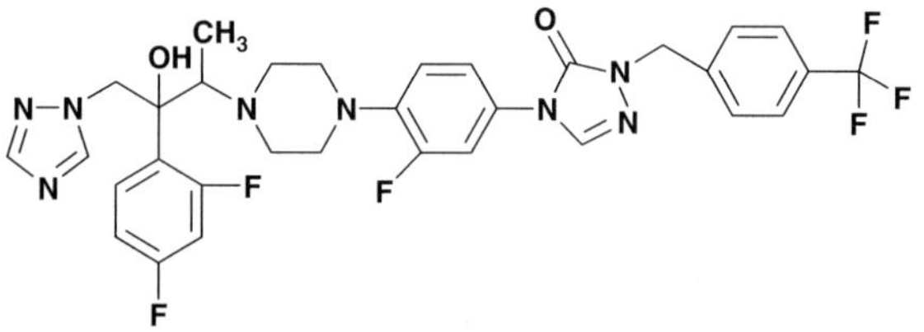

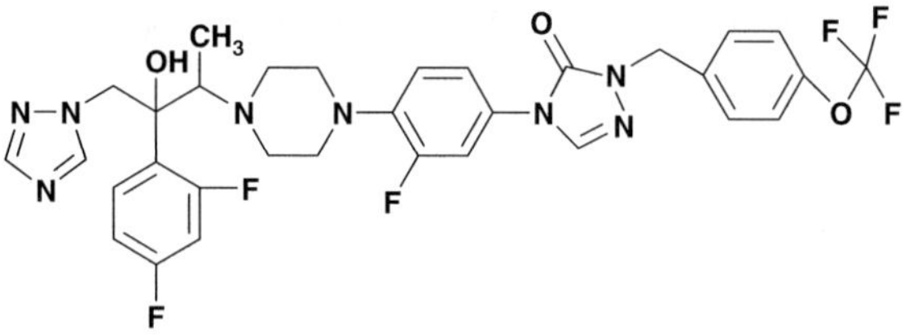

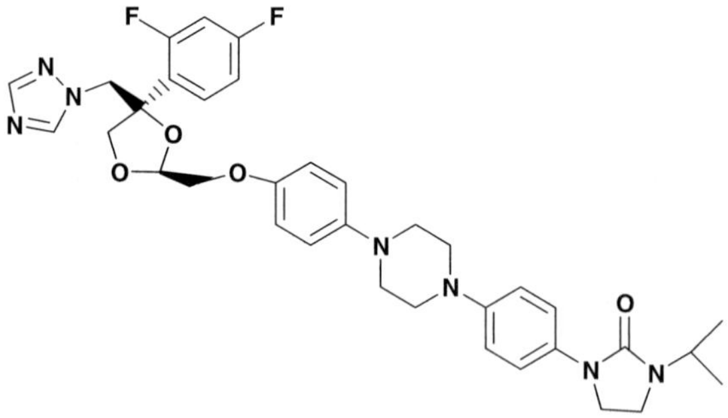

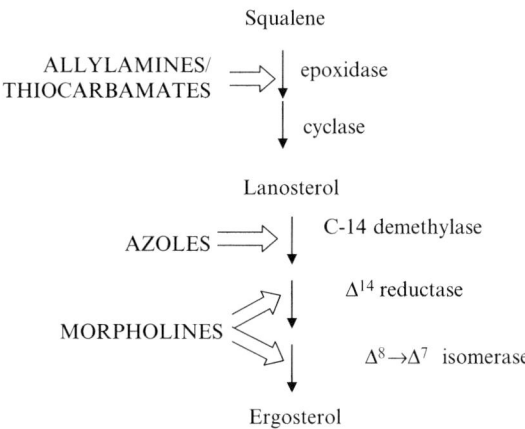

Figure 4.7 Ergosterol biosynthesis pathway, showing sites of inhibition of different antifungal agents

protein synthesis. Flucytosine is readily absorbed from the GI tract, 76–89% is bioavailable after oral administration (Groll & Kolve, 2004; Vermes et al., 2000). Small molecular size and increased hydrosolubility enables 5-FC to penetrate into CSF, vitreous and peritoneal fluids and inflamed joints. The drug is used to treat chromoblastomycosis, uncomplicated lower urinary tract candidiasis and VC (Groll & Kolve, 2004). Successful treatment of candidiasis, cryptococcosis and chromoblastomycosis using combinations of 5-FC and AMB, fluconazole or itraconazole has been reported (Schwarz et al., 2006; Johnson et al., 2004; Kontoyiannis & Lewis, 2004). The combination of 5-FC with LAMB has proven useful in clearing the CSF in non-HIV infected patients with cryptococcal meningitis (Brouwer et al., 2004; Brandt et al., 2001). 5-FC-AMB combination is recommended for treating CNS cryptococcosis and candidiasis, *Candida* endophthalmitis, renal and hepatosplenic candidiasis, *Candida* thrombophlebitis of the great veins, aspergillosis and CNS phaeohyphomycosis (Johnson et al., 2004).

Acknowledgments We thank Eman Mahmoud, Aula Abu-Halaweh, and Khawla Salem for help with figure drawing and manuscript typing. This work was supported by research grant KHA/MH-2006, Hashemite University, Jordan.

References

Abu-Elteen, K. H. & Hamad, M. A. (2005). Antifungal agents for use in human therapy. In: *Fungi: Biology and Applications*, Kavanagh, K., Ed., Wiley, London, pp. 191–217.
Abuhammour, W. & Habte-Gaber, E. (2004). *Indian J. Pediatr.*, 71:253–259.
Al-Abdely, H. M., Alkhunaizi, A. M., Al-Tawfiq, J. A., Hassounah, M., Rinaldi, M. G., & Sutton, D. A. (2005). *Med. Mycol.*, 43:91–95.

Ally, R., Schuermann, D., Kreisel, W., Carosi, G., Aguirrebengoa, K., Dupont, B., Hodges, M., Troke, P., Romero, A. J., and the Esophageal Candidiasis Study Group (2001). *Clin. Infect. Dis.*, 33:1447–1454.

Andes, D., Marchillo, K., Lowther, J., Bryskier, A., Stamstad, T., & Conklin, R. (2003a). *Antimicrob. Agents Chemother.*, 47:1187–1192.

Andes, D., Marchillo, K., Stamstad, T., & Conkline, R. (2003b). *Antimicrob. Agents Chemother.*, 47:1193–1199.

Andriole, V. T. (1998). Current and future therapy of invasive fungal infections. In: *Current Clinical Topics in Infectious Diseases*, Vol. 18, Remington, J., & Swartz, M., Eds., Blackwell Sciences, Malden, MA, pp. 19–36.

Andriole, V. T. (1999). *J. Antimicrob. Chemother.*, 44:151–162.

Arikan, S., Ostrosky-Zeichner, L., Lozano-Chiu, M., Paetznick, V., Gordon, D., Wallace, T., & Rex, J. H. (2002). *J. Clin. Microbiol.*, 40:1406–1412.

Barchiesi, F., Spreghini, E., Schimizzi, A. M., Maracci, M., Giannini, D., Carle, F., & Scalise, G. (2004). *Antimicrob. Agents Chemother.*, 48:3312–3316.

Bowden, R., Chandrasekar, P., White, M., Li, X., Pietrelli, L., Gurwith, M., vanBurik, J., Laverdiere, M., Safrin, S., & Wingard, J. R. (2002). *Clin. Infect. Dis.*, 35:359–366.

Bowden, R. A., Cays, M., Cooley, T., Mamelok, R. D., & vanBurik, J. (1996). *J. Infect. Dis.*, 173:1208–1215.

Bozza, S., Gaziano, R., Lipford, G. B., Montagnoli, C., Bacci, A., Di Francesco, P., Kurup, V. P., Wagner, H., & Romani, L. (2002). *Microbes Infect.*, 4:1281–1290.

Bozza, S., Montagnoli, C., Gaziano, R., Rossi, G., Nkwanyuo, G., Bellocchio, S., & Romani, L. (2004). *Vaccine*, 22:857–864.

Bozza, S., Perruccio, K., Montagnoli, C., Gaziano, R., Bellocchio, S., Burchielli, E., Nkwanyuo, G., Pitzurra, L., Velardi, A., & Romani, L. (2003). *Blood*, 102:3807–3814.

Brandt, M. E., Pfaller, M. A., Hajjeh, R. A., Hamill, R. J., Pappas, P. G., Reingold, A. L., Rimland, D., & Warnock, D. W. (2001). *Antimicrob. Agents Chemother.*, 43:3065–3069.

Brouwer, A. E., Rajanuwong, A., Chierakul, W., Griffin, G. E., Larsen, R. A., White, N. J., & Harrison, T. S. (2004). *Lancet*, 363:1764–1767.

Cardenas, M. E., Cruz, M. C., Del Poeta, M., Chung, N., Perfect, J. R., & Heitman, J. (1999). *Clin. Microb. Rev.*, 12:583–611.

Carrillo-Munoz, A. J., Quindos, G., Tur, C., Ruesga, M. T., Alonso, R., del Valle, O., Hernandez-Molina, J. M., McNicholas, P., Loebenberg, D., & Santos, P. (2006). *J. Antimicrob. Chemother.*, 55:317–319.

Carver, P. L. (2004). *Ann. Pharmacother.*, 38:1707–1721.

Casadevall, A. (1995). *Infect. Immun.*, 63:4211–4218.

Cenci, E., Mencacci, A., Bacci, A., Bistoni, F., Kurup, V. P., & Romani, L. (2000). *J. Immunol.*, 165:381–388.

Cenci, E., Mencacci, A., Fe d'Ostiani, C., Del Sero, G., Mosci, P., Montagnoli, C., Bacci, A., & Romani, L. (1998). *J. Infect. Dis.*, 178:1750–1760.

Chandrasekar, P. H. & Sobel, J. D. (2006). *Clin. Infect. Dis.*, 42:1171–1178.

Chaturvedi, A. K., Kavishwar, A., Shiva Keshava, G. B., & Shukla, P. K. (2005). *Clin. Diag. Lab. Immunol.*, 12(9):1063–1068.

Childs, J. L., Disney, M. D., & Turner, D. H. (2002). *Proc. Natl. Acad. Sci. USA*, 99: 11091–11096.

Chiou, C. C., Groll, A. H., & Walsh, T. J. (2000). The Oncologist, 5:120–135.

Clancy, C. J. & Nguyen, M. H. (1998). *Europ. J. Clin Microbiol. Infect. Dis.*, 17:573–575.

Clemons, K. V. & Stevens, D. A. (2000). *Antimicrob. Agents Chemother.*, 44:1874–1877.

Courchesne, W. E. (2002). *J. Pharmacol. Exp. Ther.*, 300:195–199.

Courchesne, W. E. & Oztruk, S. (2003). Mol. Microbiol., 47:223–234.

Courtney, R., Pai, S., Laughlin, M., Lim, J., & Batra, V., (2003). *Antimicrob. Agents Chemother.*, 47:2788–2795.

Courtney, R., Radwanski, E., Lim, J., & Laughlin, M. (2004a) *Antimicrob. Agents Chemother.*, 48:804–808.

Courtney, R., Wexler, D., Radwanski, E., Lim, J, & Laughlin, M. (2004b). *Br. J. Clin. Pharmacol.*, 57:218–222.

Cuenca-Estrella, M., Gomez-Lopez, A., Mellado, E., Buitrago, M. J., Monzon, A., & Rodriguez-Tudela, J. L. (2006). *Antimicrob. Agents Chemother.*, 50:917–921.

Cuenca-Estrella, M., Mellado, E., Diaz-Guerra, T. M., Monzon, A., & Rodriguez-Tudela, J. L. (2000). *J. Antimicrob. Chemother.*, 46:475–477.

Cuenca-Estrella, M., Mellado, E., Diaz-Guerra, T. M., Monzon, A., & Rodriguez-Tudela, J. L. (2001). *Antimicrob. Agents Chemother.*, 45:1905–1907.

Cupta, S. S., Ton, V. K., Beaudry, V., Rulli, S., Cunningham, K., & Rao, R. (2003). *J. Biol. Chem.*, 278:28831–28839.

De Lucca, A. J. & Walsh, T. (2000). Rev. Iberoam. Micol., 17:116–120.

De Mattos Grosso, D., de Almeida, S. R., Mariano, M., & Lopes, J. D. (2003). *Infect. Immun.*, 71:6534–6542.

De Wet, N., Bester, A. J., Viljoen, J. J., Filho, F., Suleiman, J., Ticona, E., Llanos, E. A., Fisco, C., Lau, W., & Buell, D. (2005). *Aliment. Pharmacol. Ther.*, 21:899–907.

De Wet, N., Llanos-Cuentas, A., Suleiman, J., Baraldi, E., Krantz, E. F., Della Negra, M., & Diekmann-Berndt, H. (2004). *Clin. Infect. Dis.*, 39:842–849.

Deepe, G. S. (1997). *Clin. Microb. Rev.*, 10:585–596.

Del Poeta, M., Toffaletti, D. L., Rude, T. H., Dykstra, C. C., Heitman, J., & Perfect, J. R. (1999). *Genetics* 152:167–178.

Denning, D. W., Ribaud, P., Milpied, N., Caillot, D., Herbrecht, R., Thiel, E., Hass, A., Ruhnke, M., & Lode, H. (2002). *Clin. Infect. Dis.*, 34:563–571.

Disney, M. D., Haidaris, C. G., & Turner, D. H. (2003). *Proc. Natl. Acad. Sci. USA*, 100:1530–1534.

Disney, M. D., Matray, T., Gryaznov, S. M., & Turner, D. H. (2001). *Biochemistry*, 40:6520–6526.

Dromer, F. Charreire, J., Contrepois, A., Carbon, C., & Yeni, P. (1987). *Infect. Immun.*, 55:749–752.

Endo, M., Takesako, K., Kato, I., & Yamaguchi, H. (1997). *Antimicrob. Agents Chemother.*, 41:672–676.

Ernst, E. J. (2001). *Pharmaco. Therapy*, 21(8s):165–174.

Espinel-Ingroff, A. (2003). *Rev. Iberoam Micol.*, 20:121–136.

Ezzet, F., Wexler, D., Courtney, R., Krishna, G., Lim, J., & Laughlin, M. (2005). *Clin. Pharmacol.*, 44:211–220.

Freier, S. M., Altman, K. H. (1997). *Proc. Natl. Acad. Sci. USA*, 25:4429–4443.

Galderisi, U., Cascino, A., & Giordano, A. (1999). *J. Cell Physiol.*, 181:251–257.

Ganesan, L. T., Manavathu, E. K., Cutright, J. L., Alangaden, G. J., & Chandrasekar, P. H. (2004). *Clin. Microbiol. Infect.*, 10:961–966.

Garcia-Effron, G., Gomez-Lopez, A., Mellado, E., Monzon, A., Rodriguez-Tudela, J. L., & Cuenca-Estrella, M. (2004). *J. Antimicrob. Chemother.*, 53:1086–1089.

Gargallo-Viola, D. (1999). *Curr. Opin. Anti-infect. Investig. Drugs*, 1:297–305.

Georgopapadakou, N. H. & Walsh, T. J. (1996). *Antimicrob. Agents Chemother.*, 40:279–291.

Gordon, M. A. & Lapa, E. (1964). *J. Infect. Dis.*, 114:373–378.

Gorlach, J. M., McDade, H. C., Perfect, J. R., & Cox, G. M. (2002). *Microbiology*, 148:213–219.

Graybill, J. R., Bocanegra, R., Najvar, L. K., Leobenberg, D., & Luther, M. F. (1998a). *Antimicrob. Agents Chemother.*, 42(10):2467–2473.

Graybill, J. R., Najvar, L., Fothergill, A., Bocanegra, R., & De Las Heras, F. G. (1999). *Antimicrob. Agents Chemother.*, 43:1716–1718.

Graybill, J. R., Najvar, L. K., Bocanegra, R., Hector, R. F., & Luther, M. F. (1998b). *Antimicrob. Agents Chemother.*, 42:2371–2374.

Greenberg, R. N., Mullane, K., vanBurik, J. -A. H., Raad, I., Abzug, M. J., Anstead, G., Herbrecht, R., Langston, A., Marr, K. A., Schiller, G., Schuster, M., Wingard, J. R., Gonzalez, C. E., Revankar, S. G., Corcoran, G., Kryscio, R. J., & Hare, R. (2006). *Antimicrob. Agents Chemother.*, 50:126–133.

Groll, A. H. & Kolve, H. (2004). *Eur. J. Clin. Microbiol. Infect. Dis.*, 23:256–270.

Groll, A. H. & Walsh, T. J. (2006). *Expert Rev. Anti Infect. Ther.*, 3:467–487.

Groll, A. H. & Walsh, T. J. (2002). *Swiss Med. Wkly.*, 132:303–311.

Groll, A. H., Gea-Banacloche, J. C., Glasmacher, A., Just-Nuebling, G., Maschmeyer, G., & Walsh, T. J. (2003). *Infect. Dis. Clin. North Am.*, 17:159–191.

Groll, A. H., Gonzalez, C. E., Giri, N., Kligys, K., Love, W., Peter, J., Feuerstein, E., Bacher, J., Piscitelli, S. C., & Walsh, T. J. (1999a). *J. Antimicrob. Chemother.*, 43:95–103.

Groll, A. H., Petraitis, V., Petraitiene, R., Field-Ridley, A., Calendario, M., Bacher, J., Piscitelli, S. C., & Walsh, T. J. (1999b). *Antimicrob. Agents Chemother.*, 43:2463–2467.

Gupta, A. K., Cooper, E. A., & Lynde, C. W. (2003). *Dermatol. Clin.*, 21:511–520.

Gupta, A. K., Kohli, Y., & Batra, R. (2006). *Med. Mycol.*, 43:179–185.

Gupta, A. K., Ryder, J. E., Lynch, L. E., & Tavakkol, A. (2005). *J. Drug Dermatol.*, 4:302–308.

Han, Y. & Cutler, J. E. (1995). *Infect. Immun.*, 63:2714–2719.

Harmsen, A. G. & Stankiewicz, M. (1990). *J. Exp. Med.*, 172:937–945.

Hebart, H., Bollinger, C., Fisch, P., Sarfati, J., Meisner, C., Baur, M., Loeffler, J., Monod, M., Latge, J. -P., & Einsele, H. (2002). *Blood*, 100:4521–4528.

Hector, R. F. (1993). *Clin. Microbiol. Rev.*, 6:1–21.

Hector, R. F., Zimmer, B. L., & Pappagianis, D. (1990). *Antimicrob. Agents Chemother.*, 34:587–593.

Herbrecht, R. (2004). *Int. J. Clin. Pract.*, 58:612–624.

Herreros, E., Almela, M. J., Lozano, S., Gomez De Las Heras, F., & Gargallo-Viola, D. (2001). *Antimicrob. Agents Chemother.*, 45:3132–3139.

Herreros, E., Martinez, C. M., Almela, M. J., Marriott, M. S., Gomez De Las Heras, F., & Gargallo-Viola, D. (1998). *Antimicrob. Agents Chemother.*, 42:2863–2869.

Hiemenz, J., Cagnoni, P., Simpson, D., Devine, S., Chao, N., Keirns, J., Lau, W., Facklam, D., & Buell, D. (2005). *Antimicrob. Agents Chemother.*, 49:1331–1336.

Hossain, M. A., Reyes, G. H., Long, L. A., Mukherjee, P. K., & Ghannoum, M. A. (2003). *J. Antimicrob. Chemother.*, 51:1427–1429.

Ibrahim, A. S., Spellberg, B. J., Avanesian, V., Fu, Y., & Edwards, J. E. (2006). *Infect. Immun.*, 74(5):3039–3571.

Jang, W. S., Kim, H. K., Lee, K. Y., Kim, S. A., Han, Y. S., & Lee, I. H. (2006). *FEBS Lett.*, 580:1490–1496.

Johnson, E. M., Szekely, A., & Warnock, D. W. (1999). *Antimicrob. Agents Chemother.*, 43:1260–1263.

Johnson, M. D., MacDougall, C., Ostrosky-Zeichner, L., Perfect, J. R., & Rex, J. H. (2004). *Antimicrob. Agents Chemother.*, 48:693–715.

Joly, V., Bolard, J., Saint-Julien, L., Carbon, C., & Yeni, P. (1992). *Antimicrob. Agents Chemother.*, 36:262–266.

Kacishwar, A. & Shulka, P. K. (2006). *Med. Mycol.*, 44(2):159–167.

Kahn, J. N., Hsu, M. -J., Racine, F., Giacobbe, R., & Motyl, M. (2006). *Antimicrob. Agents Chemother.*, 50:2214–2216.

Kamai, Y., Kakuta, M., Shibayama, T., Fukuoka, T., & Kuwahara, S. (2005). *Antimicrob. Agents Chemother.*, 49:52–56.

Keating, G. M. (2006). *Drugs*, 65:1553–1567.

Keating, G. M. & Jarvis, B. (2001). *Drugs*, 61:1121–1129.

Khan, J. K., Montaseri, H., Poglod, M., Bu, H. -Z., Zuo, Z., Salama, S. M., Daneshtalab, M., & Micetich, R. G. (2000). *Antimicrob. Agents Chemother.*, 44:910–915.

Kohno, S., Masaoka, T., Yamaguchi, H., Mori, T., Ito, A., Niki, Y., & Ikemoto, H. (2004). *Scand. J. Infect. Dis.*, 36:372–379.

Kontoyiannis, D. P. & Lewis, R. E. (2004). *Br. J. Haematol.*, 126:165–175.

Kovalchuke, O. & Chakraburtty, K. (1994). *Eur. J. Biochem.*, 226:133–140.

Krause, D. S., Reinhardt, J., Vazquez, J. A., Reboli, A., Goldstein, B. P., Wible, M., & Henkel, T. (2004b). *Antimicrob. Agents Chemother.*, 48:2021–2024.

Krause, D. S., Simjee, A. E., van Rensburg, C., Viljoen, J., Walsh, T. J., Goldstein, B. P., Wible, M., & Henkel, T. (2004a). *Clin. Infect. Dis.*, 39:770–775.

Krieter, P., Flannery, B., Musick, T., Gohdes, M., Martinho, M., & Courtney, R. (2004). *Antimicrob. Agents Chemother.*, 48:3543–3551.

Kurmoe, T., Inami, K., Inoue, T., Ikai, K., Takesako, K., Kato, I., & Shiba, T. (1996). *Tetrahedron*, 52:4327–4356.

Kurtz, M. B. & Douglas, C. (1997). *Antimicrob. Agents Chemother.*, 35:79–86.

Larsen, R. A., Pappas, P. G., Perfect, J., Aberg, J. A., Casadevall, A., Cloud, G. A., James, R., Filler, S., & Dismukes, W. E. (2005). *Antimicrob. Agents Chemother.*, 49:952–958.

Letscher-Bru, V. & Herbrecht, R. (2003). *J. Antimicrob. Chemother.*, 51:513–521.

Lindell, D. M.., Moore, T. A., McDonald, R. A., Toews, G. B., & Huffnagle, G. B. (2005). *J. Immunol.*, 174:7920–7928.

Maitta, R. W., Datta, K., Chang, Q., Luo, R. X., Witover, B., Subramaniam, K., & Pirofski, L. -A. (2004). *Infect. Immun.*, 72:4810–4818.

Marco, F., Pfaller, M. A., Messer, S., & Jones, R. N. (1998). *Antimicrobial Agents Chemother.*, 42:161–163.

Marr, K. A., Boeckh, M., Carter, R. A., Kim, H. W., & Corey, L. (2004). *Clin. Infect. Dis.*, 39:797–802.

Martinez, A., Aviles, P., Jimenez, E., Caballero, J., & Gargallo-Viola, D. (2000). *Antimicrob. Agents Chemother.*, 44:3389–3394.

Martinez, A., Ferrer, S., Santos, I., Jimenez, E., Sparrowe, J., Regadera, J., Gomez De Las Heras, F., & Gargallo-Viola, D. (2001). *Antimicrob. Agents Chemother.*, 45:3304–3309.

Matsumoto, S., Wakai, Y., Nakai, T., Hatano, K., Ushitani, T., Ikeda, F., Tawara, S., Goto, T., Matsumoto, F., & Kuwahara, S. (2000). Efficacy of FK463, a new lipopeptide antifungal agent, in mouse models of pulmonary aspergillosis. *Antimicrob. Agents Chemother.*, 44:619–621.

Matthews, R. C. (1992). *J. Med. Microbiol.*, 36:367–370.

Matthews, R. C. (1994). *Microbiology*, 104:1505–1511.

Matthews, R. C. & Burnie, J. (2001). *Curr. Opin. Invest. Drugs*, 2:472–476.

Matthews, R. C. & Burnie, J. P. (2004). *Vaccine*, 22:865–871.

Matthews, R. C., Rigg, G., Hodgetts, S., Cater, T., Chapman, C., Gregory, C., Illidge, C., & Burnie, J. (2003). *Antimicrob. Agents Chemother.*, 47:2208–2216.

Matthews, R., Hodgets, S., & Burnie, J. (1995). *J. Infect. Dis.*, 171:1668–1771.

McCann, P. P., & Pegg, A. E. (1992). *Pharmacol. Ther.* 54:195–215

McGinnis, M. R., Nordoff, N. G., Ryder, N. S., & Nunn, G. B. (2000). *Antimicrob. Agents Chemother.*, 44:1407–1408.

Messer, S. A., Kirby, J. T., Sader, H. S., Fritsche, T. R., & Jones, R. N. (2004). *J. Antimicrob. Chemother.*, 54:1051–1056.

Mikamo, H., Sato, Y., & Tamaya, T. (2000). *J. Antimicrob. Chemother.*, 46:485–487.

Mikamo, H., Yin, X. H., Hayasaki, Y., Shimamura, Y., Uesugi, K., Fukayama, N., Satoh, M., & Tamaya, T. (2002). *Chemotherapy*, 48:7–9.

Mora-Duarte, J., Betts, R., Rotstein, R., Lopes-Colombo, A., Thompson-Moya, L., Smietana, J., Lupinacci, R., Sable, C., Kartsonis, N., Perfect, J., and Caspofungin Invasive Candidiasis Study Group (2002). *N. Engl. J. Med.*, 347:2020–2029.

Morrison, V. A. (2006). *Expert Rev. AntiInfect. Ther.*, 4:325–342.

Mukherjee, J., Nassbaum, G., Scharf, M. D., & Casadevall, A. (1995). *J. Exp. Med.*, 181:405–409.

Mukherjee, J., Pirofski, I. A., Scharf, M. D., & Casadevall, A. (1994). *Proc. Natl. Acad. Sci. USA*, 90:3636–3640.

Mukherjee, J., Scharf, M. D., & Casadevall, A. (1992). *Infect. Immun.*, 60:4534–4541.

Murphy, J. W. (1992). *Adv. Exp. Med. Biol.*, 319:225–230.

Murphy, M., Bernard, E. M., Ishimarc, T., & Armstrong, D. (1997). *Antimicrob. Agents Chemother.*, 41:696–698.

Nagiec, M. N., Nagiec, E. E., Baltisburger, J. A., Well, G. R., Lester, R. L., & Dickson, R. L. (1997). *J. Biol. Chem.*, 272:9809–9817.

Ng, A. W. K., Wasan, K. M., & Lopez-Berestein, G. (2003). *J. Pharm. Pharmaceut.*, 6:67–83.

Nguyen, M. H. & Yu, C. Y. (1998). *Antimicrob. Agents Chemother.*, 42:471–472.

Nimrichter, L., Barreto-Bergter, E., Mendonca-Filho, R. R., Kneipp, L. F., Mazzi, M. T., Salve, P., Farias, S. E., Wait, R., Alviano, C. S., & Rodrigues, M. L. (2004). *Microbes Infect.*, 6:657–665.

Nussbaum, G., Yuan, R., Casadevall, A., & Scharff, M. D. (1996). *J. Exp. Med.*, 183:1905–1909.

Odds, F. C. (2001). *Expert Opin. Ther. Pat.*, 11:283–294.

Odds, F. C., Brown, A. J. P., & Gow, N. A. R. (2003). *Trends in Microbiol.*, 11:272–279.

Odds, F. C., Ausma, J., VanGerven, F., Woestenborghs, F., Meerpoel, L., Heeres, J., Vanden Bossche, H., & Borgers, M. (2004). *Antimicrob. Agents Chemother.*, 48:388–391.

Olli, S. & Kirti, P. B. (2006). *J. Biochem. Mol. Biol.*, 39:278–283.

Ostrosky-Zeichner, L., Rex, J. H., Pappas, P. G., Hamill, R. J., Larsen, R. A., Horowitz, H. W., Powderly, W. G., Hyslop, N., Kauffman, C. A., Cleary, J., Mangino, J. E., & Lee, J. (2003). *Antimicrob. Agents Chemother.*, 47:3149–3154.

Pappagianis, D. and the Valley Fever Study Group. (1993). *Am. Rev. Respir. Dis.*, 148:656–660.

Pearson, M. M., Rogers, P. D., Cleary, J. D., & Chapman, S. W. (2003). *Ann. Pharmacother.*, 37:420–432.

Perea, S., Gonzalez, G., Fothergill, A. W., Kirkpatrick, W. R., Rinaldi, M. G., & Patterson, T. F. (2002b). *Antimicrob. Agents Chemother.*, 46:3039–3041.

Perea, S., Gonzalez, G., Fothergill, A. W., Sutton, D. A., & Rinaldi, M. G. (2002a). *J. Clin. Microbiol.*, 40:1831–1833.

Peschen, D., Li, H. P., Fischer, R., Kreuzaler, F., Liao, Y. C. (2004). *Nature Biotechnol.*, 22(6):682–683.

Pettengell, K., Mynhardt, J., Kluyts, T., Lau, W., Facklam, D., Buell, D., and the FK463 South African Study Group. (2004). *Aliment. Pharmacol. Ther.*, 20:475–481.

Pfaller, M. A. (2004). *Expert Opin. Investig. Drugs*, 13:1183–1197.

Pfaller, M. A., Boyken, L., Hollis, R. J., Messer, S. A., Tendolkar, S., & Diekema, D. J. (2005). *J. Clin. Microbiol.*, 43:5425–5427.

Pfaller, M. A., Marco, F., Messer, S. A., & Jones, R. W. (1998b). *Diag. Microbiol. Infect. Dis.*, 30:251–255.

Pfaller, M. A., Messer, S. A., Boyken, L., Hollis, R. J., Rice, C., Tendolkar, S., and Diekema, D. J. (2004). *Diag. Microbiol. Infect. Dis.*, 48:201–205.

Pfaller, M. A., Messer, S. A., Hollis, R. J., Jones, R. N., and the Sentry Participants group. (2002). *Antimicrob. Agents Chemother.*, 46:1032–1037.

Pfaller, M. A., Messer, S. A., Hollis, R. J., Jones, R. N., Doern, G. V., Brandt, M. E., & Hajjeh, R. A. (1998a). *Antimicrob. Agents Chemother.*, 42:3242–3244.

Phair, J., Munoz, A., Detels, R., Kaslow, R., Rinaldo, C., & Saah, A. (1990). *N. Engl. J. Med.*, 322:161–165.

Polonelli, L., Magliani, W., Conti, S., Bracci, L., Lozzi, L., Neri, P., Adriani, D., De Bernardis, F., & Cassone, A. (2003). *Infect. Immun.*, 71:6205–6212.

Polonelli, L., Seguy, N., Conti, M., Gerloni, M., Bertolotti, D., Cantelli, C., Magliani, W., & Cailliez, J. C. (1997). *Clin. Diag. Lab. Immunol.*, 4:142–146.

Raad, I. I., Graybill, J. R., Bustamante, A. B., Cornely, O. A., Gaona-Flores, V., Afif, C., Graham, D. R., Greeberg, R. N., Hadley, S., Langston, A., Negroni, R., Perfect, J. R., Pitisuttithum, P., Restrepo, A., Schiller, G., Pedicone, L., & Ullmann, A. J. (2006 b). *Clin. Infect. Dis.*, 42:1726–1734.

Raad, I. I., Hachem, R. Y., Herbrecht, R., Graybill, J. R., Hare, R., Corcoran, G., & Kontoyiannis, D. R. (2006a). *Clin. Infect. Dis.*, 42:1398–1403.

Randhawa, G. K. & Sharma, G. (2004). *Indian J. Pharmacol.*, 36:65–71.

Raska, M., Belakova, J., Wudattu, N. K., Kafkova, L., Ruzickova, K., Sebestova, M., Kolar, Z., & Weigl, E. (2005). *Folia Microbiol.*, 50:77–82.

Razonable, R. R., Henault, M., Lee, L. N., Laethem, C., Johnston, P. A., Watson, H. L., & Paya, C. V. (2005a). *Antimicrob. Agents Chemother.*, 49:1617–1621.

Razonable, R. R., Henault, M., Watson, H. L., & Paya, C. V. (2005b). *Antimicrob. Agents Chemother.*, 49:3546–3549.

Reichenberger, F., Habicht, J. M., Gratwohl, A., & Tamm, M. (2002). *Eur. Respir. J.*, 19:743–755.

Ribaud, P., Chastang, C., Latge, J. -P., Baffroy-Lafitte, L., Parquet, N., Devergie, A., Esperou, H., Selimi, F., Rocha, V., Derouin, F., Socie, G., & Gluckman, E. (1999). *Clin. Infect. Dis.*, 28:322–330.

Ribedly, J. M., Christensen, J. P., Branum, K., & Doherty, P. C. (2000). J. Virol., 74:9762–9765.

Rimaroli, C. & Bruzzese, T. (1998). *Antimicrob. Agents Chemother.*, 42:3012–3013.

Rimaroli, C. & Bruzzese, T. (2000). Chemother., 46:28–35.

Rimaroli, C., Bonabello, O. A., Sala, P., & Bruzzese, T. (2002). *J. Pharm. Sci.*, 91:1252–1258.

Rivera, J. & Casadevall, A. (2005). *J. Immunol.*, 174(12):8017–8026.

Roilides, E., Holmes, A., Blake, C., Pizzo, P. A., & Walsh, T. J. (1993). *J. Infect. Dis.*, 167:905–911.

Roling, E. E., Klepser, M. E., Wasson, A., Lewis, R. E., Ernst, E. J., & Pfaller, M. A. (2002). *Diagn. Microbiol. Infect. Dis.*, 43:13–17.

Romani, L. (2002). Immunology of invasive candidiasis. In: Candida *and* Candidiasis, R. A. Calderone, Ed., ASM Press, Washington, DC, pp. 223–241.

Sandhu, P., Xu, X., Bondiskey, P. J., Balani, S. K., Morris, M. L., Tang, Y. S., Miller, A. R., & Pearson, P. G. (2004). *Antimicrob. Agents Chemother.*, 48:1272–1280.

Sanford, J. E., Lupan, D. M., Schlageter, A. M., & Kozel, T. R. (1990). *Infect. Immun.*, 58:1919–1923.

Santos, C., Rodriguez-Gabriel, M. A., Remacha, M., & Ballesta, J. P. G. (2004). *Antimicrob. Agents Chemother.*, 48:2930–2936.

Sau, K., Mambula, S. S., Latz, E., Henneke, P., Golenbock, D. T., & Levitz, S. (2003). *J. Biol. Chem.*, 278:37561–37568.

Schwarz, P., Dromer, F., Lortholary, O., & Dannaoui, E. (2006). *Antimicrob. Agents Chemother.*, 50:113–120.

Seibel, N. L., Schwartz, C., Arrieta, A., Flynn, P., Shad, A., Albano, E., Keirns, J., Lau, W. M., Facklam, D. P., Buell, D. W., & Walsh, T. J. (2005). *Antimicrob. Agents Chemother.*, 49:3317–3324.

Shellito, J., Suzara, V. V., Blumenfeld, W., Beck, J. M., Steger, H. J., & Ermak, T. H. (1990). *J. Clin. Investig.*, 85:1686–1693.

Shoham, S. & Levitz, M. L. (2005). *Br. J. Hematol.*, 129:569–582.

Stevens, D. A. (2000). *Antimicrob. Agents Chemother.*, 44:2547–2548.

Svirshchevskaya, E., Frolova, E., Alekseeva, L., Kotzareva, O., & Kurup, V. P. (2000) *Peptides*, 21:1–8.

Takakura, S., Fujihara, N., Saito, T., Kudo, T., Iinuma, Y., Ichiyama, S., and the Japan Invasive Mycosis Surveillance Study Group (2004). *J. Antimicrob. Chemother.*, 53:283–289.

Testa, S. M., Gryaznov, S. M., & Turner, D. H. (1999). *Proc. Natl. Acad. Sci. USA*, 96:2734–2739.

Thullen, T. D., Ashbaugh, A. D., Daly, K. R., Linke, M. J., Steele, P. E., & Walzer, P. D. (2003). *Infect. Immun.*, 71:6292–6297.

Torosantucci, A., Bromuro, C., Chiani, P., De Bernardis, F., Berti, F., Galli, C., Norelli, F., Bellucci, C., Polonelli, L., Costantino, P. Rappuoli, R., & Cassone, A. (2005). *J. Exp. Med.*, 202:597–606.

Torres, H. A., Hachem, R. Y., Chemaly, R. F., Kontoyiannis, D. P., & Raad, I. I. (2005). *Lancet Infect. Dis.*, 5:775–785.

Turner, M. S., Drew, R. H., & Perfect, J. R. (2006). *Expert Opin. Emerg. Drug*, 11:231–250.

Uchida, K., Nishiyama, Y., Yokota, N., & Yamaguchi, H. (2000). *J. Antibiot.* (*Tokyo*) 53:1175–1181.

Uzun, O., Kocagoz, S., Cetinkaya, Y., Arikon, S., & Unal, S. (1997). *Antimicrob. Agents Chemother.*, 41:1156–1157.

VanBurik, J. -A. H., Hare, R. S., Solomon, H. F., Corrado, M. L., & Kontoyiannis, D. P. (2006). *Clin. Infect. Dis.*, 42:61–65.

VanBurik, J. -A. H., Ratanatharathorn, V., Stepan, D. E., Miller, C. B., Lipton, J. H., Vesole, D. H., Bunin, N., Wall, D. A., Hiemenz, J. W., Satoi, Y., Lee, J. M., Walsh, T. J. and National Institute of Allergy and Infectious Diseases Mycoses Study Group (2004). *Clin. Infect. Dis.*, 39:1407–1416.

Vazquez, J. A. (2005). *Clin. Ther.*, 27:657–673.

Vazquez, J. A. & Sobel, J. D. (2006). *Clin. Infect. Dis.*, 43:215–222.

Vazquez, J. A., Skiest, D. J., Nieto, L., Northland, R., Sanne, I., Gogate, J., Greaves, W., & Isaacs, R. (2006). *Clin. Infect. Dis.*, 42:1179–1186.

Vermes, A., Gucherlaar, H. -J., & Dankert, J. (2000). *J. Antimicrob. Chemother.*, 46:171–179.

Viejo-Diaz, M., Andres, M. T., & Fierro, J. F. (2005). *Antimicrob. Agents Chemother.*, 49:2583–2588.

Villanueva, A., Arathoon, E. G., Gotuzzo, E., Berman, R. S., DiNubile, M. J., & Sable, C. A. (2001a). *Clin. Infect. Dis.*, 33:1529–1535.

Villanueva, A., Gotuzzo, E., Arathoon, E., Noriega, L. M., Kartsonis, N., Lupinacci, R. J., Smietana, J. M., DiNubile, M. J., & Sable, C. A. (2001b). *Am. J. Med.*, 113:294–299.

Vora, S., Purimetla, N., Brummer, E., & Stevens, D. A. (1998). *Antimicrob. Agents Chemother.*, 42(4):907–910.

Walsh, T. J., Lutsar, I., Driscoll, T., Dupont, B., Roden, M., Ghahramani, P., Hodges, M., Groll, A. H., & Perfect, J. R. (2002a). *Pediatr. Infect. Dis. J.*, 21:240–248.

Walsh, T. J., Pappas, P., Winston, D. J., Lazarus, H. M., Petersen, F., Raffalli, J., Yanovich, S., Stiff, P., Greenberg, R., Donowitz, G., & Lee, J. (2002b). *New Eng. J. Med.*, 346:225–234.

Wang, J. C. & Livingstone, A. M. (2003). *J. Immunol.*, 171:6339–6343.

Warn, P. A., Sharp, A., Morrissey, G., & Denning, D. W. (2005). *J. Antimicrob. Chemother.*, 56:590–593.

Warn, P. A., Sharp, A., & Denning, D. W. (2006). *J. Antimicrob. Chemother.*, 57:135–138.

Wills, E. A., Redinbo, M. R., Perfect, J., & DelPoeta, M. (2000). *Emerg. Ther. Targets*, 4:1–32.

Wingard, J., White, M., Anaissie, E., Raffalli, J., Goodman, T., Arrieta, A., and the L Amph/ABLC Collaborative Study Group (2000). *Clin. Infect. Dis.*, 31:1155–1163.

Wüthrich, M., Filutowicz, H. I., & Klein, B. S. (2000). *J. Clin. Invest.*, 106(11):1381–1389.

Yamazumi, T., Pfaller, M. A., Messer, S. A., Houston, A., Hollis, R. J., & Jones, R. N. (2000). *Antimicrob. Agents Chemother.*, 44:2883–2886.

Zhanel, G. G., Karlowsky, J. A., Harding, G. J., Balko, T. V., Zelenitsky, S. A., Friesen, M., Kabani, A., Turik, M., & Hoban, D. J. (1997). *Antimicrob. Agents Chemother.*, 41:863–865.

Zhou, Q., Gault, R. A., Kozel, T. R., & Murphy, W. J. (2006). *Infect. Immun.*, 74(4):2161–2168.

Chapter 5
Candida albicans: New Insights in Infection, Disease, and Treatment

Donna MacCallum

C. albicans: Commensal and Pathogen

Candida albicans is commonly found in the gastrointestinal tract, oral cavity, and genital area as a harmless commensal. Based on recent studies in healthy individuals, asymptomatic oral carriage of *Candida* species is estimated to occur in 24–70% of children and adults, with a reduced frequency in babies less than 1 year of age (Table 5.1). Of isolates identified, the majority are *C. albicans* (38–76% in adults and children). Again, the frequency of *C. albicans* differs across different age groups, with a far greater proportion of isolates identified as *C. albicans* in young babies and in the elderly (Table 5.1). Higher oral carriage rates are found in HIV positive patients (Sanchez-Vargas et al., 2005a; Liu et al., 2006) and diabetics (Belazi et al., 2005).

Asymptomatic vaginal carriage of *Candida* species is estimated to occur in 21–32% of healthy women, with *C. albicans* representing 20–98% of identified isolates (Agatensi et al., 1991; Holland et al., 2003; Beigi et al., 2004; Beltrame et al., 2006; Pirotta & Garland, 2006). One recent study found that within a group of women repeatedly screened over 12 months 30% were never colonised, 70% were colonised on at least 1 occasion, and 4% were persistently colonised (Beigi et al., 2004). Higher rates of vaginal carriage have been found in pregnant women, women colonised by *Lactobacillus* species (Cotch et al., 1998; Beigi et al., 2004), Type I diabetics (de Leon et al., 2002), and post-antibiotic treatment (Pirotta & Garland, 2006).

It is important, however, to note that as well as being a harmless commensal *C. albicans*, and some other *Candida* species, are opportunistic pathogens and cause a range of different infections. These cover a spectrum ranging from irritating superficial mucosal lesions to disseminated blood infections, which have an associated high mortality rate.

K. Kavanagh (ed.), *New Insights in Medical Mycology.*
© Springer 2007

Table 5.1 Oral carriage of *Candida*, and specifically *C. albicans*, in the general population. Carriage frequency of yeasts in the oral cavity of asymptomatic individuals are mean values calculated from the percentage carriage rates reported in recent studies. The range of reported values is shown below the mean value

Age group (year)	% Carriage (range)	% Isolates *C. albicans* (range)	Source references
<1	26 (8–44)	78 (56–100)	Kleinegger et al. (1996), Qi et al. (2005)
3–18	46.6 (24–70)	75.4 (38–93)	Kleinegger et al. (1996), Qi et al. (2005), Sanchez-Vargas et al. (2005a, b)
18–45	55.8 (41–64)	64.1 (36–75)	Kleinegger et al. (1996), Fanello et al. (2001), Qi et al. (2005), Sanchez-Vargas et al., (2005a, b)
>60	63.8 (59–69)	74 (72–76)	Kleinegger et al. (1996), Wang et al. (2006b)

Superficial Mucosal Infections

Superficial mucosal lesions occur in the oral and vaginal cavities and are commonly called 'thrush'. These infections can occur in both immunocompetent and immunocompromised persons. Oral thrush, or oral candidiasis, is a common problem seen in infants, the elderly, and in cancer patients, particularly those with haematological malignancies, receiving chemotherapy, or receiving head and neck radiotherapy. It is characterised by white growth on mucous membranes of the oral cavity, which have underlying red areas when the fungal growth is scraped off.

The majority of isolates associated with oral candidiasis were identified as *C. albicans* (63–84%) (Sanchez-Vargas et al., 2005a, b; Davies et al., 2006). Risk factors associated with oral candidiasis include xerostomia (dry mouth) and denture wearing (Davies et al., 2006), as well as poorly controlled diabetes mellitus (Guggenheimer et al., 2000) and immunosuppression. The frequency of oral candidiasis, but not oral carriage of *Candida* species (Sanchez-Vargas et al., 2005a), is higher in HIV positive patients with decreased CD4[+] T lymphocyte counts (Liu et al., 2006).

Vulvovaginal thrush is a relatively common problem, representing roughly a quarter of all infectious vaginitis, with a large proportion of women experiencing at least one episode of vaginal thrush during their lifetime (Sobel et al., 1998). Some women experience multiple episodes, commonly called recurrent vulvovaginal candidiasis (RVVC). Symptoms of vulvovaginal candidiasis include itching, burning, soreness, and abnormal vaginal discharge. In women presenting with vulvovaginal candidiasis 60–100% of isolates are identified as *C. albicans* (Giraldo et al., 2000; Lopes Consolaro et al., 2004; De Vos et al., 2005; Beltrame et al., 2006; Moreira & Paula, 2006; Paulitsch et al., 2006; Pirotta & Garland, 2006).

Predisposing factors for vulvovaginal candidiasis may include use of contraceptive pills (Moreira & Paula, 2006), although this remains controversial (Pirotta & Garland, 2006).

Candidaemia

Candidaemia, i.e. bloodstream infection by *Candida* species, is a significant clinical problem, with the attributable mortality estimated to be between 10% and 71% (Kibbler et al., 2003; Safdar et al., 2004; Zaoutis et al., 2005; Falagas et al., 2006; Vigouroux et al., 2006; Zaragoza & Peman, 2006). In recent years there was a gradual increase in the incidence of *Candidaemia* (Bassetti et al., 2006; Sandven et al., 2006; Sendid et al., 2006). The symptoms of *Candidaemia* generally resemble those found for bacterial sepsis, i.e. fever, and in babies includes respiratory distress (symptoms recently reviewed by Sims et al. (2005)).

Although non-*albicans Candida* species are found to cause *Candidaemia*, *C. albicans* is responsible for the majority of infections (42–100%), with some variation found in different patient groups (Schelenz & Gransden, 2003; Tortorano et al., 2004, 2006; Badiee et al., 2005; Cliff et al., 2005; Laupland et al., 2005; Bassetti et al., 2006; Fridkin et al., 2006; Pasqualotto et al., 2006; Sandven et al., 2006; Sendid et al., 2006). An example of this is found for cancer patients where, for patients with haematological malignancies, the frequency of *C. albicans* compared to non-*albicans Candida* was approximately half of that found for patients with solid tumours (Pasqualotto et al., 2006; Tortorano et al., 2006). However, even within patients with solid tumours, neutropenia permitted increased infections by non-*albicans Candida* (Pasqualotto et al., 2006).

The majority of patients who develop *Candida* bloodstream infections are those in intensive care units (ICU) and those that have undergone surgery, especially abdominal surgery (including organ transplant and surgery for gastrointestinal problems). Other patient groups at risk of *Candidaemia* are cancer patients, both solid tumours and haematological malignancies, and premature babies (<1,000 g birth weight) (Schelenz & Gransden, 2003; Sims et al., 2005; Richardson, 2005; Maschmeyer, 2006). Major predisposing factors related to *Candidaemia* are listed in Table 5.2. It should be noted that for critically ill patients several risk factors may be present in, or apply to, the same patient. For the majority of patients, a central venous catheter (CVC) had been present, and one study found that 14% of screened CVC tips were positive for micro-organisms (with *C. albicans* being the second most common microbe found after coagulase-negative *Staphylococcus* species) (Hammarskjold et al., 2006).

In order to help prevent development of serious *C. albicans* infections and to develop more effective treatments and drugs, it is essential to gain further understanding of *C. albicans* biology, the infection process, interactions that occur between host and fungus, and the behaviour of *C. albicans* cells in response to antifungal therapy.

Table 5.2 Risk factors associated with *C. albicans* bloodstream infections. Host-related and health care-related factors identified in recent publications are listed in the table. (From Schelenz & Gransden, 2003; Richardson, 2005; Sims et al., 2005; Maschmeyer, 2006.)

Central venous catheter (CVC)
Abdominal surgery, including solid organ transplants
Prolonged stay in intensive care unit (ICU)
Broad spectrum antibiotic use
Parenteral nutrition
Immunosuppression, including neutropenia
Prior heavy colonisation with *C. albicans*
Corticosteroid therapy
Very low birth weight in neonates

One Giant Step

One of the greatest steps forward in our understanding of *C. albicans* has been the release of the human annotated genome sequence of *C. albicans* strain SC5314 (Jones et al., 2004; Braun et al., 2005; d'Enfert et al., 2005; Nantel, 2006). The annotated genome data is freely available from several websites (Agabian Annotation website; *Candida albicans*@NRC-Biotechnology Research Institute website; *Candida*DB; The *Candida* Genome Database (CGD)), with the CGD resource not only providing tools for viewing, searching, analysing, and downloading data, but also collecting and organising data from published *C. albicans* literature, providing links to literature relating to genes of interest (Arnaud et al., 2005; Costanzo et al., 2006). The genome of a second *C. albicans* strain, WO-1, has been sequenced recently and the data made available to the research community (WO-1 *C. albicans* genome sequence website). Physical mapping of the *C. albicans* SC5314 genome has also been carried out (*C. albicans* physical map website).

Genome sequence data has allowed a number of significant advances, including bioinformatic identification of *C. albicans* genes (including homologues of *Saccharomyces cerevisiae* genes), application of further molecular tools for the fungus, design of a multi-locus sequence typing (MLST) scheme for characterisation of *C. albicans* strains, design of DNA microarrays for gene transcription studies, and improved identification of proteins in proteomic analyses.

An Expanding Molecular Toolbox for C. albicans

Up until recently the tools available for working with *C. albicans* were fairly limited, although gene disruption using the URA-blaster cassette has been possible since 1993 (Fonzi & Irwin, 1993). Gene disruptions have become simpler with the development of PCR knockout cassettes (de Hoogt et al., 2000; Wilson et al., 2000; Gola et al., 2003; Taneja et al., 2004) and gene function can be rapidly tested by disruption using the UAU cassette (Enloe et al., 2000).

Unfortunately, the URA-blaster technique has been found to generate mutants that can be affected in their virulence due only to the expression of the *URA3* gene at an ectopic locus (Lay et al., 1998; Sundstrom et al., 2002; Staab & Sundstrom, 2003; Brand et al., 2004). This problem can be circumvented by expressing *URA3* at a highly expressed locus, such as *ENO1* (Sundstrom et al., 2002) or *RPS1* (Brand et al., 2004) or by reintroducing *URA3* at its native locus (Cheng et al., 2003).

A further advance in gene disruption has been the development of dominant selectable markers and their use in disruption cassettes for *C. albicans* (Beckerman et al., 2001; Reuss et al., 2004; Morschhauser et al., 2005; Shen et al., 2005). Since *URA3* is no longer the selectable marker, the Ura3p effect is no longer a problem and, more importantly, gene disruptions are no longer confined to laboratory strains derived from SC5314. Dominant markers in use for *C. albicans* include resistance to mycophenolic acid (Beckerman et al., 2001; Morschhauser et al., 2005) and nourseothricin (Reuss et al., 2004; Shen et al., 2005).

The number of regulatable promoters available for use in *C. albicans* is also increasing, with the tetracycline-regulatable system (Nakayama et al., 2000; Roemer et al., 2003; Park & Morschhauser, 2005b) joining other regulatable promoters (Gerami-Nejad et al., 2004). This system not only allows phenotypes to be examined before and after the gene is switched off *in vitro*, but can also be used *in vivo*.

The development of codon-optimised fluorescent reporter genes (GFP (green), YFP (yellow), and CFP (cyan)) (Gerami-Nejad et al., 2001) allows localisation and/or co-localisation of proteins at the level of single cells (Hazan & Liu, 2002; Oberholzer et al., 2002; Gerami-Nejad et al., 2004; Soares-Silva et al., 2004; Karababa et al., 2006). In addition, GFP linked to gene promoters (Barelle et al., 2004) allows visualisation of the induction of those promoters under various conditions, both *in vitro* and *in vivo* (Strauss et al., 2001; Gaur et al., 2005; Green et al., 2005a; Barelle et al., 2006). In addition to targeted investigation of *C. albicans* genes and proteins of interest, the application of transcript profiling experiments or proteomic analyses provide unbiased, global perspectives on gene and protein expression, respectively.

DNA microarrays have been designed and transcript profiling performed to investigate changes in *C. albicans* gene expression by various groups including Eurogentec (Eurogentec website) in collaboration with the European Galar Fungail Consortium (European Galar Fungail Consortium website) and the NRC Biotechnology Research Institute, Montreal, Canada (*Candida albicans*@NRC-Biotechnology Research Institute website) as well as many others. Transcript profiling experiments and their contributions to the understanding of *C. albicans* biology and pathogenesis have recently been reviewed by Garaizar et al. (2006). One of the major uses of transcript profiling for *C. albicans* has been to examine the global effects on gene expression caused by gene knockouts (Garcera et al., 2005; Garcia-Sanchez et al., 2005; Harcus et al., 2004; Lan et al., 2004; Lane et al., 2001; Lee et al., 2004; Lotz et al., 2004; Murad et al., 2001; Sohn et al., 2003; Tournu et al., 2005; Wang et al., 2006, 2007). These analyses have allowed identification of groups of co-ordinately regulated genes, and have provided insights into pathways involved in certain cellular responses.

Proteomics studies of *C. albicans* allow greater understanding of changes that occur at the protein level, rather than at the transcription level. Several two-dimensional (2-D) polyacrylamide gel electrophoresis (PAGE) maps of the *C. albicans* proteome, both yeast and hyphal forms, have been produced (Hernandez et al., 2004; Yin et al., 2004), with the Cogeme map available at http://www.abdn.ac.uk/proteomics/ap-*Candida*-albicans-2d-map.hti. The contributions of proteomics to our understanding of *C. albicans* and disease have recently been reviewed by Pitarch et al. (2006b, c).

MLST approach for characterisation of *C. albicans* strains has also been designed (Bougnoux et al., 2002, 2004; Tavanti et al., 2003; Robles et al., 2004), which builds upon previous molecular typing systems for *C. albicans* (Schmid et al., 1990; Chowdhary et al., 2006). This system allows strains to be typed, their relatedness determined and can also be used to track the development of potential outbreaks (Bougnoux et al., 2006; Chen et al., 2006; Odds et al., 2006; Viviani et al., 2006). This has also demonstrated that the majority of *C. albicans* isolates causing systemic infection are endogenous to the host (Odds et al., 2006) and that there is some evidence of microevolution of strains in a patient over time (Bougnoux et al., 2002, 2006; Chen et al., 2006; Odds et al., 2006).

The remainder of this chapter describes recent increases in our understanding of *C. albicans*, its biology and host–fungus interactions, with emphasis on data obtained from transcript profiling and proteomic analyses. Important in the ability of *C. albicans* to infect and cause disease is its resistance to stresses, especially those experienced during interactions with host immune cells, its ability to adhere to surfaces and host cells, and the ability to form biofilms. The studies focussed upon are those concerned with increasing our understanding of life-threatening systemic infection.

Yeast–hypha Dimorphism

Yeast–hypha dimorphism is believed to be hugely important in the virulence of *C. albicans* (Kumamoto & Vinces, 2005), and has been the focus of a significant amount of research in the past. Transcript profiling studies have built upon the knowledge obtained from previous studies (Nantel et al., 2002; Sohn et al., 2003; Lee et al., 2004; Bachewich et al., 2005; Enjalbert & Whiteway, 2005; Kadosh & Johnson, 2005; Singh et al., 2005; Cao et al., 2006).

Efg1p has been confirmed as an important regulator of the yeast–hypha dimorphic switch (Nantel et al., 2002; Sohn et al., 2003; Bachewich et al., 2005; Singh et al., 2005; Cao et al., 2006). During the formation of hyphae in the presence of macrophages genes involved in galactose, protein, and lipid metabolism and stress response genes underwent changes in transcription. The promoters of differentially expressed genes were found to contain binding sites for Cph1 and Efg1 (Singh et al., 2005). Flo8p was shown to interact with Efg1, and is involved in controlling a subset of Efg1p-regulated genes, which mostly show hypha-specific expression (including *HGC1* and *IHD1*), with the *flo8* strain unable to form hyphae (Cao et al., 2006). However, Efg1p plays a major role in the induction and repression of cell wall genes in both yeast and hyphae (Sohn et al., 2003). The *SIT4* gene has also been demonstrated to be involved in induction of hypha-specific genes, with the expression of two glucanase-encoding genes, *XOG1* and *YNR67*, found to rely entirely upon Sit4p (Lee et al., 2004).

Many of the genes expressed in hyphal cells have been found to be repressed under yeast growth conditions. Repression can occur through the actions of Nrg1p and Tup1p (Kadosh & Johnson, 2005) or can involve regulation of histone genes, as found in response to the quorum sensing molecule farnesol (Enjalbert & Whiteway, 2005).

Yeast–hypha dimorphism has also been studied by proteomic analyses (Pitarch et al., 2002; Choi et al., 2003; Ebanks et al., 2006). Studies examining the cell wall proteins of yeast and hyphal forms found that there were a number of proteins upregulated in the hyphal form (Pitarch et al., 2002; Ebanks et al., 2006). Upregulated proteins included Hsp70s and Hsp90, fructose bisphosphate aldolase, glyceraldehyde-3-phosphate dehydrogenase, phosphoglycerate kinase, phosphoglycerate mutase, alcohol dehydrogenase, enolase, pyruvate kinase, agglutinin-like sequence proteins (Als family), and myo-inositol-1-phosphate synthase (Ino1p). A further study found proteins Pra1p. Phr1p, and Tsa1p to be upregulated in hyphal cells compared to yeasts (Choi et al., 2003). However, only transcription of *PHR1* was found to be increased, suggesting that the other two proteins are modulated at the post-transcriptional level. Proteomic studies have also identified some yeast-specific proteins (Ebanks et al., 2006).

C. albicans Cell Surface

The proteins displayed on the cell wall are obviously important as it is these cells that can be sensed and interact with host cells (Sohn et al., 2006a). Glycosylphosphatidylinositol (GPI)-anchored proteins in the cell wall have also been investigated using proteomic approaches. Theoretically, *C. albicans* strain SC5314 could display 104 GPI-anchored proteins on its cell surface (De Groot et al., 2003). However, proteomic analysis of exponentially growing yeast cells demonstrated that only 14 cell wall proteins were found on the cell surface (12 GPI-anchored and 2 mild-alkali sensitive proteins) (de Groot et al., 2004). Proteins identified include carbohydrate-active enzymes (chitinase, Crh11p, Pga4p, Phr1p, Scw1p), adhesins (Als1p and Als4p, Pga24), a superoxide dismutase (Sod4p), and some unknown function proteins (Ecm33.3p, Pir1p, Pga29p, Rbt5p, Ssr1p). One of these proteins, Ecm33.3p, has been demonstrated to be important for normal cell wall structure and interactions with host cells (Martinez-Lopez et al., 2006). Furthermore, two GPI-anchored proteases (Albrecht et al., 2006) and Als proteins (Hoyer, 2001; Sheppard et al., 2004) have been demonstrated to be involved in adhesion to host cells. Expression of the *ALS* gene family has been demonstrated to be differentially regulated during growth by GFP-promoter fusion experiments (Green et al., 2005b), with *ALS1* gene expression associated with transfer into fresh medium and *ALS7* showing a transient peak 2–3 h after movement into fresh medium. *ALS3* expression increases were associated with formation of germ tubes. *In vivo*, in a mouse model of systemic disease, *ALS1, 2, 3, 4*, and *9* were all found to be expressed (Green et al., 2005a).

Proteomics have also been used to identify *C. albicans* proteins that bind to blood proteins, e.g. plasminogen (Crowe et al., 2003; Jong et al., 2003). Proteins identified include phosphoglycerate mutase, alcohol dehydrogenase, glyceraldehyde-3-phosphate dehydrogenase, phosphoglycerate kinase, fructose bisphosphate aldolase (Fba1p), and enolase (Eno1p), which were also found to be upregulated in proteomic studies examining yeast–hypha dimorphism (Pitarch et al., 2002; Choi et al., 2003; Ebanks et al., 2006). *C. albicans* is able to invade and transcytose brain microvascular endothelial cells without affecting monolayer intergrity (Jong et al., 2001), with plasmin-bound cells demonstrated to do this more easily (Jong et al., 2003).

A further proteomic approach, immunoproteomics, also been applied to identify immunogenic proteins of *C. albicans* (Pitarch et al., 2001; Fernandez-Arenas et al., 2004; Pitarch et al., 2004). This involves 2-D electrophoresis followed by Western blotting.

Proteomic analysis of immunogenic proteins in a mouse model of systemic candidiasis identified more than 31 immunoreactive proteins; including glycolytic enzymes, such as fructose bisphosphate aldolase, triose phosphate isomerase (Tpi1p), glyceraldehyde-3-phosphate dehydrogenase, phosphoglycerate kinase, triose phosphate isomerase, enolase, and pyruvate kinase (Pitarch et al., 2001). Metabolic enzymes, such as methionine synthase, Imh3p, alcohol dehydrogenase and aconitase, and heat-shock proteins from the Hsp70 family were also identified as immunogenic. Different antibody profiles were identified for strains of mice with differing susceptibility to systemic *C. albicans* infection. Eno1p was the dominant immunogenic protein in BALB/c mice (most resistant), with the more susceptible mice having a stronger reaction to methionine synthase, Hsp70 proteins, and phosphoglycerate kinase (Pitarch et al., 2001).

Immunoproteomics with patient serum identified proteins including Hsp70s and Hsp90, enolase (Eno1p), pyruvate kinase (Cdc19p), pyruvate decarboxylase (Pdc11p), hexokinase, glucose-6-phosphate isomerase, glyceraldehyde-3-phosphate dehydrogenase, phosphoglycerate kinase, triose phosphate isomerase (Tpi1p), alcohol dehydrogenase, Ino1p, and fructose-bisphosphate aldolase (Fba1p) (Fernandez-Arenas et al., 2004). Two protective antigens were also identified, IMP dehydrogenase (Imh3p) and acetyl CoA synthetase (Acs2p) (Fernandez-Arenas et al., 2004; Pitarch et al., 2004). The agreement between the immunogenic proteins found for the murine model and for human patients reinforces the usefulness of the mouse model for studying immune response to *C. albicans* (Pitarch et al., 2001).

In surviving and non-surviving patients with candidiasis differing antibody profiles were found (Pitarch et al., 2004), with recovering patients maintaining relatively high levels of antibodies to the following proteins: Eno1p, Pgk1p, and Fba1p. In an attempt to identify diagnostic or prognostic biomarkers for systemic candidiasis a systematic proteomic approach has been used (Pitarch et al., 2006a). Statistical analyses demonstrated that only high levels of antibodies against 1,3-β-glucosidase (Bgl2p) and phosphoglycerate kinase (Pgk1p) were independent predictors of systemic candidiasis. High levels of anti-Bgl2 antibodies, or seropositivity for enolase antibodies, are associated with a reduced risk of death (Pitarch et al., 2006a).

C. albicans and Stress

The ability of the *C. albicans* to resist various stresses is of great importance in its ability to infect and cause disease (Kraus & Heitman, 2003; Kruppa & Calderone, 2006; Monge et al., 2006). Stresses influencing *C. albicans* behaviour during infection and disease include oxidative and nitrosative stresses and nutrient limitation.

Transcript profiling experiments have shown that *C. albicans* has both a core stress response and stress-specific responses (Enjalbert et al., 2006). The Hog1p pathway was found to be involved in responses to osmotic and heavy metal stresses, but not to oxidative stress (Enjalbert et al., 2006). However, the transcription factor Cap1p has been demonstrated to be involved in the oxidative stress response via multiple pathways, including the cellular antioxidant defence system (Enjalbert et al., 2006; Wang et al., 2006). Transcript profiling experiments demonstrated that 76 out of 89 genes differentially expressed in response to oxidative stress were Cap1p-dependent (Wang et al., 2006). Cap1p has also been shown to regulate expression of genes related to intracellular redox in response to oxidative stress (Wang et al., 2007).

Homologues of *S. cerevisiae* stress response zinc-finger transcription factor genes, *MSN4* and *MNL1*, have been shown to have little or no effect upon resistance of *C. albicans* to heat, osmotic, ethanol, nutrient, oxidative, or heavy metal stresses (Nicholls et al., 2004), suggesting divergence of stress responses in these fungi.

The response of *C. albicans* to nitric oxide and nitrosative stress has also been examined (Hromatka et al., 2005). Many genes were transiently altered in their expression levels, but a subset of nine genes remained at elevated levels during exposure to nitric oxide. These genes include *YHB1* (encoding a flavohaemoglobin, proteins involved in detoxifying nitric oxide in a variety of pathogens), *AOX1* and *AOX 2* (encoding proteins suggested to function to reduce nitric oxide-induced oxidative stress), *SSU1, CTR2*, and *RBT5*. Deletion of the *YHB1* gene resulted in several changes in transcription pattern, including the inappropriate expression of hypha-associated genes under non-inducing conditions (Hromatka et al., 2005). This *C. albicans* strain was also attenuated in virulence, which may be due either to the nitric oxide or the filamentous phenotypes.

Proteomic studies of nutrient limitation in *C. albicans* have included responses to amino acid starvation (Yin et al., 2004) and reduced Ura3p (Brand et al., 2004). Amino acid starvation was found to induce not only amino acid biosynthetic proteins, but proteins involved in carbon metabolism (Yin et al., 2004).

Proteins affected by Ura3p deficiency included those involved in purine and pyrimidine biosynthesis, heme biosynthesis, and aromatic amino acid turnover (Brand et al., 2004).

C. albicans and the Host Immune System

When *C. albicans* infects a host it enters into a battle with the host immune system, particularly cells of the innate immune system. Neutrophils and macrophages are involved in mopping up fungal cells found in the bloodstream, and it is important to know how both host and fungus respond during these interactions. The adaptive immune response is involved in determining the outcome of systemic infection, so interactions of *C. albicans* with the major antigen processing and presenting cells, dendritic cells, are also important. One of the major areas of recent improved understanding is the identification of host receptors and their corresponding fungal ligands. *C. albicans*–host cell receptor interactions have been recently reviewed by Filler (2006).

Dendritic cells stimulate *C. albicans*-specific lymphocyte proliferation, with recognition of *C. albicans* occurring mainly via the mannose receptor (Newman & Holly, 2001). The glycosylated portion of *C. albicans* mannoprotein 65 (MP65) was demonstrated to stimulate production of TNF-α and IL-6 by dendritic cells via the mannose receptor, while the protein portion of the same protein was found to stimulate dendritic cell maturation and T cell activation via toll-like receptors (TLRs) and the MyD88-dependent signalling pathway (Pietrella et al., 2006).

An elegant study by Romani et al. (2004) demonstrated that different macrophage surface receptors were involved in phagocytosis of unopsonised yeast and hyphal cells (Romani et al., 2004). Blocking the mannose receptor had the greatest effect on phagocytosis of yeast cells; with CR3 and dectin-1 also having significant effects. However, the mannose receptor had no effect on phagocytosis of hyphae, which was mediated mostly by CR3, dectin-1, and the FcγRII/III receptors. Binding of the mannose receptor was associated with a Type I cytokine responses, whereas entry via the FcγR receptors produced a Type II response (Romani et al., 2004). The mannose receptor, however, does not appear to be essential for host defence or phagocytosis in an intraperitoneal (i.p.) infection model of candidiasis (Lee et al., 2003).

The MyD88-dependent signalling pathway is essential for resistance to *C. albi cans* infection (Bellocchio et al., 2004; Villamon et al., 2004a). MyD88-deficient mice infected with *C. albicans* had significantly reduced survival and higher organ burdens compared to control mice (Villamon et al., 2004a). This was associated with reduced neutrophil infiltrates *in vivo*, and reduced proinflammatory cytokines *in vitro*. Innate and adaptive immune responses against *C. albicans* requires coordinated action of members of the TLR super family, which signal via the MyD88 adaptor (Bellocchio et al., 2004; Villamon et al., 2004a).

TLRs have been recently reviewed (Gil & Gozalbo, 2006). Macrophages deficient in either TLR2 or TLR4 demonstrated that TLR2, but not TLR4, is involved in anti-candidal defence (Blasi et al., 2005). TRL2-deficient mice were also shown to be more susceptible to *C. albicans* infections (Villamon et al., 2004c), with macrophages producing less TNF-α and macrophage inhibitory protein 2 (MIP2). However, a second study demonstrated that TLR2-deficent mice were more resistant to systemic disease (Netea et al., 2004). Differences found between the two studies may be due to different mouse strain backgrounds, or may reflect differences in the *C. albicans* strains used to infect the mice. Little research has examined differences in virulence between strains of *C. albicans*. *C. albicans* is thought to induce immunosuppression of mice via TLR2 and MyD88-mediated signals, which increase IL10 and survival of T regulatory cells (Tregs) (Netea et al., 2004; Sutmuller et al., 2006). However, TLR2 has been shown to be dispensable for cell-mediated immune responses (Villamon et al., 2004b).

TLR4-deficient mice showed no significant difference in survival and no differences were found in neutrophil recruitment for i.p. infection or macrophage TNF-α production, suggesting that TLR4 was dispensable for murine immune responses to *C. albicans* (Murciano et al., 2006). Different morphological forms of *C. albicans* are found to signal via either TLR2 or TLR4 (Blasi et al., 2005). Interaction of hyphal cells with peripheral blood mononuclear cells (PBMCs) or murine splenic lymphocytes stimulated IL10 production in a TLR2-dependent manner (Blasi et al., 2005), while yeast cells induced IFN-γ in a TLR4-dependent mechanism (Blasi et al., 2005), with the switch from yeast to hyphae losing the TLR4-mediated signal. *C. albicans* mannosylation mutants suggest that TLR4 recognises *O*-linked mannoses on the cell surface, TLR2 is involved in recognition of β-glucan and the mannose receptor recognises mannosyl groups (Table 5.3) (Netea et al., 2006). This suggests that yeast and hyphae display very different epitopes on their cell surfaces. Differential signalling through the TLRs produces differing immune responses, with TLR4 signalling producing proinflammatory cytokines, and TLR2 signalling producing anti-inflammatory cytokines (van der Graaf et al., 2005).

Dectin-1 is a receptor shown to bind β-glucan (Table 5.3) and is important for macrophage phagocytosis (Lee et al., 2003; Gantner et al., 2005; Heinsbroek et al., 2005). β-glucan is a key molecular pattern recognised by human polymorphonuclear leukocytes (PMNs) (Lavigne et al., 2006). In differing studies dectin-1 deficient mice either showed no difference in their susceptibility to *C. albicans* infections

Table 5.3 Receptor–ligand interactions between host cells and *C. albicans*

Host receptor	*C. albicans* ligand	References
TLR4	*O*-linked mannoses	Netea et al. (2006)
TLR2/Dectin-1	β-glucan	Netea et al. (2006)
Dectin-1	β-glucan	Lee et al. (2003), Gantner et al. (2005)
Dectin-2	High mannose structures (man$_9$)	McGreal et al. (2006)
Galectin-3	β-1,2-oligomannosides	Kohatsu et al. (2006)
Mannose receptor	Mannosyl groups	Netea et al. (2006)

(Saijo et al., doi:10.1038/ni1425), or were more susceptible, with reduced lymphocyte infiltration and higher organ burdens (Taylor et al., 2006). Again, these differences may reflect different mouse strain backgrounds or *C. albicans* strains.

β-glucan is largely protected in yeast cells, but is exposed during cell separation and then induces antimicrobial responses via dectin-1 (Gantner et al., 2005). Hyphal cells, which do not expose their β-glucan, do not stimulate dectin-1 mediated responses (Gantner et al., 2005; Wheeler & Fink, 2006). Card9 has recently been identified as one of the key transducers of dectin-1 signalling, controlling dectin-1 mediated myeloid activation, cytokine production and innate antifungal activity (Gross et al., 2006). A mutant screen recently identified a number of genes involved in β-glucan masking (Wheeler & Fink, 2006). Genes identified included those encoding proteins involved in protein mannosylation (*MNN10, MNN11, OCH1, OST3*, and *OST4*) and several transcription factors (*ASF1, IES6, NOT4*, and *SSN8*) (Wheeler & Fink, 2006). Disruption of several of these genes has been demonstrated to affect host immune responses and/or virulence (The *Candida* Genome Database).

C. albicans can also bind other receptors; including dectin-2, galectin-3, and DC-SIGN (Table 5.3). Dectin-2 recognises high mannose structures (McGreal et al., 2006) and, like dectin-1, preferentially binds hyphal cells (Sato et al., 2006). Binding of dectin-2 leads to phosphorylation of FcγRs, which mediate the signal from dectin-2, inducing innate immune responses (Sato et al., 2006). Galectin-3, expressed on epithelial cells, macrophages, and dendritic cells, binds to β-1,2-oligomannosides (Kohatsu et al., 2006). Binding to galectin-3 directly induces death of *C. albicans* (Kohatsu et al., 2006). Although a synthetic analogue of β-1,2-oligomannosides prevented colonisation of mouse gut epithelium (Dromer et al., 2002), galectin-3 was not required for recognition and endocytosis by endothelial cells, which is TLR2-mediated (Jouault et al., 2006). DC-SIGN (CD209) on human monocyte-derived dendritic cells is able to bind *C. albicans*, and in immature dendritic cells internalises the yeasts in specific DC-SIGN-enriched vesicles (Cambi et al., 2003). *C. albicans* also binds complement regulators, including C4b-binding protein (C4BP), a classical pathway inhibitor (Meri et al., 2004), factor H, and FHL-1 (Meri et al., 2002). Both yeast and hyphal forms bound C4BP, with a prominent binding site being the tip of germ tubes. C4BP was found to bind to the same ligand as FHL-1 (an alternative pathway inhibitor) (Meri et al., 2004). Binding of the complement regulators is suggested to inhibit complement activation at the fungal cell surface, but also enhances binding to endothelial cells (Meri et al., 2004). *C. albicans* hyphae are also able bind to epithelial cells via vitronectin (Santoni et al., 2001) and to induce endocytosis into endothelial cells via *N*-cadherin (Phan et al., 2005). Endocytosis may also be stimulated via *C. albicans* phosphorylating two endothelial cell proteins (Belanger et al., 2002).

C. albicans mutant strains unable to damage endothelial cells *in vitro* were found to be attenuated in virulence *in vivo* (Sanchez et al., 2004). This may be due, in part, to differential inflammatory responses noted for *C. albicans* strains with different invasive potentials in epithelial and endothelial cells, with highly invasive strains triggering higher levels of proinflammatory cytokines (Villar et al., 2005).

Interactions with the different host cell receptors explain the various observations found for *C. albicans* influencing cytokine responses. *C. albicans* cells with defects in mannosylation, hence differential signalling via the mannose receptor and TLR4, were shown to stimulate lower levels of cytokine production in mononuclear cells or murine macrophages (Netea et al., 2006). *C. albicans* yeast cells were found to stimulate large amounts of IFN-γ in PBMCs or murine splenic lymphocytes, but hyphal cells did not (van der Graaf et al., 2005). This again reflects the differences in signalling found for the different morphological forms.

C. albicans–host Cell Interaction from the Host Point of View

A number of studies have utilised host DNA microarrays to examine transcriptional changes occurring when host cells interact with *C. albicans* or with *C. albicans*-derived cell wall components (Huang et al., 2001; Ishibashi et al., 2004; McLaren et al., 2004; Mullick et al., 2004; Barker et al., 2005; Kim et al., 2005; Fradin et al., 2006). These studies have been performed for a human monocytic cell line (Barker et al., 2005), blood cells (Ishibashi et al., 2004; McLaren et al., 2004; Kim et al., 2005; Fradin et al., 2006), cell line-derived granulocytes (Mullick et al., 2004), and dendritic cells (Huang et al., 2001). Unsurprisingly, for the monocytic cell line, PBMCs, granulocytes and dendritic cells, upregulation of genes involved in the pro-inflammatory response occurred, including TNFA (encodes TNF-α), MIP1A and MIP1B (encoding macrophage inflammatory proteins), and IL1B (encoding IL1-β) (Huang et al., 2001; Mullick et al., 2004; Barker et al., 2005; Kim et al., 2005; Fradin et al., 2006). These studies define the transcript profile of monocytic cells in the early response to *C. albicans* (Barker et al., 2005), and demonstrated regulation of genes over several hours of interaction (Kim et al., 2005). In addition, it was shown that *C. albicans* also induced genes associated with apoptosis in granulocytic cells (Mullick et al., 2004). Therefore, *C. albicans* is not only an inducer of genes involved in recruitment and activation of neutrophils and monocytes, but also affects genes involved in cell survival. The expression profile of neutrophils suggested that transcription was not required to attack and kill microbes, although interaction with *C. albicans* did induce expression of genes associated with immune cell communication (Fradin et al., 2006).

Proteomics have also been used to examine changes in the proteome occurring when macrophages are infected by *C. albicans* (Shin et al., 2005; Shin et al., 2006). The most prominent changes were to glycolytic enzymes, proteins involved in cell integrity and in nitric oxide production (Shin et al., 2005). Galectin-3 was also found to be significantly downregulated in infected macrophages (Shin et al., 2006). As binding to galectin-3 directly induces death of *C. albicans* (Kohatsu et al., 2006), this may represent a method of evading the immune system.

Transcript profiling has also shown that PBMCs differentially respond to soluble and particulate glucans from *C. albicans* (Ishibashi et al., 2004). Although common genes were regulated by the two glucans (many encoding proinflammatory molecules), glucan-specific transcription patterns were also noted. It is suggested that the different glucans stimulate different biological activities via differing activation mechanisms (Ishibashi et al., 2004).

One of the major findings from transcript profiling experiments was that the transcript profile obtained for the whole population of PBMC was not reflective of the major cells in the population, the CD4$^+$ or CD8$^+$ lymphocytes (McLaren et al., 2004). This suggests that results obtained for transcript profiling cannot always be extrapolated to the whole population. Transcript profiling has also demonstrated that dendritic cells have pathogen-specific transcript profiles, as well as a core response, suggested to produce pathogen-specific responses (Huang et al., 2001).

C. albicans–host Interaction from the Fungal Point of View

Transcript profiling experiments have been carried out to investigate gene expression changes occurring when *C. albicans* interacts with macrophages (Lorenz et al., 2004), neutrophils (Rubin-Bejerano et al., 2003; Fradin et al., 2005), or whole blood (Fradin et al., 2005). Lorenz et al. (2004) demonstrated changes in transcription occurring when *C. albicans* is phagocytosed by macrophages (Lorenz et al., 2004). In the early phase, the cells demonstrated starvation responses, but in later stages switched to hyphal growth and switched to glycolytic growth. *C. albicans* cells also demonstrated induction of oxidative stress responses (Lorenz et al., 2004). In a murine model of systemic candidiasis, GFP-promoter fusions confirmed that the glyoxylate cycle and gluconeogenic genes were induced following phagocytosis by macrophages and neutrophils (Barelle et al., 2006). However, the majority of cells in an infected kidney were of hyphal morphology and expressed glycolytic genes, not glyoxylate and gluconeogenic genes (Barelle et al., 2006).

C. albicans incubated with different blood fractions showed differing behaviours (Fradin et al., 2005). Transcript profiling showed that the profile obtained for *C. albicans* exposed to whole blood that was not reflective of all of the other different blood fractions. *C. albicans* exposed to red blood cells, mononuclear cells, plasma, or blood lacking neutrophils were active and rapidly switched to filamentous growth. In mononuclear cells *C. albicans* again showed induction of genes associated with hyphal growth. However, when incubated with neutrophils, *C. albicans* showed growth arrest and changes in gene expression to overcome nitrogen and carbon starvation (Rubin-Bejerano et al., 2003; Fradin et al., 2005). Genes involved in overcoming oxidative stress were also induced. *C. albicans* incubated in whole blood showed similar transcript profiles to those incubated with neutrophils, suggesting that neutrophils play a key role in systemic candidiasis (Fradin et al., 2005).

Expression profiling has also been carried out during infection of Hep2 epithelial cells, which resulted in upregulation of *ALS2 & 5*, among others (Sandovsky-Losica et al., 2006). The expression profile during adherence to human epithelia has also been determined, with profiles obtained during adherence to epithelium being very different to that obtained for plastics (Sohn et al., 2006b). Genes show-ing adhesion-dependent induction of expression included the cell surface genes *PRA1, PGA23, PGA7,* and *HWP1* (Sohn et al., 2006b).

Methodology for the removal of contaminating host material has recently been published, allowing transcript profiling of *C. albicans* cells from infected tissue to be carried out (Andes et al., 2005). Induced genes *in vivo* in the kidney of infected mice included several previously shown to be important for pathogenesis; secreted enzymes and morphology-associated genes (Andes et al., 2005).

Mating in C. albicans

The *C. albicans* Mating Type-Like (MTL) locus was originally described by Hull & Johnson (1999). Homologues of *S. cerevisiae* mating pathway genes have also been identified in the *C. albicans* genome (Tzung et al., 2001). The majority of *C. albicans* isolates are heterozygous at the MTL locus, with only a minority (10–11%) homozygous a or α (Legrand et al., 2004; Tavanti et al., 2005). It has subse-quently been shown that these homozygous strains are capable of switching between white and opaque forms (Lockhart et al., 2002), and it is opaque cells that are highly competent for mating (Miller & Johnson, 2002). Mating in *C. albicans* has been the subject of several reviews (Hull & Heitman, 2002; Johnson, 2003; Soll et al., 2003; Soll, 2004).

The cell biology of the mating process is beautifully described by Lockhart et al. (2003a). Analysis of gene expression changes occurring during mating has shown that there are several different categories of genes regulated during mating, some opaque switch-dependent and some opaque switch-independent (Lockhart et al., 2003b).

Opaque cells were previously shown to possess a unique morphology; elongated cells with pimples on the surface (Anderson & Soll, 1987). Opaque cells have since also been shown to have unique transcription patterns (Lan et al., 2002; Lockhart et al., 2003b; Zhao et al., 2005), with differences in metabolic patterns; opaque cells expressing genes associated with oxidative metabolism and white cells expressing a more fermentative profile (Lan et al., 2002). During mating genes involved in filamentation are upregulated and opaque-specific genes downregu-lated in *C. albicans* (Zhao et al., 2005).

Histone deacetylases have been shown to have a role in phase-specific gene expression (Srikantha et al., 2001), with transcription of two deacetylases, *HDA1* and *RPD3* being reduced in opaque phase cells (Srikantha et al., 2001). *HDA1* deletion increased switching from white to opaque and reduced *EFG1* expression, while *RPD3* deletion resulted in increased switching frequencies in both directions, and was associated with reduced expression of a number of opaque-specific genes and reduced *EFG1* expression (Srikantha et al., 2001). Expression of *EFG1* in opaque phase cells induced conversion back to the white form (Sonneborn et al., 1999). *TUP1* has also been implicated in white-opaque switching (Zhao et al., 2002), although *tup1* null mutants still undergo switching and mate successfully (Park & Morschhauser, 2005a).

A master regulator of white-opaque switching was recently identified, *WOR1/TOS9* (Huang et al., 2006; Srikantha et al., 2006). *WOR1* was found to be expressed at low levels in white cells, but in opaque cells forms a positive feedback loop, binding elements in its own promoter to maintain high levels of *WOR1* expression (Zordan et al., 2006). Misexpression of *WOR1* induced white cells to become stably opaque (Huang et al., 2006; Srikantha et al., 2006). This effect occurs not only in a and α cells, but also in a/α heterozygotes (Huang et al., 2006).

An essential *C. albicans* gene, *HBR1*, identified due to its induced expression in response to haemoglobin (Pendrak et al., 2004), has also been found to affect white-opaque switching. Deletion of one copy of the gene allowed opaque-phase switching and mating competence in an *MTL* heterozygous strain, leading to the suggestion that this may allow mating to occur without allelic deletion at the *MTL* locus (Pendrak et al., 2004). However, it should be noted that *C. albicans* strains can become homozygous at the *MTL* locus via loss of one chromosome 5, followed by duplication of the remaining copy (Legrand et al., 2004; Wu et al., 2005).

Mating in *C. albicans* is thought to occur on skin, where the lower temperature aids the switch to opaque cells, and responses to the surface topography facilitate mating (Lachke et al., 2003). The result of sex between two *C. albicans* cells of opposite mating type is a tetraploid cell (Lockhart et al., 2003a). These cells have been shown to lose chromosomes to become diploid once again (Bennett & Johnson, 2003). Tetraploids are less virulent than diploids in a mouse model of infection and have been shown to undergo ploidy changes during infection (Ibrahim et al., 2005). However, *MTL* homozygosity does not appear to affect virulence (Ibrahim et al., 2005), although this may depend upon *C. albicans* strain (Lockhart et al., 2005). There may be a possible immunological advantage to being able to switch between white and opaque cells, as opaque cells, unlike white cells, do not secrete a chemoattractant for PMNs (Geiger et al., 2004). Recent research also points to a role of *C. albicans* opaque cells in biofilm formation (see below) (Daniels et al., 2006). It has been suggested that the mating process in *C. albicans* may have evolved to allow survival in the host (Bennett & Johnson, 2005), although there is some evidence from MLST studies that sex may allow some recombination in *C. albicans* (Tavanti et al., 2004; Odds et al., 2006).

C. albicans Biofilms

C. albicans-associated biofilms can be found on catheter tips and on dentures. As discussed earlier, catheters are one of the major risk factors for systemic candidiasis. A number of reviews of *C. albicans* biofilms have been recently published (Lopez-Ribot, 2005; Mukherjee et al., 2005; Nett & Andes, 2006; Nobile & Mitchell, 2006; Ramage et al., 2006). Biofilms are composed of *C. albicans* cells embedded in an extracellular matrix material composed of carbohydrate, protein, hexosamine, phosphorus, and uronic acid, with the major component of the *C. albi cans* matrix being glucose (32%) (Al-Fattani & Douglas, 2006). Understanding of the behaviour of *C. albicans* growing as a biofilm has been enhanced by the development of an *in vivo* rat CVC biofilm model (Andes et al., 2004). The development of a biofilm over time has been characterised, with yeast cells being densely embedded in extracellular matrix on the catheter surface. The outermost surface of the biofilm contained both yeast and hyphal cells in a more fibrous extracellular material. Host cells are also found within the matrix (Ramage et al., 2001; Andes et al., 2004). While the presence of serum and saliva conditioning films increased initial adherence, there was little effect on overall biofilm formation (Ramage et al., 2004).

Biofilms were demonstrated to have very different transcript profiles when compared to planktonic grown cells (Garcia-Sanchez et al., 2004). Transcript profiling demonstrated that transcriptional changes in biofilm formation begin within 30 min of contact with the substrate, and involve genes associated with sulphur metabolism and amino acid biosynthesis being upregulated (Garcia-Sanchez et al., 2004; Murillo et al., 2005). Some gene expression changes are restricted to the earliest stages of biofilm formation (Murillo et al., 2005).

ALS genes were found to be differentially expressed in biofilms (Nailis et al., 2006). An *ALS3* mutant was found to produce defective biofilms, weakened structurally and much reduced in biomass (Zhao et al., 2006). GFP-promoter fusion with the *ALS3* promoter demonstrated that GFP was produced in hyphal cells throughout the biofilm. Overexpression of *ALS3* resulted in biofilms with similar mass to wild-type cells, but the cells had a yeast-like morphology (Zhao et al., 2006). The role in biofilm formation did not appear to be purely an adhesion role. Bcr1p was identified as a protein required for biofilm formation, but not hyphal formation, although several of the regulated genes are induced during hyphal formation (Nobile & Mitchell, 2005). Overexpression of *ALS3*, one of the targets for Brc1p, rescued the biofilm defect, with overexpression of other Bcr1p targets (*ALS1, ECE1*, and *HWP1*) only partially rescuing the biofilm phenotype (Nobile et al., 2006).

Other biofilm-deficient mutants have been identified and include insertions in *NUP85, MDS3, KEM1*, and *SUV3* (Richard et al., 2005). All except *kem1* were blocked at early stages of biofilm development, with *kem1* at an intermediate stage. The mutants were all defective in hypha formation in several different media, leading to the suggestion that hyphae provide an adherent scaffold to stabilise the structure (Richard et al., 2005). Protein mannosylation (Pmt) mutants were also found to be defective for biofilm formation (Peltroche-Llacsahuanga et al., 2006). A Pmt inhibitor was blocked early stages of biofim formations, suggesting that surface anchoring and adherence to the substrate may be affected (Peltroche-Llacsahuanga et al., 2006).

Proteomics have been applied to biofilm formation, demonstrating that alcohol dehydrogenase (Adh1p) is downregulated in *C. albicans* biofilms (Mukherjee et al., 2006). Subsequently it was shown that Adh1p restricts biofilm formation through an ethanol-dependent mechanism, and that ethanol treatment of a rabbit model of catheter-associated biofilm actually inhibited biofilm formation (Mukherjee et al., 2006).

Further studies (Thomas et al., 2006) investigated changes in proteins between planktonic and biofilm growing *C. albicans*. In biofilms the following proteins were upregulated: Hsp70, pyruvate decarboxylase, inositol-1-phosphate synthase (Ino1p), enolase (Eno1p), and inosine 5' monophosphate dehydrogenase (Imh3p) (Thomas et al., 2006). Many of the upregulated proteins are the same as those found to be upregulated in hyphae, and those that are immunogenic in patients.

The quorum-sensing molecules tyrosol and farnesol are produced by *C. albi cans*, accelerating and blocking yeast–hypha transition, respectively. Both appear to have roles to play in biofilm formation, with tyrosol's action most significant in the earlier stages of biofilm formation. Biofilms secrete more tyrosol compared to planktonic cells relative to dry weight, with addition of farnesol inhibiting biofilm formation (Alem et al., 2006). DNA microarrays have been carried out for biofilms exposed to farnesol, demonstrating that hypha formation-associated genes (*TUP1, CRK1*, and *PDE2*), genes associated with drug resistance (*FCR1* and *PDR16*), cell wall maintenance genes (*CHT2* and *CHT3*), and several heat-shock protein (HSP70s and 90) genes were differentially regulated (Cao et al., 2005).

A surprising finding was that *C. albicans* opaque cells can influence the overall structure of a biofilm (Daniels et al., 2006). Pheromone produced by opaque cells actually selectively upregulates mating-associated genes in white cells, and produces more cohesive white cells (Daniels et al., 2006). The resulting biofilm formed, when there are only occasional opaque cells in a population, tends to be thicker than those produced with white cell only biofilms (Daniels et al., 2006). However, for this to occur would require a host to be infected by both *MTL* a and α simultaneously, or for strains to become *MTL* homozygous within the host.

One of the major problems with *C. albicans* biofilms associated with patients is that they are inherently more resistant to antifungal agents (Ramage et al., 2001; Kuhn et al., 2002; Lewis et al., 2002; Andes et al., 2004; Ramage et al., 2004; Cocuaud et al., 2005; Al-Fattani & Douglas, 2006; Khot et al., 2006; Seidler et al., 2006; Shuford et al., 2006a). Biofilm have been shown to be more resistant to amphotericin B (Ramage et al., 2001; Kuhn et al., 2002; Lewis et al., 2002; Al- Fattani & Douglas, 2006; Khot et al., 2006; Shuford et al., 2006a), azoles (Ramage et al., 2001; Kuhn et al., 2002; Lewis et al., 2002; Al-Fattani & Douglas, 2006; Shuford et al., 2006a), and echinocandins (Cocuaud et al., 2005; Seidler et al., 2006; Shuford et al., 2006a). However, at therapeutic levels echinocandins and lipid-formulation amphotericin B were shown to significantly reduce biofilm metabolism (Kuhn et al., 2002; Lewis et al., 2002; Cocuaud et al., 2005; Seidler et al., 2006; Shuford et al., 2006b). Fluconazole and voriconazole were also shown to have some antifungal effects against biofilms, but never completely eradicated colonisation (Lewis et al., 2002). Antifungal agents do experience some problems in penetrating the biofilm (Samaranayake et al., 2005), but this does not appear to a major mechanism of drug resistance (Al-Fattani & Douglas, 2004). It has been suggested however, that β-glucans found within the matrix may have a role in biofilm drug resistance (Nett et al., 2006).

The biofilm layer of filamentous cells and yeasts were only slightly more antifungal resistant compared to planktonic cells (Khot et al., 2006). However, the substratum layer of yeasts was much more resistant to amphotericin B, which was associated with differential regulation of ergosterol and β-1,6-glucan synthesis pathways, again implicating β-glucans (Khot et al., 2006). It has been suggested that within this layer there is a subpopulation of persister cells usually associated with attachment (LaFleur et al., 2006). These cells are switch variants, rather than mutants as detachment of the cells leads to the cells eventually becoming susceptible again (Andes et al., 2004). Within 2 h of attaching to a silicone surface, cells had increased tolerance to fluconazole (Mateus et al., 2004). GFP-promoter fusions demonstrated that enhanced tolerance of attached cells was partially due to increased expression of the drug pump genes *CDR1* and *MDR1* (Mateus et al., 2004). *CDR1* and *CDR2* expression was also shown to be upregulated in biofilms (Andes et al., 2004). Two proteins previously associated with drug resistance, Grp2p and orf19.822p, were also found in greater abundance when *C. albicans* cells formed a biofilm on catheter material (Vediyappan & Chaffin, 2006). These studies suggest that attachment to a substrate induce changes in *C. albicans* which incidentally also increase their resistance to antifungal agents. This also suggests that genes involved in drug resistance also have roles in normal *C. albicans* biology, including adaptation to being adhered to a surface.

Antifungals and C. albicans

A huge number of studies have focussed on transcript profiling of gene expression changes occurring in response to anti-infective or antifungal treatment/exposure (De Backer et al., 2001; Kontoyiannis & May, 2001; Barker & Rogers, 2005; Copping et al., 2005; Lee et al., 2005; Lepak et al., 2006; Neuhof et al., 2006), as well as studying strains that have become resistant to antifungals (Rogers & Barker, 2002; Barker et al., 2004; Xu et al., 2006). These studies demonstrate that treatment with different antifungal agent classes induce different transcriptional responses (Table 5.4).

The antifungal glycosylated lipopeptide Hassallidin A (Neuhof et al., 2005) was found to induce metabolic and mitotic genes (Neuhof et al., 2006). In addition, both fluconazole and caspofungin were found to modulate expression of *SAP2* and *SAP9* (Copping et al., 2005). The anti-infective agent ciclopirox olamine, used to treat superficial mycoses, thought to work as an iron chelator, was found to induce genes involved in iron uptake as well as virulence-associated genes in *C. albicans* (Niewerth et al., 2003; Lee et al., 2005).

Azole-resistant strains, either experimentally or clinically induced, were found to have upregulated genes involved in cell stress (*DDR48* and *RTA2*), as well as ergosterol biosynthetic genes (Rogers & Barker, 2002; Barker et al., 2004; Karababa et al., 2004). *CDR1* and *CDR2*, previously demonstrated to be associated with azole resistance, have also been found to be upregulated in azole resistant strains when compared with azole sensitive strains (Rogers & Barker, 2002; Karababa et al., 2004; Xu et al., 2006). The *CDR* genes and some of the stress genes

Table 5.4 *C. albicans* genes regulated in response to antifungal treatment found by transcript profiling experiments

Antifungal drug	Genes regulated in *C. albicans* during exposure to antifungal	References
Amphotericin B	Small molecule (ion) transport genes	Liu et al. (2005)
	Stress genes (including oxidative stress genes)	
	Lipid, fatty acid and sterol synthesis genes	
Azoles	Lipid, fatty acid and sterol synthesis genes	De Backer et al. (2001)
	Small molecule transport	Liu et al. (2005)
	Carbohydrate metabolism	Lepak et al. (2006)
	Stress genes	
Echinocandins	Cell wall maintenance genes	Liu et al. (2005)
	Small molecule transport	
	Stress genes	
	Lipid, fatty acid and sterol synthesis genes	
5-flucytosine	Protein synthesis	Liu et al. (2005)
	Purine/ pyrimidine synthesis genes	
Hassallidin A	Metabolic genes	Neuhof et al. (2006)
	Mitotic genes	

induced during antifungal treatment contain DRE elements in their promoters and have been found to be co-ordinately regulated by the transcription factor Tac1p (Coste et al., 2004). It is also interesting to note that among *C. albicans* isolates changes in transcription of different groups of genes were associated with drug resistance (Xu et al., 2006).

Proteomic analysis was carried out to examine the changes that occur in response to treatment with either azoles or echinocandins (Bruneau et al., 2003). The response to azoles could be clearly differentiated from the response to echinocandins, and the authors suggest that proteomic analyses may provide clues to the mechanism of action for drugs of unknown action. Azole resistance-associated proteins include Grp2p, Ifd1p, Ifd4p, Ifd5p, and Erg10p, a protein involved in the ergosterol biosynthesis pathway (Hooshdaran et al., 2004).

Diagnosis and Treatment

Molecular diagnosis of disseminated candidiasis has been reviewed recently (Bretagne & Costa, 2005). Methods for detecting and quantifying *C. albicans* in biological samples include a real-time PCR for the mannoprotein 65 (MP65) gene, which demonstrated specificity for *C. albicans* and was highly sensitive (1 genome for sera and urine and 10 genomes for blood) (Arancia et al., 2006). Nucleic acid sequence-based amplification has been shown to improve detection rates of *C. albicans* in blood cultures (Borst et al., 2001). A multianalyte profiling system has also been developed, where DNA probes specific for six medically important *Candida* species are linked to beads. Biotinylated PCR products from samples were then hybridised to the probes and bound amplicons detected fluorometrically (Das et al., 2006). It is suggested that a sample could be processed and analysed within 1 h post-PCR amplification.

Other PCR methods based upon rRNA sequences to identify *Candida* species have also been devised (Evertsson et al., 2000; Goldenberg et al., 2005; Leinberger et al., 2005; Nazzal et al., 2005; Klingspor & Jalal, 2006). However, problems are found with DNA extraction from biological samples, such as whole blood (Fredricks et al., 2005). This may be less of a problem if plasma is used, since free fungal DNA has been detected in plasma samples (Kasai et al., 2006). The PCR methods are designed to be pan-fungal, amplifying DNA from a number of fungal species, with the species identified either through use of hemi-nested PCR (Nazzal et al., 2005), through fungal-specific probes (Evertsson et al., 2000), or by high-resolution separation of PCR products by HPLC (Goldenberg et al., 2005). The PCR products have also been hybridised to a DNA microarray designed to allow species identification (Leinberger et al., 2005). An attempt is being been made to produce a consensus PCR test for use in detection of systemic fungal infections, including those caused by *C. albicans* (White et al., 2005, 2006). These molecular methods are designed to reduce the time taken to identify the fungus causing a systemic infection. This is especially important as it has been demonstrated, both in a mouse model and in patients, that any delay in the time taken to initiate antifungal therapy significantly affects mortality (MacCallum & Odds, 2004; Morrell et al., 2005; Garey et al., 2006).

An immunoassay has also been developed and marketed (Unimedi *Candida* monotest) which detects serum mannan antigens. This is highly sensitive and specific and is a promising tool for diagnosis, especially as the test can be carried out in 1 h (Fujita et al., 2006). The optimisation of these tests and development of new molecular tests should, hopefully, improve diagnosis of systemic fungal infection.

A huge amount of literature describes the antifungal agents currently, and soon to be, available for the treatment of systemic *C. albicans* infections, with a number of reviews recently published (Potter, 2005; Aperis et al., 2006; Chamilos & Kontoyiannis, 2006; Deck & Guglielmo, 2006; Enoch et al., 2006; Kauffman, 2006; Munoz et al., 2006; Spellberg et al., 2006; Turner et al., 2006).

One of the exciting areas of treatment of systemic *C. albicans* infection involves use of antibody therapy. Mice immunised with a specific epitope of *C. albicans* Hsp90 were found to have significantly lower kidney burdens compared to control mice (Wang et al., 2006a). An antibody against this fungal Hsp90 (Mycograb) has since been shown to reduce fungal counts and improve mortality rates in invasive candidiasis when used in combination with amphotericin B (compared to amphotericin alone) (Matthews et al., 2003; Pachl et al., 2006). It is suggested that this, and other antibody therapies, are promising treatments for the future.

Which Questions Still Require Answering?

C. albicans systemic infections appear to be most common in patients in ICU and those that have undergone surgery, yet it is still difficult to predict which patients are at greatest risk. The improvements in early diagnosis of fungal infections, without the need to culture the causative organism, will allow earliest initiation of antifungal therapy, and greatly increase the probability of a positive outcome. It is also important to consider the huge effect of biofilms on colonised catheters (another of the major contributing factors towards the likelihood of developing systemic candidiasis). Further understanding of the biology of biofilm formation may lead to drug targets for prevention of their formation, or may suggest coatings for plastics to prevent biofilms. Biofilm formation by *C. albicans* can be affected by surface modification of plastics commonly used for medical devices, with 6% polyethylene oxide modification of polyetherurethane preventing biofilm formation (Chandra et al., 2005). The application of coated plastics for medical devices requires future research. However, the next big step in our understanding of systemic *C. albicans* infections will be investigation of how gene expression in the fungus changes during infection development in patients, which should now be possible by transcript profiling. It should also be possible to link changes in immune responses to the fungus with changes in fungal gene expression, or protein expression. It will also be possible to consider whether all isolates of *C. albicans* behave in a similar manner to cause infection and disease.

References

Agabian Annotation. http://agabian.ucsf.edu/canoDB/anno.php.

Agatensi, L., Franchi, F., Mondello, F., Bevilacqua, R. L., Ceddia, T., De Bernardis, F., & Cassone, A. (1991). *J. Clin. Pathol.*, 44:826–830.

Albrecht, A., Felk, A., & Pichova, I., et al. (2006). *J. Biol. Chem.*, 281:688–694.

Alem, M. A., Oteef, M. D., Flowers, T. H., & Douglas, L. J. (2006). *Eukaryot. Cell*, 5:1770–1779.

Al-Fattani, M. A. & Douglas, L. J. (2006). *J. Med. Microbiol.*, 55:999–1008.

Al-Fattani, M. A. & Douglas, L. J. (2004). *Antimicrob. Agents Chemother.*, 48:3291–3297.

Anderson, J. M. & Soll, D. R. (1987). *J. Bacteriol.*, 169:5579–5588.

Andes, D., Lepak, A., Pitula, A., Marchillo, K., & Clark, J. (2005). *J. Infect. Dis.*, 192:893–900.

Andes, D., Nett, J., Oschel, P., Albrecht, R., Marchillo, K., & Pitula, A. (2004). *Infect. Immun.*, 72:6023–6031.

Aperis, G., Myriounis, N., Spanakis, E. K., & Mylonakis, E. (2006). *Expert Opin. Investig. Drugs*, 15:1319–1336.

Arancia, S., Carattoli, A., La Valle, R., Cassone, A., & De Bernardis, F. (2006). *Mol. Cell Probes*, 20:263–268.

Arnaud, M. B., Costanzo, M. C., Skrzypek, M. S., Binkley, G., Lane, C., Miyasato, S. R., & Sherlock, G. (2005). *Nucleic Acids Res.*, 33:358–363.

Bachewich, C., Nantel, A., & Whiteway, M. (2005). *Mol. Microbiol.*, 57:942–959.

Badiee, P., Kordbacheh, P., Alborzi, A., Zeini, F., Mirhendy, H., & Mahmoody, M. (2005). *Exp. Clin. Transplant*, 3:385–389.

Barelle, C. J., Manson, C. L., MacCallum, D. M., Odds, F. C., Gow, N. A., & Brown, A. J. (2004). *Yeast*, 21:333–340.

Barelle, C. J., Priest, C. L., Maccallum, D. M., Gow, N. A., Odds, F. C., & Brown, A. J. (2006). *Cell Microbiol.*, 8:961–971.

Barker, K. S., Crisp, S., Wiederhold, N., Lewis, R. E., Bareither, B., Eckstein, J., Barbuch, R., Bard, M., & Rogers, P. D. (2004). *J. Antimicrob. Chemother.*, 54:376–385.

Barker, K. S., Liu, T., & Rogers, P. D. (2005). *J. Infect. Dis.*, 192:901–912.

Barker, K. S., & Rogers, P. D. (2005). *Methods Mol. Med.*, 118:45–56.

Bassetti, M., Righi, E., Costa, A., Fasce, R., Molinari, M. P., Rosso, R., Pallavicini, F. B., & Viscoli, C. (2006). *BMC Infect. Dis.*, 6:21.

Beckerman, J., Chibana, H., Turner, J., & Magee, P. T. (2001). *Infect. Immun.*, 69:108–114.

Beigi, R. H., Meyn, L. A., Moore, D. M., Krohn, M. A., & Hillier, S. L. (2004). *Obstet. Gynecol.*, 104:926–930.

Belanger, P. H., Johnston, D. A., Fratti, R. A., Zhang, M., & Filler, S. G. (2002). *Cell Microbiol.*, 4:805–812.

Belazi, M., Velegraki, A., Fleva, A., Gidarakou, I., Papanaum, L., Baka, D., Daniilidou, N., & Karamitsos, D. (2005). *Mycoses*, 48:192–196.

Bellocchio, S., Montagnoli, C., Bozza, S., Gaziano, R., Rossi, G., Mambula, S. S., Vecchi, A., Mantovani, A., Levitz, S. M., & Romani, L. (2004). *J. Immunol.*, 172:3059–3069.

Beltrame, A., Matteelli, A., Carvalho, A. C., Saleri, N., Casalini, C., Capone, S., Patroni, A., Manfrin, M., & Carosi, G. (2006). *Int. J. STD AIDS*, 17:260–266.

Bennett, R. J. & Johnson, A. D. (2005). *Annu. Rev. Microbiol.*, 59:233–255.

Bennett, R. J. & Johnson, A. D. (2003). *EMBO J.*, 22:2505–2515.

Blasi, E., Mucci, A., & Neglia, R., et al. (2005). *FEMS Immunol. Med. Microbiol.*, 44:69–79.

Borst, A., Leverstein-Van Hall, M. A., Verhoef, J., & Fluit, A. C. (2001). *Diagn. Microbiol. Infect. Dis.*, 39:155–160.

Bougnoux, M. E., Aanensen, D. M., Morand, S., Theraud, M., Spratt, B. G., & d'Enfert, C. (2004). *Infect. Genet. Evol.*, 4:243–252.

Bougnoux, M. E., Diogo, D., Francois, N., Sendid, B., Veirmeire, S., Colombel, J. F., Bouchier, C., Van Kruiningen, H., d'Enfert, C., & Poulain, D. (2006). *J. Clin. Microbiol.*, 44:1810–1820.

Bougnoux, M. E., Morand, S., & d'Enfert, C. (2002). *J. Clin. Microbiol.*, 40:1290–1297.

Brand, A., MacCallum, D. M., Brown, A. J., Gow, N. A., & Odds, F. C. (2004). *Eukaryot. Cell*, 3:900–909.

Braun, B. R., van Het Hoog, M., & d'Enfert, C., et al. (2005). *PLoS Genet.*, 1:36–57.

Bretagne, S. & Costa, J. M. (2005). *FEMS Immunol. Med. Microbiol.*, 45:361–368.

Bruneau, J. M., Maillet, I., Tagat, E., Legrand, R., Supatto, F., Fudali, C., Caer, J. P., Labas, V., Lecaque, D., & Hodgson, J. (2003). *Proteomics*, 3:325–336.

C. albicans Physical Map. http://albicansmap.ahc.umn.edu/.

Cambi, A., Gijzen, K., de Vries, J. M., Torensma, R., Joosten, B., Adema, G. J., Netea, M. G., Kullberg, B. J., Romani, L., & Figdor, C. G. (2003). *Eur. J. Immunol.*, 33:532–538.

Candida albicans@NRC-Biotechnology Research Institute. http://*Candida*.bri.nrc.ca/.

CandidaDB. http://genolist.pasteur.fr/*Candida*DB/.

Cao, F., Lane, S., Raniga, P. P., Lu, Y., Zhou, Z., Ramon, K., Chen, J., & Liu, H. (2006). *Mol. Biol. Cell*, 17:295–307.

Cao, Y. Y., Cao, Y. B., Xu, Z., Ying, K., Li, Y., Xie, Y., Zhu, Z. Y., Chen, W. S., & Jiang, Y. Y. (2005). *Antimicrob. Agents Chemother.*, 49:584–589.

Chamilos, G. & Kontoyiannis, D. P. (2006). *Curr. Opin. Infect. Dis.*, 19:380–385.

Chandra, J., Patel, J. D., Li, J., Zhou, G., Mukherjee, P. K., McCormick, T. S., Anderson, J. M., & Ghannoum, M. A. (2005). *Appl. Environ. Microbiol.*, 71:8795–8801.

Chen, K. W., Chen, Y. C., Lo, H. J., Odds, F. C., Wang, T. H., Lin, C. Y., & Li, S. Y. (2006). *J. Clin. Microbiol.*, 44:2172–2178.

Cheng, S., Nguyen, M. H., Zhang, Z., Jia, H., Handfield, M., & Clancy, C. J. (2003). *Infect. Immun.*, 71:6101–6103.

Choi, W., Yoo, Y. J., Kim, M., Shin, D., Jeon, H. B., & Choi, W. (2003). *Yeast*, 20:1053–1060.

Chowdhary, A., Lee-Yang, W., Lasker, B. A., Brandt, M. E., Warnock, D. W., & Arthington-Skaggs, B. A. (2006). *Med. Mycol.*, 44:405–417.

Cliff, P. R., Sandoe, J. A., Heritage, J., & Barton, R. C. (2005). *J. Med. Microbiol.*, 54:391–394.

Cocuaud, C., Rodier, M. H., Daniault, G., & Imbert, C. (2005). *J. Antimicrob. Chemother.*, 56:507–512.

Copping, V. M., Barelle, C. J., Hube, B., Gow, N. A., Brown, A. J., & Odds, F. C. (2005). *J. Antimicrob. Chemother.*, 55:645–654.

Costanzo, M. C., Arnaud, M. B., Skrzypek, M. S., Binkley, G., Lane, C., Miyasato, S. R., & Sherlock, G. (2006). *FEMS Yeast Res.*, 6:671–684.

Coste, A. T., Karababa, M., Ischer, F., Bille, J., & Sanglard, D. (2004). *Eukaryot. Cell*, 3:1639–1652.

Cotch, M. F., Hillier, S. L., Gibbs, R. S., & Eschenbach, D. A. (1998). *Am. J. Obstet. Gynecol.*, 178:374–380.

Crowe, J. D., Sievwright, I. K., Auld, G. C., Moore, N. R., Gow, N. A., & Booth, N. A. (2003). *Mol. Microbiol.*, 47:1637–1651.

Daniels, K. J., Srikantha, T., Lockhart, S. R., Pujol, C., & Soll, D. R. (2006). *EMBO J.*, 25:2240–2252.

Das, S., Brown, T. M., Kellar, K. L., Holloway, B. P., & Morrison, C. J. (2006). *FEMS Immunol. Med. Microbiol.*, 46:244–250.

Davies, A. N., Brailsford, S. R., & Beighton, D. (2006). *Oral. Oncol.*, 42:698–702.

De Backer, M. D., Ilyina, T., Ma, X. J., Vandoninck, S., Luyten, W. H., & Vanden Bossche, H. (2001). *Antimicrob. Agents. Chemother.*, 45:1660–1670.

de Groot, P. W., de Boer, A. D., Cunningham, J., Dekker, H. L., de Jong, L., Hellingwerf, K. J., de Koster, C., & Klis, F. M. (2004). *Eukaryot. Cell*, 3:955–965.

De Groot, P. W., Hellingwerf, K. J., & Klis, F. M. (2003). *Yeast*, 20:781–796.

de Hoogt, R., Luyten, W. H., Contreras, R., & De Backer, M. D. (2000). *BioTechniques*, 28:1112–1116.

de Leon, E. M., Jacober, S. J., Sobel, J. D., & Foxman, B. (2002). *BMC Infect. Dis.*, 2:1.

De Vos, M. M., Cuenca-Estrella, M., Boekhout, T., Theelen, B., Matthijs, N., Bauters, T., Nailis, H., Dhont, M. A., Rodriguez-Tudela, J. L., & Nelis, H. J. (2005). *Clin. Microbiol. Infect.*, 11:1005–1011.

Deck, D. H. & Guglielmo, B. J. (2006). *Expert Rev. Anti. Infect. Ther.*, 4:137–149.

d'Enfert, C., Goyard, S., & Rodriguez-Arnaveilhe, S., et al. (2005). *Nucleic Acids Res.*, 33:353–357.

Dromer, F., Chevalier, R., Sendid, B., Improvisi, L., Jouault, T., Robert, R., Mallet, J. M., & Poulain, D. (2002). *Antimicrob. Agents Chemother.*, 46:3869–3876.

Ebanks, R. O., Chisholm, K., McKinnon, S., Whiteway, M., & Pinto, D. M. (2006). *Proteomics*, 6:2147–2156.

Enjalbert, B., Smith, D. A., Cornell, M. J., Alam, I., Nicholls, S., Brown, A. J., & Quinn, J. (2006). *Mol. Biol. Cell*, 17:1018–1032.

Enjalbert, B. & Whiteway, M. (2005). *Eukaryot. Cell*, 4:1203–1210.

Enloe, B., Diamond, A., & Mitchell, A. P. (2000). *J. Bacteriol.*, 182:5730–5736.

Enoch, D. A., Ludlam, H. A., & Brown, N. M. (2006). *J. Med. Microbiol.*, 55:809–818.

Eurogentec. http://www.eurogentec.com.

European Galar Fungail Consortium. http://www.pasteur.fr/recherche/unites/Galar_Fungail/.

Evertsson, U., Monstein, H. J., & Johansson, A. G. (2000) *APMIS*, 108:385–392.

Falagas, M. E., Apostolou, K. E., & Pappas, V. D. (2006). *Eur. J. Clin. Microbiol. Infect. Dis.*, 25:419–425.

Fanello, S., Bouchara, J. P., Jousset, N., Delbos, V., & LeFlohic, A. M. (2001). *J. Hosp. Infect.*, 47:46–52.

Fernandez-Arenas, E., Molero, G., Nombela, C., Diez-Orejas, R., & Gil, C. (2004). *Proteomics*, 4:3007–3020.

Filler, S. G. (2006). *Curr. Opin. Microbiol.*, 9:333–339.

Fonzi, W. A. & Irwin, M. Y. (1993). *Genetics*, 134:717–728.

Fradin, C., De Groot, P., MacCallum, D., Schaller, M., Klis, F., Odds, F. C., & Hube, B. (2005). *Mol. Microbiol.*, 56:397–415.

Fradin, C., Mavor, A. L., Weindl, G., Schaller, M., Hanke, K., Kaufmann, S. H., Mollenkopf, H., & Hube, B. (2006). *Infect. Immun.* PMID: 17145939.

Fredricks, D. N., Smith, C., & Meier, A. (2005). *J. Clin. Microbiol.*, 43:5122–5128.

Fridkin, S. K., Kaufman, D., Edwards, J. R., Shetty, S., & Horan, T. (2006). *Pediatrics*, 117:1680–1687.

Fujita, S., Takamura, T., Nagahara, M., & Hashimoto, T. (2006). *J. Med. Microbiol.*, 55:537–543.

Gantner, B. N., Simmons, R. M., & Underhill, D. M. (2005). *EMBO J.*, 24:1277–1286.

Garaizar, J., Brena, S., Bikandi, J., Rementeria, A., & Ponton, J. (2006). *FEMS Yeast Res.*, 6:987–998.

Garcera, A., Castillo, L., Martinez, A. I., Elorza, M. V., Valentin, E., & Sentandreu, R. (2005). *Res. Microbiol.*, 156:911–920.

Garcia-Sanchez, S., Aubert, S., Iraqui, I., Janbon, G., Ghigo, J. M., & d'Enfert, C. (2004). *Eukaryot. Cell*, 3:536–545.

Garcia-Sanchez, S., Mavor, A. L., Russell, C. L., Argimon, S., Dennison, P., Enjalbert, B., & Brown, A. J. (2005). *Mol. Biol. Cell*, 16:2913–2925.

Garey, K. W., Rege, M., Pai, M. P., Mingo, D. E., Suda, K. J., Turpin, R. S., & Bearden, D. T. (2006). *Clin. Infect. Dis.*, 43:25–31.

Gaur, N. A., Manoharlal, R., Saini, P., Prasad, T., Mukhopadhyay, G., Hoefer, M., Morschhauser, J., & Prasad, R. (2005). *Biochem. Biophys. Res. Commun.*, 332:206–214.

Geiger, J., Wessels, D., Lockhart, S. R., & Soll, D. R. (2004). *Infect. Immun.*, 72:667–677.

Gerami-Nejad, M., Berman, J., & Gale, C. A. (2001). *Yeast*, 18:859–864.

Gerami-Nejad, M., Hausauer, D., McClellan, M., Berman, J., & Gale, C. (2004). *Yeast*, 21:429–436.

Gil, M. L. & Gozalbo, D. (2006). *Microbes Infect.*, 8:2299–2304.

Giraldo, P., von Nowaskonski, A., Gomes, F. A., Linhares, I., Neves, N. A., & Witkin, S. S. (2000). *Obstet. Gynecol.*, 95:413–416.

Gola, S., Martin, R., Walther, A., Dunkler, A., & Wendland, J. (2003). *Yeast*, 20:1339–1347.

Goldenberg, O., Herrmann, S., Adam, T., Marjoram, G., Hong, G., Gobel, U. B., & Graf, B. (2005). *J. Clin. Microbiol.*, 43:5912–5915.

Green, C. B., Zhao, X., & Hoyer, L. L. (2005a). *Infect. Immun.*, 73:1852–1855.

Green, C. B., Zhao, X., Yeater, K. M., & Hoyer, L. L. (2005b). *Microbiology*, 151:1051–1060.

Gross, O., Gewies, A., Finger, K., Schafer, M., Sparwasser, T., Peschel, C., Forster, I., & Ruland, J. (2006). *Nature*, 442:651–656.

Guggenheimer, J., Moore, P. A., Rossie, K., Myers, D., Mongelluzzo, M. B., Block, H. M., Weyant, R., & Orchard, T. (2000). *Oral Surg. Oral Med. Oral Pathol. Oral Radiol. Endod.*, 89:570–576.

Hammarskjold, F., Wallen, G., & Malmvall, B. E. (2006). *Acta Anaesthesiol. Scand.*, 50:451–460.

Harcus, D., Nantel, A., Marcil, A., Rigby, T., & Whiteway, M. (2004). *Mol. Biol. Cell*, 15:4490–4499.

Hazan, I. & Liu, H. (2002). *Eukaryot. Cell*, 1:856–864.

Heinsbroek, S. E., Brown, G. D., & Gordon, S. (2005). *Trends Immunol.*, 26:352–354.

Hernandez, R., Nombela, C., Diez-Orejas, R., & Gil, C. (2004). *Proteomics*, 4:374–382.

Holland, J., Young, M. L., Lee, O., & Chen, S. C. -A. (2003). *Sex Transm. Infect.*, 79:249–250.

Hooshdaran, M. Z., Barker, K. S., Hilliard, G. M., Kusch, H., Morschhauser, J., & Rogers, P. D. (2004). *Antimicrob. Agents Chemother.*, 48:2733–2735.

Hoyer, L. L. (2001). *Trends Microbiol.*, 9:176–180.

Hromatka, B. S., Noble, S. M., & Johnson, A. D. (2005). *Mol. Biol. Cell*, 16:4814–4826.

Huang, G., Wang, H., Chou, S., Nie, X., Chen, J., & Liu, H. (2006). *Proc. Natl. Acad. Sci. USA*, 103:12813–12818.

Huang, Q., Liu, D., Majewski, P., Schulte, L. C., Korn, J. M., Young, R. A., Lander, E. S., & Hacohen, N. (2001). *Science*, 294:870–875.

Hull, C. M. & Heitman, J. (2002). *Curr. Biol.*, 12:R782–R784.

Hull, C. M. & Johnson, A. D. (1999). *Science*, 285:1271–1275.

Ibrahim, A. S., Magee, B. B., Sheppard, D. C., Yang, M., Kauffman, S., Becker, J., Edwards, J. E., Jr., & Magee, P. T. (2005). *Infect. Immun.*, 73:7366–7374.

Ishibashi, K., Miura, N. N., Adachi, Y., Ogura, N., Tamura, H., Tanaka, S., & Ohno, N. (2004). *Int. Immunopharmacol.*, 4:387–401.

Johnson, A. (2003). *Nat. Rev. Microbiol.*, 1:106–116.

Jones, T., Federspiel, N. A., & Chibana, H., et al. (2004). *Proc. Natl. Acad. Sci. USA*, 101:7329–7334.

Jong, A. Y., Chen, S. H., Stins, M. F., Kim, K. S., Tuan, T. L., & Huang, S. H. (2003). *J. Med. Microbiol.*, 52:615–622.

Jong, A. Y., Stins, M. F., Huang, S. H., Chen, S. H., & Kim, K. S. (2001). *Infect. Immun.*, 69:4536–4544.

Jouault, T., El Abed-El Behi, M., Martinez-Esparza, M., Breuilh, L., Trinel, P. A., Chamaillard, M., Trottein, F., & Poulain, D. (2006). *J. Immunol.*, 177:4679–4687.

Kadosh, D. & Johnson, A. D. (2005). *Mol. Biol. Cell*, 16:2903–2912.

Karababa, M., Coste, A. T., Rognon, B., Bille, J., & Sanglard, D. (2004). *Antimicrob. Agents Chemother.*, 48:3064–3079.

Karababa, M., Valentino, E., Pardini, G., Coste, A. T., Bille, J., & Sanglard, D. (2006). *Mol. Microbiol.*, 59:1429–1451.

Kasai, M., Francesconi, A., Petraitiene, R., Petraitis, V., Kelaher, A. M., Kim, H. S., Meletiadis, J., Sein, T., Bacher, J., & Walsh, T. J. (2006). *J. Clin. Microbiol.*, 44:143–150.

Kauffman, C. A. (2006). *Proc. Am. Thorac. Soc.*, 3:35–40.

Khot, P. D., Suci, P. A., Miller, R. L., Nelson, R. D., & Tyler, B. J. (2006). *Antimicrob. Agents Chemother.*, 50:3708–3716.

Kibbler, C. C., Seaton, S., Barnes, R. A., Gransden, W. R., Holliman, R. E., Johnson, E. M., Perry, J. D., Sullivan, D. J., & Wilson, J. A. (2003). *J. Hosp. Infect.*, 54:18–24.

Kim, H. S., Choi, E. H., & Khan, J., et al. (2005). *Infect. Immun.*, 73:3714–3724.

Kleinegger, C. L., Lockhart, S. R., Vargas, K., & Soll, D. R. (1996). *J. Clin. Microbiol.*, 34:2246–2254.

Klingspor, L. & Jalal, S. (2006). *Clin. Microbiol. Infect.*, 12:745–753.

Kohatsu, L., Hsu, D. K., Jegalian, A. G., Liu, F. T., & Baum, L. G. (2006). *J. Immunol.*, 177:4718–4726.

Kontoyiannis, D. P. & May, G. S. (2001). *Antimicrob. Agents Chemother.*, 45:3674–3676.

Kraus, P. R. & Heitman, J. (2003). *Biochem. Biophys. Res. Commun.*, 311:1151–1157.

Kruppa, M. & Calderone, R. (2006). *FEMS Yeast Res.*, 6:149–159.

Kuhn, D. M., George, T., Chandra, J., Mukherjee, P. K., & Ghannoum, M. A. (2002). *Antimicrob. Agents Chemother.*, 46:1773–1780.

Kumamoto, C. A. & Vinces, M. D. (2005). *Cell Microbiol.*, 7:1546–1554.

Lachke, S. A., Lockhart, S. R., Daniels, K. J., & Soll, D. R. (2003). *Infect. Immun.*, 71:4970–4976.

LaFleur, M. D., Kumamoto, C. A., & Lewis, K. (2006). *Antimicrob. Agents Chemother.*, 50:3839–3846.

Lan, C., Newport, G., Murillo, L. A., Jones, T., Scherer, S., Davis, R. W., & Agabian, N. (2002). *Proc. Natl. Acad. Sci. USA*, 99:14907–14912.

Lan, C. Y., Rodarte, G., Murillo, L. A., Jones, T., Davis, R. W., Dungan, J., Newport, G., & Agabian, N. (2004). *Mol. Microbiol.*, 53:1451–1469.

Lane, S., Birse, C., Zhou, S., Matson, R., & Liu, H. (2001). *J. Biol. Chem.*, 276:48988–48996.

Laupland, K. B., Gregson, D. B., Church, D. L., Ross, T., & Elsayed, S. (2005). *J. Antimicrob. Chemother.*, 56:532–537.

Lavigne, L. M., Albina, J. E., & Reichner, J. S. (2006). *J. Immunol.*, 177:8667–8675.

Lay, J., Henry, L. K., Clifford, J., Koltin, Y., Bulawa, C. E., & Becker, J. M. (1998). *Infect. Immun.*, 66:5301–5306.

Lee, C. M., Nantel, A., Jiang, L., Whiteway, M., & Shen, S. H. (2004). *Mol. Microbiol.*, 51:691–709.

Lee, R. E., Liu, T. T., Barker, K. S., Lee, R. E., & Rogers, P. D. (2005). *J. Antimicrob. Chemother.*, 55:655–662.

Lee, S. J., Zheng, N. Y., Clavijo, M., & Nussenzweig, M. C. (2003). *Infect. Immun.*, 71:437–445.

Legrand, M., Lephart, P., Forche, A., Mueller, F. M., Walsh, T., Magee, P. T., & Magee, B. B. (2004). *Mol. Microbiol.*, 52:1451–1462.

Leinberger, D. M., Schumacher, U., Autenrieth, I. B., & Bachmann, T. T. (2005). *J. Clin. Microbiol.*, 43:4943–4953.

Lepak, A., Nett, J., Lincoln, L., Marchillo, K., & Andes, D. (2006). *Antimicrob. Agents Chemother.*, 50:1311–1319.

Lewis, R. E., Kontoyiannis, D. P., Darouiche, R. O., Raad, I. I., & Prince, R. A. (2002). *Antimicrob. Agents Chemother.*, 46:3499–3505.

Liu, T. T., Lee, R. E., Barker, K. S., Lee, R. E., Wei, L., Homayouni, R., & Rogers, P. D. (2005). *Antimicrob. Agents Chemother.*, 49:2226–2236.

Liu, X., Liu, H., Guo, Z., & Luan, W. (2006). *Oral. Dis.*, 12:41–44.

Lockhart, S. R., Daniels, K. J., Zhao, R., Wessels, D., & Soll, D. R. (2003a). *Eukaryot. Cell*, 2:49–61.

Lockhart, S. R., Pujol, C., Daniels, K. J., Miller, M. G., Johnson, A. D., Pfaller, M. A., & Soll, D. R. (2002). *Genetics*, 162:737–745.

Lockhart, S. R., Wu, W., Radke, J. B., Zhao, R., & Soll, D. R. (2005). *Genetics*, 169:1883–1890.

Lockhart, S. R., Zhao, R., Daniels, K. J., & Soll, D. R. (2003b). *Eukaryot. Cell*, 2:847–855.

Lopes Consolaro, M. E., Aline Albertoni, T., Shizue Yoshida, C., Mazucheli, J., Peralta, R. M., & Estivalet Svidzinski, T. I. (2004). *Rev. Iberoam. Micol.*, 21:202–205.

Lopez-Ribot, J. L. (2005). *Curr. Biol.* 15:R453–R455.

Lorenz, M. C., Bender, J. A., & Fink, G. R. (2004). *Eukaryot. Cell*, 3:1076–1087.

Lotz, H., Sohn, K., Brunner, H., Muhlschlegel, F. A., & Rupp, S. (2004). *Eukaryot. Cell*, 3:776–784.

MacCallum, D. M. & Odds, F. C. (2004). *Antimicrob. Agents Chemother.*, 48:4911–4914.

Martinez-Lopez, R., Park, H., Myers, C. L., Gil, C., & Filler, S. G. (2006). *Eukaryot. Cell*, 5:140–147.

Maschmeyer, G. (2006). *Int. J. Antimicrob. Agents*, 27:3–6.

Mateus, C., Crow, S. A., Jr., & Ahearn, D. G. (2004). *Antimicrob. Agents Chemother.*, 48:3358–3366.

Matthews, R. C., Rigg, G., Hodgetts, S., Carter, T., Chapman, C., Gregory, C., Illidge, C., & Burnie, J. (2003). *Antimicrob. Agents. Chemother.*, 47:2208–2216.

McGreal, E. P., Rosas, M., Brown, G. D., Zamze, S., Wong, S. Y., Gordon, S., Martinez-Pomares, L., & Taylor, P. R. (2006). *Glycobiology*, 16:422–430.

McLaren, P. J., Mayne, M., Rosser, S., Moffatt, T., Becker, K. G., Plummer, F. A., & Fowke, K. R. (2004). *Clin. Diagn. Lab. Immunol.*, 11:977–982.

Meri, T., Blom, A. M., Hartmann, A., Lenk, D., Meri, S., & Zipfel, P. F. (2004). *Infect. Immun.*, 72:6633–6641.

Meri, T., Hartmann, A., Lenk, D., Eck, R., Wurzner, R., Hellwage, J., Meri, S., & Zipfel, P. F. (2002). *Infect. Immun.*, 70:5185–5192.

Miller, M. G. & Johnson, A. D. (2002). *Cell*, 110:293–302.

Monge, R. A., Roman, E., Nombela, C., & Pla, J. (2006). *Microbiology*, 152:905–912.

Moreira, D. & Paula, C. R. (2006). *Int. J. Gynaecol. Obstet.*, 92:266–267.

Morrell, M., Fraser, V. J., & Kollef, M. H. (2005). *Antimicrob. Agents Chemother.*, 49:3640–3645.

Morschhauser, J., Staib, P., & Kohler, G. (2005). *Methods Mol. Med.*, 118:35–44.

Mukherjee, P. K., Mohamed, S., & Chandra, J., et al. (2006). *Infect. Immun.*, 74:3804–3816.

Mukherjee, P. K., Zhou, G., Munyon, R., & Ghannoum, M. A. (2005). *Med. Mycol.*, 43:191–208.

Mullick, A., Elias, M., & Harakidas, P., et al. (2004). *Infect. Immun.*, 72:414–429.

Munoz, P., Singh, N., & Bouza, E. (2006). *Curr. Opin. Infect. Dis.*, 19:365–370.

Murad, A. M., d'Enfert, C., Gaillardin, C., Tournu, H., Tekaia, F., Talibi, D., Marechal, D., Marchais, V., Cottin, J., & Brown, A. J. (2001). *Mol. Microbiol.*, 42:981–993.

Murciano, C., Villamon, E., Gozalbo, D., Roig, P., O'Connor, J. E., & Gil, M. L. (2006). *Med. Mycol.*, 44:149–157.

Murillo, L. A., Newport, G., Lan, C. Y., Habelitz, S., Dungan, J., & Agabian, N. M. (2005). *Eukaryot. Cell*, 4:1562–1573.

Nailis, H., Coenye, T., Van Nieuwerburgh, F., Deforce, D., & Nelis, H. J. (2006). *BMC Mol. Biol.*, 7:25.

Nakayama, H., Mio, T., Nagahashi, S., Kokado, M., Arisawa, M., & Aoki, Y. (2000). *Infect. Immun.*, 68:6712–6719.

Nantel, A. (2006). *Fungal Genet. Biol.*, 43:311–315.

Nantel, A., Dignard, D., & Bachewich, C., et al. (2002). *Mol. Biol. Cell*, 13:3452–3465.

Nazzal, D., Yasin, S., & Abu-Elteen, K. (2005). *New Microbiol.*, 28:245–250.

Netea, M. G., Gow, N. A., & Munro, C. A., et al. (2006). *J. Clin. Invest.*, 116:1642–1650.

Netea, M. G., Sutmuller, R., Hermann, C., Van der Graaf, C. A., Van der Meer, J. W., van Krieken, J. H., Hartung, T., Adema, G., & Kullberg, B. J. (2004). *J. Immunol.*, 172:3712–3718.

Nett, J. & Andes, D. (2006). *Curr. Opin. Microbiol.*, 9:340–345.

Nett, J., Lincoln, L., Marchillo, K., Massey, R., Holoyda, K., Hoff, B., Vanhandel, M., & Andes, D. (2006). *Antimicrob. Agents Chemother.*, 49:473.

Neuhof, T., Schmieder, P., Preussel, K., Dieckmann, R., Pham, H., Bartl, F., & von Dohren, H. (2005). *J. Nat. Prod.*, 68:695–700.

Neuhof, T., Seibold, M., Thewes, S., Laue, M., Han, C. O., Hube, B., & von Dohren, H. (2006). *Biochem. Biophys. Res. Commun.*, 349:740–749.

Newman, S. L. & Holly, A. (2001). *Infect. Immun.*, 69:6813–6822.

Nicholls, S., Straffon, M., Enjalbert, B., Nantel, A., Macaskill, S., Whiteway, M., & Brown, A. J. (2004). *Eukaryot. Cell*, 3:1111–1123.

Niewerth, M., Kunze, D., Seibold, M., Schaller, M., Korting, H. C., & Hube, B. (2003). *Antimicrob. Agents Chemother.*, 47:1805–1817.

Nobile, C. J., Andes, D. R., Nett, J. E., Smith, F. J., Yue, F., Phan, Q. T., Edwards, J. E., Filler, S. G., & Mitchell, A. P. (2006). *PLoS Pathog.*, 2:e63.

Nobile, C. J. & Mitchell, A. P. (2006). *Cell Microbiol.*, 8:1382–1391.

Nobile, C. J. & Mitchell, A. P. (2005). *Curr. Biol.*, 15:1150–1155.

Oberholzer, U., Marcil, A., Leberer, E., Thomas, D. Y., & Whiteway, M. (2002). *Eukaryot. Cell*, 1:213–228.

Odds, F. C., Davidson, A. D., Jacobsen, M. D., Tavanti, A., Whyte, J. A., Kibbler, C. C., Ellis, D. H., Maiden, M. C., Shaw, D. J., & Gow, N. A. (2006). *J. Clin. Microbiol.*, 4:3647–3658.

Pachl, J., Svoboda, P., & Jacobs, F., et al. (2006). *Clin. Infect. Dis.*, 42:1404–1413.

Park, Y. N. & Morschhauser, J. (2005a). *Mol. Microbiol.*, 58:1288–1302.

Park, Y. N. & Morschhauser, J. (2005b). *Eukaryot. Cell*, 4:1328–1342.

Pasqualotto, A. C., Rosa, D. D., Medeiros, L. R., & Severo, L. C. (2006). *BMC Infect. Dis.*, 6:50.

Paulitsch, A., Weger, W., Ginter-Hanselmayer, G., Marth, E., & Buzina, W. (2006). *Mycoses*, 49:471–475.

Peltroche-Llacsahuanga, H., Goyard, S., d'Enfert, C., Prill, S. K., & Ernst, J. F. (2006). *Antimicrob. Agents Chemother.*, 50:3488–3491.

Pendrak, M. L., Yan, S. S., & Roberts, D. D. (2004). *Eukaryot. Cell*, 3:764–775.

Phan, Q. T., Fratti, R. A., Prasadarao, N. V., Edwards, J. E., Jr., & Filler, S. G. (2005). *J. Biol. Chem.*, 280:10455–10461.

Pietrella, D., Bistoni, G., Corbucci, C., Perito, S., & Vecchiarelli, A. (2006). *Cell Microbiol.*, 8:602–612.

Pirotta, M. V., & Garland, S. M. (2006). *J. Clin. Microbiol.*, 44:3213–3217.

Pitarch, A., Abian, J., Carrascal, M., Sanchez, M., Nombela, C., & Gil, C. (2004). *Proteomics*, 4:3084–3106.

Pitarch, A., Diez-Orejas, R., Molero, G., Pardo, M., Sanchez, M., Gil, C., & Nombela, C. (2001). *Proteomics*, 1:550–559.

Pitarch, A., Jimenez, A., Nombela, C., & Gil, C. (2006a). *Mol. Cell Proteomics*, 5:79–96.

Pitarch, A., Nombela, C., & Gil, C. (2006b). *Methods Biochem. Anal.*, 49:285–330.

Pitarch, A., Nombela, C., & Gil, C. (2006c). *Methods Biochem. Anal.*, 49:331–361.

Pitarch, A., Sanchez, M., Nombela, C., & Gil, C. (2002). *Mol. Cell Proteomics*, 1:967–982.

Potter, M. (2005). *J. Antimicrob. Chemother.*, 56 (Suppl 1):i49–i54.

Qi, Q. G., Hu, T., & Zhou, X. D. (2005). *J. Oral Pathol. Med.*, 34:352–356.

Ramage, G., Martinez, J. P., & Lopez-Ribot, J. L. (2006). *FEMS Yeast Res.*, 6:979–986.

Ramage, G., Tomsett, K., Wickes, B. L., Lopez-Ribot, J. L., & Redding, S. W. (2004). *Oral Surg. Oral Med. Oral Pathol. Oral Radiol. Endod.*, 98:53–59.

Ramage, G., Vandewalle, K., Wickes, B. L., & Lopez-Ribot, J. L. (2001). *Rev. Iberoam. Micol.*, 18:163–170.

Reuss, O., Vik, A., Kolter, R., & Morschhauser, J. (2004). *Gene*, 341:119–127.

Richard, M. L., Nobile, C. J., Bruno, V. M., & Mitchell, A. P. (2005). *Eukaryot. Cell*, 4:1493–1502.

Richardson, M. D. (2005). *J. Antimicrob. Chemother.*, 56:5–11.

Robles, J. C., Koreen, L., Park, S., & Perlin, D. S. (2004). *J. Clin. Microbiol.*, 42:2480–2488.

Roemer, T., Jiang, B., & Davison, J., et al. (2003). *Mol. Microbiol.*, 50:167–181.

Rogers, P. D. & Barker, K. S. (2002). *Antimicrob. Agents Chemother.*, 46:3412–3417.

Romani, L., Montagnoli, C., Bozza, S., Perruccio, K., Spreca, A., Allavena, P., Verbeek, S., Calderone, R. A., Bistoni, F., & Puccetti, P. (2004). *Int. Immunol.*, 16:149–161.

Rubin-Bejerano, I., Fraser, I., Grisafi, P., & Fink, G. R. (2003). *Proc. Natl. Acad. Sci. USA*, 100:11007–11012.

Safdar, A., Bannister, T. W., & Safdar, Z. (2004). *Int. J. Infect. Dis.*, 8:180–186.

Saijo, S., Fujikado, N., & Furuta, T., et al. *Nat. Immunol.*, doi:10.1038/ni1425

Samaranayake, Y. H., Ye, J., Yau, J. Y., Cheung, B. P., & Samaranayake, L. P. (2005). *J. Clin. Microbiol.*, 43:818–825.

Sanchez, A. A., Johnston, D. A., Myers, C., Edwards, J. E., Jr., Mitchell, A. P., & Filler, S. G. (2004). *Infect. Immun.*, 72:598–601.

Sanchez-Vargas, L. O., Ortiz-Lopez, N. G., Villar, M., Moragues, M. D., Aguirre, J. M., Cashat-Cruz, M., Lopez-Ribot, J. L., Gaitan-Cepeda, L. A., & Quindos, G. (2005a). *J. Clin. Microbiol.*, 43:4159–4162.

Sanchez-Vargas, L. O., Ortiz-Lopez, N. G., Villar, M., Moragues, M. D., Aguirre, J. M., Cashat-Cruz, M., Lopez-Ribot, J. L., Gaitan-Cepeda, L. A., & Quindos, G. (2005b). *Rev. Iberoam. Micol.*, 22:83–92.

Sandovsky-Losica, H., Chauhan, N., Calderone, R., & Segal, E. (2006). *Med. Mycol.*, 44:329–334.

Sandven, P., Bevanger, L., Digranes, A., Haukland, H. H., Mannsaker, T., Gaustad, P., and Norwegian Yeast Study Group (2006). *J. Clin. Microbiol.*, 44:1977–1981.

Santoni, G., Spreghini, E., Lucciarini, R., Amantini, C., & Piccoli, M. (2001). *Microb. Pathog.*, 31:159–172.

Sato, K., Yang, X. L., Yudate, T., Chung, J. S., Wu, J., Luby-Phelps, K., Kimberly, R. P., Underhill, D., Cruz, P. D., Jr., & Ariizumi, K. (2006). *J. Biol. Chem.*, 281:38854–38866.

Schelenz, S. & Gransden, W. R. (2003). *Mycoses*, 46:390–396.

Schmid, J., Voss, E., & Soll, D. R. (1990). *J. Clin. Microbiol.*, 28:1236–1243.

Seidler, M., Salvenmoser, S., & Muller, F. M. (2006). *Int. J. Antimicrob. Agents*, 28:568–573.

Sendid, B., Cotteau, A., Francois, N., D'Haveloose, A., Standaert, A., Camus, D., & Poulain, D. (2006). *BMC Infect. Dis.*, 6:80.

Shen, J., Guo, W., & Kohler, J. R. (2005). *Infect. Immun.*, 73:1239–1242.

Sheppard, D. C., Yeaman, M. R., Welch, W. H., Phan, Q. T., Fu, Y., Ibrahim, A. S., Filler, S. G., Zhang, M., Waring, A. J., & Edwards, J. E., Jr. (2004). *J. Biol. Chem.*, 279:30480–30489.

Shin, Y. K., Kim, K. Y., & Paik, Y. K. (2005). *Mol. Cells*, 20:271–279.

Shin, Y. K., Lee, H. J., Lee, J. S., & Paik, Y. K. (2006). *Proteomics*, 6:1143–1150.

Shuford, J. A., Piper, K. E., Steckelberg, J. M., & Patel, R. (2006a). *Diagn. Microbiol. Infect. Dis.*, 57(3):277–281.

Shuford, J. A., Rouse, M. S., Piper, K. E., Steckelberg, J. M., & Patel, R. (2006b). *J. Infect. Dis.*, 194:710–713.

Sims, C. R., Ostrosky-Zeichner, L., & Rex, J. H. (2005). *Arch. Med. Res.*, 36:660–671.

Singh, V., Sinha, I., & Sadhale, P. P. (2005). *Biochem. Biophys. Res. Commun.*, 334:1149–1158.

Soares-Silva, I., Paiva, S., Kotter, P., Entian, K. D., & Casal, M. (2004). *Mol. Membr. Biol.*, 21:403–411.

Sobel, J. D., Faro, S., Force, R. W., Foxman, B., Ledger, W. J., Nyirjesy, P. R., Reed, B. D., & Summers, P. R. (1998). *Am. J. Obstet. Gynecol.*, 178:203–211.

Sohn, K., Schwenk, J., Urban, C., Lechner, J., Schweikert, M., & Rupp, S. (2006a). *Curr. Drug Targets*, 7:505–512.

Sohn, K., Senyurek, I., Fertey, J., Konigsdorfer, A., Joffroy, C., Hauser, N., Zelt, G., Brunner, H., & Rupp, S. (2006b). *FEMS Yeast Res.*, 6:1085–1093.

Sohn, K., Urban, C., Brunner, H., & Rupp, S. (2003). *Mol. Microbiol.*, 47:89–102.

Soll, D. R. (2004). *Bioessays*, 26:10–20.

Soll, D. R., Lockhart, S. R., & Zhao, R. (2003). *Eukaryot. Cell*, 2:390–397.

Sonneborn, A., Tebarth, B., & Ernst, J. F. (1999). *Infect. Immun.*, 67:4655–4660.

Spellberg, B. J., Filler, S. G., & Edwards, J. E., Jr. (2006). *Clin. Infect. Dis.*, 42:244–251.

Srikantha, T., Borneman, A. R., Daniels, K. J., Pujol, C., Wu, W., Seringhaus, M. R., Gerstein, M., Yi, S., Snyder, M., & Soll, D. R. (2006). *Eukaryot. Cell*, 5:1674–1687.

Srikantha, T., Tsai, L., Daniels, K., Klar, A. J., & Soll, D. R. (2001). *J. Bacteriol.*, 183:4614–4625.

Staab, J. F. & Sundstrom, P. (2003). *Trends Microbiol.*, 11:69–73.

Strauss, A., Michel, S., & Morschhauser, J. (2001). *J. Bacteriol.*, 183:3761–3769.

Sundstrom, P., Cutler, J. E., & Staab, J. F. (2002). *Infect. Immun.*, 70:3281–3283.

Sutmuller, R. P., den Brok, M. H., Kramer, M., Bennink, E. J., Toonen, L. W., Kullberg, B. J., Joosten, L. A., Akira, S., Netea, M. G., & Adema, G. J. (2006). *J. Clin. Invest.*, 116:485–494.

Taneja, V., Paul, S., & Ganesan, K. (2004). *FEMS Yeast Res.*, 4:841–847.

Tavanti, A., Davidson, A. D., Fordyce, M. J., Gow, N. A., Maiden, M. C., & Odds, F. C. (2005). *J. Clin. Microbiol.*, 43:5601–5613.

Tavanti, A., Gow, N. A., Maiden, M. C., Odds, F. C., & Shaw, D. J. (2004). *Fungal Genet. Biol.*, 41:553–562.

Tavanti, A., Gow, N. A., Senesi, S., Maiden, M. C., & Odds, F. C. (2003). *J. Clin. Microbiol.*, 41:3765–3776.

Taylor, P. R., Tsoni, S. V., & Willment, J. A., et al. (2006). *Nat. Immunol.*, 8:31–38.

The *Candida* Genome Database. http://www.Candidagenome.org/.

Thomas, D. P., Bachmann, S. P., & Lopez-Ribot, J. L. (2006). *Proteomics*, 6:5795–5804.

Tortorano, A. M., Caspani, L., Rigoni, A. L., Biraghi, E., Sicignano, A., & Viviani, M. A. (2004). *J. Hosp. Infect.*, 57:8–13.

Tortorano, A. M., Kibbler, C., Peman, J., Bernhardt, H., Klingspor, L., & Grillot, R. (2006). *Int. J. Antimicrob. Agents*, 27:359–366.

Tournu, H., Tripathi, G., Bertram, G., Macaskill, S., Mavor, A., Walker, L., Odds, F. C., Gow, N. A., & Brown, A. J. (2005). *Eukaryot. Cell*, 4:1687–1696.

Turner, M. S., Drew, R. H., & Perfect, J. R. (2006). *Expert Opin. Emerg. Drugs*, 11:231–250.

Tzung, K. W., Williams, R. M., & Scherer, S., et al. (2001). *Proc. Natl Acad. Sci. USA*, 98:3249–3253.

van der Graaf, C. A., Netea, M. G., Verschueren, I., van der Meer, J. W., & Kullberg, B. J. (2005). *Infect Immun.*, 73:7458–7464.

Vediyappan, G. & Chaffin, W. L. (2006). *Mycopathologia*, 161:3–10.

Vigouroux, S., Morin, O., Moreau, P., Harousseau, J. L., & Milpied, N. (2006). *Haematologica*, 91:717–718.

Villamon, E., Gozalbo, D., Roig, P., Murciano, C., O'Connor, J. E., Fradelizi, D., & Gil, M. L. (2004a). *Eur. Cytokine Netw.*, 15:263–271.

Villamon, E., Gozalbo, D., Roig, P., O'Connor, J. E., Ferrandiz, M. L., Fradelizi, D., & Gil, M. L. (2004b). *Microbes Infect.*, 6:542–548.

Villamon, E., Gozalbo, D., Roig, P., O'Connor, J. E., Fradelizi, D., & Gil, M. L. (2004c). *Microbes Infect.*, 6:1–7.

Villar, C. C., Kashleva, H., Mitchell, A. P., & Dongari-Bagtzoglou, A. (2005). *Infect. Immun.*, 73:4588–4595.

Viviani, M. A., Cogliati, M., Esposto, M. C., Prigitano, A., & Tortorano, A. M. (2006). *J. Clin. Microbiol.*, 44:218–221.

Wang, G., Sun, M., Fang, J., Yang, Q., Tong, H., & Wang, L. (2006a). *Vaccine*, 24:6065–6073.

Wang, J., Ohshima, T., Yasunari, U., Namikoshi, S., Yoshihara, A., Miyazaki, H., & Maeda, N. (2006b). *Gerodontology*, 23:157–163.

Wang, Y., Cao, Y. Y., & Cao, Y. B., et al. (2007). *Front Biosci.*, 12:145–153.

Wang, Y., Cao, Y. Y., Jia, X. M., Cao, Y. B., Gao, P. H., Fu, X. P., Ying, K., Chen, W. S., & Jiang, Y. Y. (2006). *Free Radic. Biol. Med.*, 40:1201–1209.

Wheeler, R. T. & Fink, G. R. (2006). *PLoS Pathog.* 2:e35.

White, P. L., Archer, A. E., & Barnes, R. A. (2005). *J. Clin. Microbiol.*, 43:2181–2187.

White, P. L., Barton, R., & Guiver, M., et al. (2006). *J. Mol. Diagn.*, 8:376–384.

Wilson, R. B., Davis, D., Enloe, B. M., & Mitchell, A. P. (2000). *Yeast*, 16:65–70.

WO-1 *C. albicans* genome sequence. http://www.broad.mit.edu/annotation/genome/Candida_albicans/Home.html.

Wu, W., Pujol, C., Lockhart, S. R., & Soll, D. R. (2005). *Genetics*, 169:1311–1327.

Xu, Z., Zhang, L. X., Zhang, J. D., Cao, Y. B., Yu, Y. Y., Wang, D. J., Ying, K., Chen, W. S., & Jiang, Y. Y. (2006). *Int. J. Med. Microbiol.*, 296:421–434.

Yin, Z., Stead, D., Selway, L., Walker, J., Riba-Garcia, I., McLnerney, T., Gaskell, S., Oliver, S. G., Cash, P., & Brown, A. J. (2004). *Proteomics*, 4:2425–2436.

Zaoutis, T. E., Argon, J., Chu, J., Berlin, J. A., Walsh, T. J., & Feudtner, C. (2005). *Clin. Infect. Dis.*, 41:1232–1239.

Zaragoza, R. & Peman, J. (2006). *Rev. Iberoam. Micol.*, 23:59–63.

Zhao, R., Daniels, K. J., Lockhart, S. R., Yeater, K. M., Hoyer, L. L., & Soll, D. R. (2005). *Eukaryot. Cell*, 4:1175–1190.

Zhao, R., Lockhart, S. R., Daniels, K., & Soll, D. R. (2002). *Eukaryot. Cell*, 1:353–365.

Zhao, X., Daniels, K. J., Oh, S. H., Green, C. B., Yeater, K. M., Soll, D. R., & Hoyer, L. L. (2006). *Microbiology*, 152:2287–2299.

Zordan, R. E., Galgoczy, D. J., & Johnson, A. D. (2006). *Proc. Natl. Acad. Sci. USA*, 103:12807–12812.

Chapter 6
Pathogenesis of *Cryptococcus neoformans*

Erin E. McClelland, Arturo Casadevall, and Helene C. Eisenman

Introduction

Cryptococcus neoformans (*Cn*) is a fungal pathogen, commonly found in urban environments (Tampieri, 2006) that primarily affects immunocompromised individuals through inhalation of spores. In healthy individuals *Cn* infection is usually cleared, or can remain in a latent form for prolonged periods of time. However, in individuals with impaired immune function, the infection may spread to the central nervous system (CNS), causing life-threatening meningitis (Casadevall & Perfect, 1998; Hull & Heitman, 2002). Thus, the disease is relatively common in AIDS patients. A recent study shows that the prevalence of cryptococcosis has declined with the increasing availability of highly active retroviral therapy (HAART) to treat HIV (Lortholary et al., 2006; Mirza et al., 2003). However, the disease continues to be a problem for those with limited access to HAART, especially in the developing world (Banerjee et al., 2001; Marques et al., 2000). Another group of individuals who are susceptible to cryptococcosis are organ transplant recipients receiving immunosuppressive therapy (Husain et al., 2001; Vilchez et al., 2002). However, cryptococcosis is not limited to immunocompromised persons, as shown by the recent outbreak in Vancouver among healthy individuals (Hoang et al., 2004).

Cn is a basidiomycete that normally grows as a haploid-budding yeast. Opposite mating types exist and *Cn* can undergo sexual reproduction and meiosis to produce spores (Idnurm et al., 2005). *Cn* strains manifest antigenic differences that allow them to be grouped into five different serotypes (A, B, C, D, and an AD hybrid) as well as different varieties. *C. neoformans* var. *neoformans* includes serotypes D and AD while var. *grubii* includes serotype A and var. *gatti* includes serotypes B and C. *Cn* var. *neoformans* and var. *grubii* are responsible for the majority of clinical infections in immunocompromised hosts while var. *gatti* causes disease primarily in immunocompetent hosts (Casadevall & Perfect, 1998; Fraser et al., 2005). Since there are considerable genetic differences among the varieties, it has been suggested that they should be considered separate species (Kwon-Chung & Varma, 2006).

The *Cn* genome is 19 Mb and contains 6572 genes on 14 chromosomes. Ten percent of the genes have no homologs among sequenced fungi. Some of the key features found in analyzing the genome are a large amount of transposons and

K. Kavanagh (ed.), *New Insights in Medical Mycology.*
© Springer 2007

Figure 6.1 Diagram illustrating a *Cn* cell and the various factors that contribute to virulence in this fungus. The polysaccharide capsule (blue) is the main virulence factor. It varies on both a macroscopic and molecular level. See text for details on each of these factors

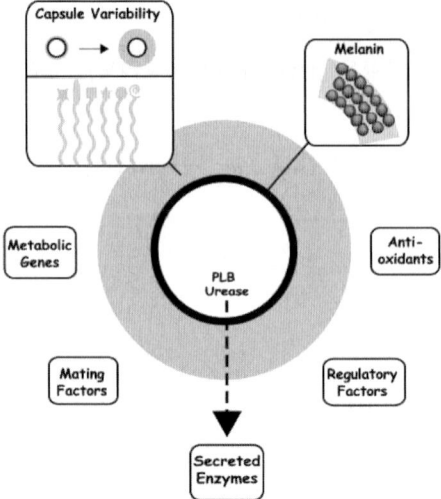

alternative splice variants (Loftus et al., 2005). The genome sequences of the serotype D strains JEC21 and B3501, the serotype A strain H99 and the serotype B strain WM276 involved in the Vancouver outbreak are available through a variety of sources including NCBI, Stanford University (http://www-sequence.stanford.edu/group/C.neoformans/), Oklahoma University (http://www.genome.ou.edu/cneo.html), Duke University (http://fungal.genome.duke.edu/), Broad Institute (http://www.broad.mit.edu/annotation/genome/cryptococcus_neoformans/Home.html), and Canada's Michael Smith Genome Sciences Centre (http://www.bcgsc.ca/project/cryptococcus).

There are three major sections in this chapter. We first introduce all of the virulence factors of *Cn* and the factors that affect and regulate virulence (Figure 6.1). Next, we describe the host immune response to *Cn*. Finally, we discuss possible new treatments for cryptococcosis in patients and future directions of research on *Cn*.

Pathogen Virulence: Virulence Factors

Polysaccharide Capsule

The capsule of *Cn* is one of its major virulence factors as evidenced by the fact that acapsular strains are avirulent (Bulmer et al., 1967; Kwon-Chung & Rhodes, 1986). The capsule contains three major components: glucuronoxylomannan (GXM), galactoxylomannan (GalXM), and mannoproteins. GXM and GalXM are polysaccharides and together make up most of the capsule with mannoproteins being a minor component on a mass basis. Mannoproteins are contained in the cell wall or secreted into the extracellular space. They are the subject of many immunological studies as they are involved in triggering the host immune response, primarily through T cell stimulation (Mansour et al., 2002; Chaka et al., 1997; Huang et al., 2002; Pietrella et al., 2001a, 2002). There are a large number of different manno-proteins in the capsule (Levitz et al., 2001) that all contain several common motifs, including a signal sequence, a functional domain, a serine/threonine-rich region and an attachment site for a glycosylphosphatidylinositol anchor (Levitz & Specht, 2006). Many of these proteins are identified using sequence homology searches from genomic data. Numerous efforts are underway to characterize their function and location in the capsule (Huang et al., 2002).

The two polysaccharide components of the capsule are well characterized. The structure of GXM and GalXM was deduced from sugar composition analysis, chemical medication, and NMR spectroscopy (Cherniak et al., 1998; Vaishnav et al., 1998). The structure of GXM consists of a mannan backbone with β-D-xylopyranosyl, β-D-glucuronosyl, and 6-O-acetyl branches. Six different structural reporter groups (M1–M6, Figure 6.2) are defined to describe the various repeat elements of GXM with a specific reporter group predominating in each cryptococcal serotype (Cherniak et al., 1998). The structure of GalXM is more complex and consists of a branched α(1,6)-galactan backbone with the following possible branches on alternate galactose residues: β(1,3)-linked galactose, α(1,4)-linked mannose, α(1,3)-linked mannose, β(1,2)-linked xylose or β(1,3)-linked xylose (Figure 6.2). O-acetyl groups are also found in GalXM, but their precise location is unknown (Vaishnav et al., 1998).

The structure of the capsule is heterogenous (McFadden et al., 2006) and under-goes microevolution during infection (Charlier et al., 2005) or *in vitro* passaging (Garcia-Hermoso et al., 2004). This capacity to change its structure may help the fungus avoid immune clearance (Fries & Casadevall, 1998) and could help explain the large differences in capsule structure between clinical and environmental strains

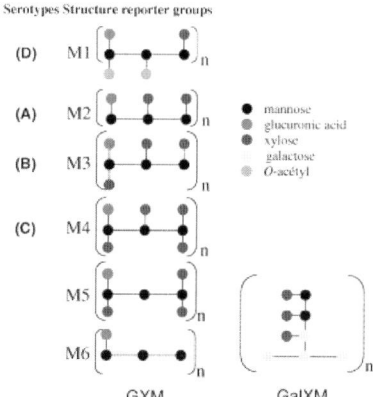

Figure 6.2 Structural reporter groups of *Cn*. (Reprinted with permission from Janbon, 2004. With permission from Blackwell Publishing.)

(Brandt et al., 1996), since *in vivo* passage may select for new polysaccharide variants.

The process of capsule synthesis and assembly is a lively topic of debate (Bose et al., 2003; Janbon, 2004; Zaragoza et al., 2006; McFadden et al., 2006). Several models of capsule synthesis are proposed, but most groups agree on a broad outline for capsule synthesis. First, the precursors are made in the intracellular space. GXM requires GDP-mannose, UDP-xylose, and UDP-glucuronic acid as precursors. GDP-mannose is made in a three-step process involving three different enzymes, a phosphomannose isomerase (Wills et al., 2001), a phosphomannomutase, and a GDP-mannose pyrophosphorylase. The last two enzymes have not yet been characterized in *Cn*. UDP-xylose is made in a two-step process in which UDP-glucose is converted to UDP-glucuronic acid via UDP-glucose dehydrogenase (Bar-Peled et al., 2004; Moyrand & Janbon, 2004) and then UDP-glucuronic acid is converted to UDP-xylose via UDP-glucuronic acid decarboxylase (Bar-Peled et al., 2001). Next, the precursors are localized to the site of polysaccharide construction and the monosaccharides are assembled into polymers using an array of sugar transferases (Sommer et al., 2003) and at least one acetyl transferase (Moyrand et al., 2002, 2004; Janbon et al., 2001). The presence of antibody-reactive GXM components in the cytoplasm of *Cn* by immunoelectron microscopy suggests that much of the synthesis and assembly of capsular polysaccharide occur in the intracellular space (Feldmesser et al., 2000; Garcia-Rivera et al., 2004). The polysaccharide is then secreted from the plasma membrane via exocytosis (Yoneda & Doering, 2006) and then secreted through the cell wall via vesicle secretion to the extracellular space (Rodrigues et al., 2006). In fact, vesicular transport of capsular polysaccharide appears a solution to the problem of how to export large molecules through the fungal cell wall. Finally, the individual capsule components are assembled into a mature capsule through a mechanism that may involve self-association of polysaccharide molecules exported to the outside of the cell.

One capsule synthesis model proposed envisions capsule biosynthesis involving the addition of new capsular material to the inner capsule near the cell wall, thus displacing older capsular material outwards (Bose et al., 2003) (Figure 6.3A). However,

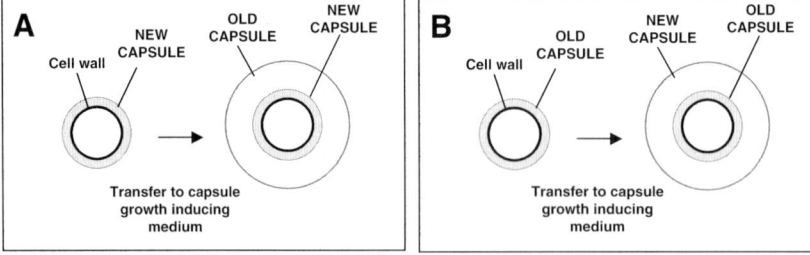

Figure 6.3 Two models of capsular growth. See text for details. (Adapted from Zaragoza et al., 2006. With permission from Blackwell Publishing.)

a more recent model has proposed that capsule enlargement occurs by apical growth, with new capsular material added onto the outer edges of the capsule and the older capsular material closest to the cell wall (Zaragoza et al., 2006) (Figure 6.3B). This model proposes that smaller polysaccharides travel through the old capsular material to the capsule surface where they are assembled into the enlarging structure. This is consistent with another recent model based primarily on the physical properties of GXM (McFadden et al., 2006). In this recent model, the authors suggest that capsule assembly occurs by self-association of small GXM fibers that are released from the cell, since a single GXM molecule is not long enough to span the distance from the cell wall to the outer edge of the capsule. According to this scheme, shed polysaccharide are GXM fibers that are unable to find a suitable attachment point (McFadden et al., 2006). Ongoing research in this field will hopefully unify these models and increase our understanding of capsule structure and assembly.

The capsule is one of the major virulence factors for *Cn*, primarily because it is antiphagocytic, required for intracellular replication and the shed polysaccharide functions as a major modulator of the host immune response. The release of GXM into tissue adversely affects immune function and interferes with host defense mechanisms. GXM inhibits migration and activation of antigen-presenting cells and also inhibits a Th1 response by inducing large amounts of proinflammatory cytokines. In addition, GXM seems to be required for dissemination from macrophages. Mouse macrophages infected with three different varieties of *Cn* show phagosomal fusion and extrusion of *Cn*, resulting in extracellular dissemination of *Cn* and survival of the macrophage to undergo phagocytosis of additional *Cn* (Alvarez & Casadevall, 2006). Examples of the large number of deleterious effects the capsule has on host defenses are listed in Table 6.1.

In *Cn*, a number of capsular phenotypic switch variants have been reported that manifest as changes in colony morphology, including wrinkled, pseudohyphal, smooth, and mucoid (Fries et al., 1999). The phenomenon of phenotypic switching occurs in numerous pathogens and is defined as reversible phenotypic changes that arise in a population of cells at rates higher than the background mutation rate (Guerrero et al., 2006). Phenotypic switching occurs at rates of 10^{-3}–10^{-5} in *Cn* and is reversible (Fries et al., 1999, 2001). Much of the work in *Cn* phenotypic switching has focused on a set of smooth (SM) and mucoid (MC) switch variants derived from a standard serotype D strain (Fries et al., 2001). The cells in MC colonies produce excess polysaccharide making the colonies appear shiny. Phenotypic switching between SM and MC occurs both *in vivo* and *in vitro* (Fries et al., 2001). In murine infections, MC is more virulent than SM, indicating that phenotypic switching has a role in virulence, possibly helping the fungus rapidly adapt to the host environment (Fries et al., 2001). Furthermore, in infected mice, SM *Cn* is more susceptible to antifungal therapies (Fries et al., 2005). Analyses of SM and MC *Cn* strains show differences in the structure of GXM, with different repeat structures between the variants. Other differences include an increase in viscosity of MC GXM and increased size (Fries et al., 1999, 2001). Phenotypic switching between SM and MC has also been observed in *Cn* var. *gattii* strains and clinical isolates of both variants have been isolated from patients (Jain et al., 2006).

Table 6.1 List of capsule effects on host

Immunomodulatory Effect	References
Inhibition of phagocytosis	Bulmer and Sans (1968), Kozel and Mastroianni (1976)
Promotion of phagosome fusion and extrusion	Alvarez and Casadevall (2006)
Interference with antigen presentation	Retini et al. (1998), Monari et al. (2005a)
Downregulation of MHC class II and B7 expression on monocytes	Monari et al. (1999)
Induces differentiation towards a non-protective Th2 response	Almeida et al. (2001)
Induction of IL-10 production	Chiapello et al. (2004), Vecchiarelli et al. (1996)
Induction of IL-6 production	Delfino et al. (1997)
Induction of pro-inflammatory cytokines (IL-1β, IL-6, IL-8, and TNF-α)	Retini et al. (1996)
Inhibition of IL-12	Retini et al. (2001)
Downregulation of TNF-α and IL-1β production	Vecchiarelli et al. (1995)
Induction of immune unresponsiveness	Kozel et al. (1977), Murphy and Cozad (1972)
Reduction of delayed-type hypersensitivity response and Th1 response	Retini et al. (1999)
Induction of suppressor T cells	Murphy and Cozad (1972), Blackstock et al. (1987)
Inhibition of lymphoproliferation	Syme et al. (1999)
Inhibition of T cell activation and proliferation	Mody and Syme (1993), Yauch et al. (2006)
Inhibition of leukocyte migration	Dong and Murphy (1995b)
Inhibition of chemokine production in human EC	Mozaffarian et al. (2000)
Interference with chemotaxis and chemokinesis	Coenjaerts et al. (2001), Dong and Murphy (1995a)
Induction of premature L-selectin shedding from neutrophils	Dong and Murphy (1996)
Blockade of the neutrophil CD18 receptor	Dong and Murphy (1997)
Reduction of C5aR expression on the surface of neutrophils	Monari et al. (2002)
Inhibition of neutrophil adhesion and rolling	Ellerbroek et al. (2002, 2004)
Reduction of neutrophil and macrophage killing activity	Monari et al. (2003), Vecchiarelli (2000)
Induction of apoptosis in spleen mononuclear cells and T cells	Chiapello et al. (2003, 2004), Monari et al. (2005b)
Induction of human alveolar cell lysis	Barbosa et al. (2006)
Interference with human DC maturation	Vecchiarelli et al. (2003)
Contributes to cerebral edema and increased intracranial pressure	Denning et al. (1991), Hirano et al. (1964a, b)
Induces enhanced infectivity of HIV	Pettoello-Mantovani et al. (1992, 1994)

Abbreviations: IL – interleukin; TNF-α – tumor necrosis factor alpha; EC – endothelial cell; DC – dendritic cell.

Melanin

Melanization is associated with virulence in *Cn* and other fungi. Melanins are dark pigments composed of a polymer of indole and phenol subunits of unknown secondary structure. They absorb light across the UV and visible spectrum, have high physical and chemical strength and can resist degradation, even by strong acids. Lastly, melanins have redox properties and can scavenge free radicals (Hill, 1992; Nosanchuk & Casadevall, 2003b).

In *Cn*, melanin is observed as an electron-dense layer in the cell wall (Nosanchuk & Casadevall, 2003a). Treatment of melanized cells with strong acids and denaturants yields particles of melanin that retain the original cell shape. Because of this they are referred to as melanin 'ghosts' (Wang et al., 1996). High-resolution microscopy of melanin architecture in *Cn* shows a granular surface and multiple layers (Eisenman et al., 2005). Melanin in *Cn* has pores that presumably allow melanized cells to obtain nutrients and export products. NMR cryporometry measurements show that melanin ghosts contain pores of 10–50 Å, with the porosity decreasing with time of melanization (Eisenman et al., 2005). Using size exclusion chromatography, Jacobsen and Ikeda (2005) found that *Cn* melanin has pores with a maximum size of 4–16 nm. Together, these studies show that rather than being a solid layer, melanin in *Cn* has a complex architecture that includes pores large enough to accommodate nutrients.

The development of monoclonal antibodies to *Cn* melanin demonstrates that melanin is a target of the immune response (Nosanchuk et al., 1998). In addition, the antibodies are useful in studying *Cn* melanin. The antibodies bind to yeast in the lung and brain tissue of animals infected with *Cn*, indicating that *Cn* is melanized in these tissues (Rosas et al., 2000). Interestingly, the antibodies are able to inhibit growth of *Cn in vitro* under melanizing conditions and they also protect mice from *Cn* infection (Rosas et al., 2001).

In *Cn*, melanin polymerization is catalyzed by the enzyme laccase, encoded by the *CNLAC1* and *LAC2* genes (Missall et al., 2005; Pukkila-Worley et al., 2005; Williamson, 1994). Of the two genes, *LAC1* is the major contributor to melanin production. As shown by insertional mutagenesis, a number of other genes are required for melanization, including genes involved in metal transport and metabolism and cell wall biosynthesis (Walton et al., 2005; Zhu & Williamson, 2003).

Melanization of *Cn in vitro* requires the addition of exogenous laccase substrate. *In vitro* melanization can be induced by the addition of catecholamine neurotransmitters to the culture medium. These include L-dopa, norepinephrine, and dopamine, for example. Comparison of melanin particles produced from different substrates reveals significant differences between the pigments produced by *Cn*. The structure and amount of incorporation of the different substrates into pigment varies. Furthermore, not all of the pigments have all the hallmarks of melanins, such as a stable free radical signal, or specific oxidative breakdown products (Garcia-Rivera et al., 2005a). It is assumed that, *in vivo*, *Cn* uses host neurotransmitters to produce melanin. However, this has yet to be shown. The substrates used by *Cn* in the environment to produce melanin are likewise unknown. However, recent work suggests that catecholamines produced by other microbes may be a source of substrates as *Cn* produces melanin when cocultured with the bacterium, *Klebsiella aerogenes* (Frases et al., 2006a). In addition, *Cn* has been shown to be able to use the bacterial melanin precursor homogentisic acid for melanization (Frases et al., 2006b).

The question of *in vivo* melanization and its importance to virulence is the subject of some debate and recent studies address this issue. Laccase is expressed during infection, consistent with *in vivo* melanization (Garcia-Rivera et al., 2005b; Waterman et al., 2006). Melanized *Cn* is detectable in tissue samples from infected patients. Furthermore, melanin particles isolated from these tissues by chemical extraction are similar to *in vitro* melanized *Cn* when viewed by electron microscopy (Nosanchuk et al., 2000). Similar results are obtained in experimental mouse infections (Rosas et al., 2000). The results of other *in vivo* studies highlight the importance of melanization to virulence. Experimental cryptococcosis in a mouse model can be improved by treatment with glyphosate, an inhibitor of melanization (Nosanchuk et al., 2001). Together, these data show that melanization of *Cn* occurs *in vivo*.

A number of *in vitro* and *in vivo* studies focus on the role of melanin in virulence. Melanin is a negatively charged, hydrophobic surface capable of binding many substances including antibiotics and heavy metals (Nosanchuk & Casadevall, 2003b). *Cn* melanin binds to a number of proteins, as well antimicrobial peptides normally produced by the host. Melanin protects *Cn* from the antimicrobial activity of the peptides protegrin and defensin, by binding and sequestering them (Doering et al., 1999). When wild type and *CNLAC1* deletion strains are used to infect mice, there is no difference in lung clearance between the wild type and deletion strains. However, the fungal burden in the brain and spleen is higher for wild type *Cn*, showing that laccase, and likely melanization, is required for dissemination and/or survival outside the lung (Noverr et al., 2004).

Alternate functions for laccase in virulence, including protection from oxidative stress are proposed. One study found that strains lacking *LAC1* are more susceptible to killing by alveolar macrophages, even in the absence of L-dopa, the substrate for melanization (Liu et al., 1999). Oxidation of iron from Fe^{2+} to Fe^{3+} by laccase may protect *Cn* from free radicals generated by Fe^{2+}, suggesting an alternate function for laccase. Also, the *LAC1* and *LAC2* genes have altered transcription in response to oxidative and nitrosative stress, consistent with the view that laccase has an important antioxidant function in *Cn*. Finally, *LAC2* has genetic interactions with the thiol-specific antioxidant *TSA1* (Missall et al., 2005). The role of antioxidant genes in *Cn* virulence is discussed in depth below. A second connection between laccase and response to stress exists as expression of laccase is regulated in part by the heat-shock proteins Hsp70 and HSF (Zhang et al., 2006).

Phospholipase B

Phospholipase B (PLB) is a secreted protein that is found in all *Cn* serotypes (Chen et al., 1997; Wright et al., 2004). This enzyme removes acyl chains from phospholipids. The mechanism of action of PLB in virulence could involve damage to host tissues, nutrient acquisition, and immune modulation through alteration of lipid signaling molecules. Infection of mice with *plb* mutant strains shows that PLB is necessary for dissemination but not survival in the lungs (Noverr et al., 2003). The inflammatory response is also reduced in the lungs of mice infected with the mutant strain. Additionally, lower levels of eicosanoids are found in the lungs of mice infected with the *plb* mutants compared to WT. These lipids are known to modulate the immune system, suggesting that PLB contributes to virulence by reducing the host immune response (Noverr et al., 2003). The failure of *plb* mutants to disseminate from the lungs is described in other studies. When mice are infected with *Cn* by the endotracheal route, *plb* mutants are found in interstitial lung and alveolar macrophages, but not peripheral blood monocytes, brain or lymph nodes. However, in an adoptive transfer experiment with infected monocytes, PLB is not required for infection of the central nervous system (CNS), further supporting a role for PLB in dissemination from the lungs by default (Santangelo et al., 2004).

Urease

Urease catalyzes the hydrolysis of urea to ammonia and carbamate and is produced in large quantities by *Cn* (Cox et al., 2000). Mice infected intravenously with a strain lacking urease and a urease-reconstituted strain show profound differences in sequestration of *Cn* within the microcapillaries of the brain, a key step in dissemination of *Cn* to the CNS (Olszewski et al., 2004). These data, in combination with data showing laccase is also involved in yeast survival and replication within the brain indicates that urease and laccase play synergistic roles in CNS invasion (Olszewski et al., 2004). Thus, it is not surprising that strains lacking urease are significantly less virulent in mice (Cox et al., 2000; Olszewski et al., 2004).

Factors Affecting Virulence

Metabolic Genes

Cn strains lacking genes involved in amino acid biosynthesis (Met3, Met6, Lys9, and Ilv2) are also avirulent in mice. Most of these genes result in an auxotrophic phenotype when they are disrupted and thus their non-pathogenicity in mice is not surprising (Yang et al., 2002; Pascon et al., 2004; Kingsbury et al., 2004a, b).

In addition, genes that are involved in cellular transport processes or signaling networks that regulate different virulence factors are obviously involved in *Cn* virulence. Two examples of these higher order virulence genes are Vph1 (encoding a vesicular proton pump (Erickson et al., 2001)) and Vad1 (encoding a DEAD-box RNA helicase (Panepinto et al., 2005). Strains lacking these genes are avirulent in mice and show numerous defects in melanin production, capsule formation, urease production or growth at 37°C.

Antioxidant Genes

The ability to resist oxidative and nitrosative stress is important to the virulence of *Cn*. Numerous studies with oxidative stress factor mutants reveal a connection between susceptibility to oxidative/nitrosative stress and loss of virulence. Deletion of these factors manifests *in vitro* as sensitivity to oxidative agents such as hydrogen peroxide and t-butylhydroperoxide and/or nitrosative compounds like sodium nitrite (Missall et al., 2004; Missall & Lodge, 2005; Wormley et al., 2005). Table 6.2 lists some of the oxidative stress factors contributing to virulence. It is not surprising that oxidative and nitrosative stress resistance affects virulence considering the importance of survival inside host phagocytic cells to the pathogenesis of *Cn* (see section on innate immunity). However, this interpretation is oversimplified as some deletion mutants have secondary effects on other virulence traits, such as the ability to grow at 37°C (Missall et al., 2004; Giles et al., 2005). Furthermore, while some of the mutants are hypersensitive to killing by macrophages or neutrophils *in vitro*, not all are (Missall et al., 2005; Wormley et al., 2005; Narasipura et al., 2003).

Regulation of Virulence

A major research interest in the *Cn* field is signal transduction and the regulation of virulence in *Cn*. Components of the mitogen-activated protein kinase (MAPK) pathway, the protein kinase A (PKA) pathway, the calcineurin and sphingolipid pathways contribute to virulence in *Cn*. These pathways regulate multiple virulence attributes, as well as other aspects of *Cn* biology, such as mating.

Pathways Regulating Mating and Virulence

During infection, *Cn* cells grow as haploid-budding yeast. However, like other members of the fungal kingdom, *Cn* can also undergo mating and meiosis to produce spores. This process is not observed *in vivo*, but occurs *in vitro* under specific conditions, and presumably occurs in the environment. The mating process involves complex morphological transitions and regulatory mechanisms. Interestingly, some

Table 6.2 Oxidative stress factors that affect virulence

Factor	References
Thiol-specific Antioxidant	Missall et al. (2004)
Thioredoxin	Missall and Lodge (2005)
Skn7	Coenjaerts et al. (2006), Wormley et al. (2005)
Superoxide dismutase	Giles et al. (2005), Narasipura et al. (2003)
Alternative oxidase	Akhter et al. (2003)

of the factors controlling mating also contribute to virulence. For example, signal transduction factors involved in mating regulation also influence virulence. These include factors involved in MAPK signaling, such as components of the HOG (high osmolarity glycerol) pathway, which controls the stress response (Bahn et al., 2005, 2006). Members of the PAK (p21 GTPase activated kinase) family, such as STE20α, STE20a, and PAK1 also play a role in virulence (Nichols et al., 2004; Wang et al., 2002). Lastly, the *Cn* Ras1 homolog, a GTP-binding protein, contributes to virulence and regulates mating filamentation (Alspaugh et al., 2000).

Other examples of mating factors that contribute to virulence are the mating pheromone receptor and the mating pheromone (MFα). Deletion of *CPR*a, encoding the receptor, results in total loss of virulence in mice and smaller capsule size *in vivo* (Chang et al., 2003). However, deletion of the *MFα* gene has only a slight effect on virulence in mice (Shen et al., 2002). *STE12α* and *STE12a* are transcription factor genes located within the mating loci (*MATa* and *MATα*) that control mating in *Cn*. Deletion of these genes results in either attenuation or total loss of virulence in mouse models of infection. Furthermore, known virulence attributes of *Cn* are reduced in the deletion mutants, including melanin, capsule, and PLB activity. However, the degree of attenuation and virulence factor effects is strain dependent (Chang et al., 2001; Davidson et al., 2003; Ren et al., 2006). In addition, not all of the factors regulating mating are also required for virulence (e.g. Sxi1α and Sxi2a) (Hull et al., 2005). The variability of effects on virulence shows that the relationship between mating and virulence is complex.

Besides the role of specific mating factors in virulence, mating has another significant influence on virulence: generation of virulent *Cn* strains through genetic recombination. A prominent example of this is the *Cn* strains causing the current outbreak of cryptococcosis in Vancouver. Genetic analyses of the outbreak strains suggest they arose from recombination between an endemic strain and more virulent strain from Australia (Fraser et al., 2005).

cAMP and PKA Signaling

Heterotrimeric G proteins, consisting of the α, β, and γ subunits, are primary components of many signal transduction cascades and they, and the signal transduction pathways, are conserved across many different organisms. In *Cn*, the PKA pathway that regulates cyclic AMP (cAMP) is an example of a signaling pathway involving G proteins that is important for fungal development and virulence (Pukkila-Worley & Alspaugh, 2004). The PKA pathway is activated during nutrient deprivation via a 7-transmembrane domain G protein-coupled receptor (Gpr4) that senses amino acids (methionine) (Xue et al., 2006). Once methionine binds Gpr4, the G proteins are activated and the α subunit (Gpa1, Alspaugh et al., 1997) is released from the Gβγ dimer (D'Souza & Heitman, 2001). In *Cn*, the Gβ subunit is encoded by the Gib2 gene while there are two Gγ subunits encoded by the genes Gpg1/2 (Palmer et al., 2006). Gpa1 in combination with the adenylyl cyclase associated protein

(Aca1, Bahn et al., 2004) can bind to the enzyme adenylyl cyclase (Cac1) inducing it to produce cAMP (Alspaugh et al., 2002). cAMP levels are negatively regulated by the phosphodiesterase Pde1 (and possibly by Pde2) (Hicks et al., 2005).

One of the primary targets of cAMP is PKA, which is a tetrameric protein composed of dimeric regulatory subunits (Pkr1) and two monomeric catalytic subunits (Pka1/2) (D'Souza et al., 2001). Increased levels of cAMP result in the release of the regulatory subunits and activation of the kinase. This results in the phosphorylation of downstream targets of the signaling cascade by Pka1/2 (Pukkila-Worley & Alspaugh, 2004) ultimately leading to transcription of genes involved in capsule biosynthesis (Cramer et al., 2006) and melanin production (Pukkila-Worley et al., 2005).

Because strains deficient in PKA are sterile (D'Souza et al., 2001), there is thought to be cross talk between the MAPK and PKA pathways. Thus, one hypothesis is that for *Cn* to undergo mating, both pathways are needed: the PKA pathway signals nutrient deprivation while the MAPK pathway signals the presence of an appropriate mating partner (Pukkila-Worley & Alspaugh, 2004).

Calcineurin Signaling

Calcineurin is a conserved calcium–calmodulin-regulated protein phosphatase that consists of a catalytic A subunit and a regulatory B subunit. Both subunits are required for protein function, growth at high temperatures, and virulence in mice (Odom et al., 1997; Fox et al., 2001). The phosphatase function of calcineurin is inhibited by the immunosuppressive drugs cyclosporine A and FK506, which bind cyclophilin A and FKBP12, respectively (Kraus & Heitman, 2003). *Cn* contains two homologous genes for cyclophilin A, both of which are required for cell growth, mating, virulence in mice, and toxicity to cyclosporine A (Wang et al., 2001). Due to these toxic effects of cyclosporine A, research is ongoing to find analogs that are nonimmunosuppressive as potential antifungals against *Cn* (Cruz et al., 2000). Calcineurin affects *Cn* virulence in a variety of ways as it is also involved in cell wall integrity (mutants lacking the MAP kinase mpk1 are defective in growth at high temperatures (Kraus et al., 2003), hyphal elongation during mating (Fox & Heitman, 2005), and haploid fruiting (Cruz et al., 2001).

Lipid Signaling

Sphingolipid signaling through protein kinase C (PKC) also regulates virulence in *Cn*. The lipid signaling molecule diacylglycerol (DAG) is generated by the enzyme inositol phosphoryl ceramide synthase (Ipc1) (Shea & Del Poeta, 2006). DAG regulates the activity of laccase, possibly through activation of PKC. Thus, strains with reduced Ipc1 have corresponding defects in melanization (Heung et al., 2004). Ipc1 is a fungal-specific enzyme that is essential in *Cn*. Reducing the expression of Ipc1 leads to defects in virulence, such as decreased survival in macrophages *in vitro* and attenuated virulence *in vivo* (Luberto et al., 2001). A second mechanism of action of Ipc1 (through DAG) is regulation of the expression of App1 (antiphagocytic protein), a secreted protein that inhibits phagocytosis by macrophages (Luberto et al., 2003; Mare et al., 2005).

The phosphoryl inositol group of IPC can be removed by inositol phosphosphingolipid phospholipase C (Isc1) (Shea et al., 2006). *Cn* strains lacking Isc1 have decreased fungal burden in the brain of immunocompetent mice, suggesting a role for this gene in dissemination to the brain. In addition, the *isc1* strain is highly susceptible to killing by macrophages *in vitro* and hypersensitive to oxidative and nitrosative stresses compared to wild type (Shea et al., 2006).

Host Response: Innate Immunity

The host response to a *Cn* infection is usually quite efficient and effective, as evident by the fact that cryptococcal infection is common but disease is rare. If the host is immunocompromised, *Cn*-induced disease is more likely. This section will highlight the advances in the host immune response to a *Cn* infection, including both innate and adaptive immunity, and introduce some alternate models for the study of *Cn* infection.

Macrophages

When *Cn* spores are first inhaled, the host cells that initially phagocytose the yeast are presumably alveolar macrophages with neutrophil and dendritic cell involvement shortly thereafter. These important innate antigen-presenting cells are critical to initiating an effective immune response. Macrophages are thought to be one of the primary defenses against *Cn* infection. They help control the immunomodulatory effects of *Cn* by sequestering shed polysaccharide (GXM) in a receptor-dependent manner through phagocytosis or pinocytosis (He et al., 2003; Chang et al., 2006). However, the efficiency of phagocytosis is dependent on the cell cycle (Luo et al., 2006) and the activation state of the macrophage (He et al., 2003). In addition, a recent study found that macrophages derived from the B-1 subset of B lymphocytes show increased fungal killing via a NO-mediated mechanism (Ghosn et al., 2006).

Neutrophils

Human polymorphonuclear neutrophils kill *Cn* through both oxidative (Chaturvedi et al., 1996; Aratani et al., 2006) and nonoxidative mechanisms (Mambula et al., 2000). Despite their ability to kill *Cn*, neutrophil influx into mouse tissues is not associated with protective responses to pulmonary *Cn* infection (Kawakami et al., 1999). Depletion of neutrophils before infection with *Cn* results in increased survival times in BALB/c mice (Mednick et al., 2003). However, a recent study found that C57Bl/6 mice deficient in myeloperoxidase (an enzyme important in the oxidative systems of neutrophils) show decreased survival times, increased fungal burden, and more severe pathology (Aratani et al., 2006). These apparently discordant results may be explained by the different mouse strains used in each study.

Dendritic Cells

Dendritic cells (DC) inhibit growth or kill *Cn* through both oxidative and nonoxidative mechanisms (Kelly et al., 2005). In addition, DCs induce antigen-specific T cell activation in both mice (Wozniak et al., 2006) and humans (Syme et al., 2002; Mansour et al., 2006). Human DCs phagocytose (or pinocytose) *Cn* via the binding of mannoproteins to mannose receptors, DC-SIGN (Mansour et al., 2006), and FCγRII receptors (Syme et al., 2002) on the cell surface. Once inside the cell, mannoproteins are degraded in an endo-lysosomal compartment and then loaded onto MHC class II proteins to activate mannoprotein-specific T cells (Mansour et al., 2006).

NKT Cells

Natural killer T cells (NKT) are important innate and adaptive immune bridging cells. They are thought to be involved in the regulation of Th1 and Th2 cell development. Evidence for this regulation is seen in mice lacking NKT cells as Th1 responses are significantly reduced in mice infected with *Cn* (Kawakami et al., 2001a). During *Cn* infection, both the number and percentage of NKT cells in the lungs increase, as well as the expression of NKT specific gene Vα14. Cells lacking Vα14, and therefore NKT cells, are less able to fight *Cn* infection, as indicated by reduced IFN-γ in lungs, reduced delayed type hypersensitivity (DTH) response to *Cn*, and higher fungal burden in the lung (Kawakami et al., 2001a). Conversely, activation of NKT cells improves the host response to *Cn*. α-Galactoceramide, a synthetic lipid that is recognized by Vα14 and activates NKT cells, activates NKT cells against *Cn* during infection, resulting in lower fungal burdens in the lung, but has no effect in mice lacking NKT cells (Kawakami et al., 2001b). The ability of NKT cells to inhibit *Cn* is also observed in passive transfer experiments in which thymocytes from older, resistant mice are transferred to younger, susceptible mice. The increased resistance in the recipient mice is due to NKT cells (Blackstock & Murphy, 2004a).

γδ T Cells

γδ T cells are also important innate immune effector cells that are found primarily in non-lymphoid tissues. In a *Cn* infection, γδ T cells are thought to play a down-regulatory role of host defenses (Kawakami, 2004) due to their production of TGF-β, which is known to suppress host defenses to infectious pathogens (Reed, 1999). Similar to NKT cells, both the number and percentage of γδ cells increases during *Cn* infection. However, mice lacking γδ cells have fewer CFUs in lung and more IFN-γ in serum (Uezu et al., 2004). Further studies are needed to elucidate the function of these cells in *Cn* infection.

Host Response: Adaptive Immunity

Humoral Immunity

Because of the general acceptance that cell-mediated immunity is crucial for clearance of *Cn*, the contribution of an antibody-mediated response has historically been uncertain. With the advent of monoclonal antibodies (mAb), there are now numerous studies showing that antibodies can provide protection or prolong survival in *Cn* infection (Parra et al., 2005; Yuan et al., 1995; Rivera & Casadevall, 2005). In fact, one mAb to GXM is in use an adjunctive therapy for cryptococcosis (Casadevall et al., 1998). However, it has also become clear that not all antibodies are protective against *Cn* (Lendvai & Casadevall, 1999) and the mechanism of Ab protection is quite complex (Casadevall & Pirofski, 2005). Ab protection can vary depending on the concentration of Ab, the innoculum size (Taborda et al., 2003), the strain of mouse (Rivera & Casadevall, 2005; Lovchik et al., 1999), the specificity (Mukherjee et al., 1995) and isotype of the Ab (Yuan et al., 1995; Yuan et al., 1998) and the cellular response to the infection, specifically the role of inducible nitric oxide and CD4[+] and CD8[+] T cells (Yuan et al., 1997; Rivera et al., 2002). The current view is that the antibody response to a *Cn* infection acts to enhance the cellular response by reducing inflammatory damage and modulating the immune response to contain the infection (Rivera et al., 2005; Feldmesser et al., 2002).

Cell-mediated Immunity

Cell-mediated immunity (CMI) has an important role in host defense against cryptococcosis. Not only is CMI important to initial infection with *Cn*, it is maintained during persistent, secondary infections as well (Lindell et al., 2005, 2006). Comparison of high- and low-virulent *Cn* strains *in vivo* shows that infection with the more virulent strain induces a rapid DTH response to the cryptococcal antigen CneF, indicative of CMI activation (Blackstock et al., 1999). T cells proliferate *in vitro* in response to *Cn*, especially CD4[+], but also CD8[+] T cells. Furthermore, depletion experiments with mAbs against CD4 or CD8 demonstrated that CD4[+] T cells are required for CD8[+] T cell proliferation, but not *vice versa* (Syme et al., 1997). A more detailed analysis of the T cell response to *Cn* infection shows that a large diversity of T-cell receptor Vβ subsets proliferate, especially for CD4[+] T cells (Lindell et al., 2006).

CD8[+] cells are also involved in CMI to *Cn*. Aβ-/-mice (C57BL6/J background) that lack major histocompatibility complex (MHC) II and therefore CD4[+] T cells can be vaccinated with a low-virulent *Cn* strain and subsequently survive challenge with a low dose of *Cn* cells. When the mice are also given mAb to CD8, the effect of immunization is somewhat blocked, indicating resistance was partially CD8 dependent (Aguirre et al., 2004).

The role of cell-mediated immunity in host response to *Cn* is further illustrated by the finding that MHC genotype affects the host response to *Cn*. Strains of mice differing in the MHC locus and infected with *Cn* have different susceptibilities to the fungus, as assessed by fungal burden in the liver (McClelland et al., 2003b). Furthermore, when *Cn* is passaged in these mice, the time to death decreases at different rates in the different MHC backgrounds, showing the influence of MHC on host response to *Cn* (McClelland et al., 2004b).

Another component of CMI, co-stimulatory T cell surface proteins such as CD40, play a role in immunity to *Cn*. CD40-/-mice infected with *Cn* have higher lung fungal burdens and reduced lung inflammatory cell infiltration compared to wild type. Cultured splenocytes from these mice produce lower levels of Th1 cytokines (Pietrella et al., 2004). The role of CD40 signaling is further demonstrated by treating infected animals with a combination therapy of anti-CD40 Ab and interleukin (IL)-2. The treated mice have increased survival and fewer *Cn* in the brain and kidney (Zhou et al., 2006). The co-inhibitory receptor CTLA4 also functions in the T cell response to *Cn* as CTLA4 expression increases on the surface of T cells in response to *Cn* (Pietrella et al., 2001b). Anti-CTLA4 antibodies administered to infected mice increase the DTH response to *Cn* and reduce the fungal burden. When combined with CneF immunization, the antibodies increase the survival of mice infected with *Cn* (McGaha & Murphy, 2000).

A strong Th1 response is clearly required for effective host defense against *Cn*. When mice are infected with high- or low-virulent *Cn* strains and the cytokines produced by cultured spleen cells in response to CneF are measured, the low-virulent strains elicit more Th1 cytokines (IFN-γ and IL-2) later in infection compared to the high-virulent strains, showing the importance of a sustained Th1 response against *Cn* (Blackstock et al., 1999). Measurement of cytokines in bronchoalveolar lavage fluid of such mice indicates that the high virulent strain induces higher levels of IL-4, suggesting a Th2 response is associated with a negative outcome of infection (Abe et al., 2000). Additionally, experimental treatment of mice that promotes a strong Th1 response correlates with reduced cryptococcal burden and inflammation (Edwards et al., 2005).

Studies with knockout mice show the importance of specific Th1 cytokines. IL-18-/-mice infected with *Cn* have reduced lung and brain fungal clearance, DTH response to *Cn*, and Th1 cytokines, including IFN-γ (Kawakami et al., 2000). Other host factors are required for the induction of the Th1 response, including urokinase-type plasminogen activator (uPA). Mice lacking uPA are unable to mount an adequate Th1 response against *Cn* infection, and lung histopathology shows that the alveolar spaces are filled with macrophages containing *Cn*. In addition, these mice have smaller lymph nodes and less T cell proliferation (Gyetko et al., 2002).

In contrast, a Th2 response is less effective against *Cn*. In the absence of IL-10, a Th2-promoting cytokine, mice survive *Cn* infection longer (Blackstock et al., 1999). Results with another Th2 cytokine, IL-4, are more complex. IL-4 -/-mice have reduced DTH to CneF with *Cn* infection, indicating a need for IL-4 in CMI against *Cn*. However, cultured spleen cells from infected mice stimulated with CneF produce more IL-2 and IFN-γ and less IL-5 and IL-10, indicating an enhanced Th1 response in the knockout mice compared to wild type. Despite this, there are no survival differences in the mouse strains when infected with *Cn*, except that the knockout mice survive better when primed with CneF before infection (Blackstock & Murphy, 2004b).

Since *Cn* usually infects the CNS, the role of CMI in the CNS is also studied using the intracerebral (i.c.) infection model, in which mice are infected by injecting fungal cells directly into the brain, eliminating the complicating factor of dissemination in interpreting results. When mice are immunized with CneF and subsequently infected i.c., the fungal burden in the lung, brain, and spleen is lower than the control. In addition, more leukocytes are present in the brain, especially CD4+ T cells and macrophages. Moreover, antibodies to CD4 and IFN-γ increase the fungal burden in the brain (Buchanan & Doyle, 2000) and CneF-immunized mice have increased expression of a number of chemokines and Th1 cytokines in the brain after i.c. infection (Uicker et al., 2005). Using mouse strains that favored either Th1 or Th2 response, Huffnagle and McNeil (1999) found that Th1-polarized mice have reduced infection of the CNS. Other studies show the importance of MHC in fighting *Cn* in the CNS. Adoptive transfer experiments in MHC class II deficient mice show that CD4+ T cells require interaction with MHC II on perivascular microglial cells in the CNS (Aguirre & Miller, 2002). Together, these data show that CMI is important in fighting *Cn* in the CNS.

Alternate Hosts for *Cryptococcus neoformans*

There are a few *in vivo* models for the study of *Cn* infection and each has its own advantages and disadvantages. *Acanthameoba castellanii* is a good nonmammalian host for *Cn* infection (Steenbergen et al., 2001) as many of the hallmarks of *Cn* infection in macrophages are also seen in amoeba and amoeba are viable at 37°C. *Dictyostelium discoideum* (Steenbergen et al., 2003) is another important nonmammalian host because of the many similarities seen between infection of *Cn* and macrophages. There are a few major advantages of this system in that *Dictyostelium* is a genetically malleable host and there are numerous mutant lines available, unlike *A. castellanii*. However, slime mold amoeboid cells are smaller than amoebae and are less efficient at phagocytosis. Thus, both systems have advantages and disadvantages for studying the genetics of the interaction between *Cn* and its potential environmental host.

Another genetically malleable non-mammalian host is the worm *Caenorhabditis elegans* (Mylonakis et al., 2002, 2003). This host model is probably the most extensively used to study *Cn* infection thus far because there are many genetic tools available, such as a variety of mutant lines, RNA interference (RNAi) libraries and microarrays. In addition, its genome is sequenced and annotated. Research using *C. elegans* has already identified two different genes involved in virulence of *Cn*. Kin1 is a protein kinase homologue that binds macrophages and is associated with virulence in mice (Mylonakis et al., 2004). Rom2 is a guanyl-nucleotide exchange factor homologue that has a significant role in growth at 37°C, mating and the maintenance of the stability of the cell wall under osmotic pressures. It is also associated with virulence in mice (Tang et al., 2005).

Another model host for *Cn* infections is *Drosophila melanogaster* (Apidianakis et al., 2004). This host is useful for the study of innate immune responses to *Cn* infections as there is a high degree of conservation between mammals and insects. In addition, *Drosophila* is also genetically malleable with a large number of mutant lines available. It has a sequenced and annotated genome and RNAi libraries and microarrays are available. A recently studied model for *Cn* infection is the greater wax moth caterpillar *Galleria mellonella* (Mylonakis et al., 2005). This is a particularly useful host model because some of the disadvantages with the other models are not a problem. For example, with *G. mellonella* it is possible to administer exact fungal inocula by injection, unlike *C. elegans* and *D. melanogaster*, which have to ingest *Cn* to be infected. In addition, this host can be studied at higher temperatures (37°C) and can be used to study the effects of antifungal compounds. One major disadvantage, however, is the lack of genetic tools available for *G. mellonella*.

New Treatments

Much of the research on developing treatments for *Cn* focuses on immune-based therapies including antibody administration and vaccine development. GXM, a component of the polysaccharide capsule, is a main target of immune-based therapy. In addition, peptide and drug based treatments are being explored.

Antibody Therapy

A monoclonal antibody to GXM, 18B7, underwent preliminary testing for treatment of cryptococcosis (Casadevall et al., 1998). In phase I clinical trials with HIV-infected patients, 18B7 administration correlates with reduced antigen titers in serum. This study is the first to use mAbs as therapy against a fungal infection (Larsen et al., 2005). Another groundbreaking therapy, the use of radioimmunotherapy against microbial infection, utilizes 18B7. In such a therapy, antibody is radiolabeled with the β-emitter Rhenium-188 and administered to *Cn*-infected mice. The radiolabeled antibodies kill *Cn in vitro*, and more importantly, increase the survival time of the mice and reduce fungal burden in the lung and brain (Dadachova et al., 2003).

Vaccine Possibilities

Vaccination against *Cn* is another strategy to fight infection. One approach is the use of attenuated strains. Mice injected with an avirulent *Cn* strain that is unable to produce melanin are able to survive longer than control mice when subsequently challenged with more virulent strains (Barluzzi et al., 2000). More specific antigens may be useful vaccine candidates. Culture supernatants from *Cn* can elicit an immune response in mice. In combination with adjuvant, the supernatant can immunize T cell positive mice to better survive challenge with *Cn* infection. This is likely due to a combination of factors, since fractionation of the supernatant into mannoprotein and non-mannoprotein components results in partial protection by each component (Mansour et al., 2004). A third vaccination strategy is the use of peptide mimotope of GXM. The rationale for using such peptides is that protective antibodies against GXM have been isolated, but the peptide has fewer limitations than GXM for use as a vaccine. Peptide P13, isolated by screening a phage display library with an anti-GXM antibody, is one such candidate peptide (Zhang et al., 1997). Mice vaccinated with P13 have increased serum titers against GXM. Moreover, they are able to survive *Cn* infection for a longer time, showing that protective immunity could be induced through molecular mimicry (Fleuridor et al., 2001). Human antibodies to GXM are also produced by P13 immunization of transgenic mice expressing human immunoglobulins (Maitta et al., 2004).

Other Therapies

Nonimmune treatments are also being tested against cryptococcosis. One such experimental treatment is the use of a naturally occurring antimicrobial peptide, killer peptide (KP), produced by *Pichia anomala*. KP binds to a common cell wall component – β1–3 glucan, and is active against many microorganisms. KP has *in vitro* effects against *Cn* including growth inhibition, reduced capsule size and increased susceptibility to macrophage and neutrophil killing. *In vivo*, KP treatment increases survival time of mice infected with *Cn* and reduces fungal CFU in the brain (Cenci et al., 2004). In addition, a surprising new drug treatment for *Cn* is the HIV protease inhibitor Indinavir. *In vitro*, indinavir inhibits some virulence characteristics, including capsule size. It also increases killing of *Cn* by phagocytic cells (Monari et al., 2005c). Furthermore, it can increase the survival time of mice infected with *Cn* (Pericolini et al., 2006).

Conclusion

Tremendous progress has been made in the last two decades in understanding the biology, ecology, and pathogenesis of *C. neoformans*. The emerging picture is that *Cn* virulence factors for mammals are selected for survival in the soil. In particular, attributes such as the capsule, melanin, and phospholipase are dual use virulence factors that allow the property of pathogenicity for animals and the ability to survive in the harsh environment of soils. In addition, *Cn* presents new cellular biological problems that are unique to this organism. For example, understanding the construction and assembly of the polysaccharide capsule will provide new insights into eukaryotic cell biology. Similarly, this organism may provide some insights into the emergence of sexual types in biology. The realization that *Cn* has a unique biology is catalyzing a transition from the study of pathogenesis to an increasing focus on understanding the cellular biology of this remarkable organism.

References

Abe, K., Kadota, J., Ishimatsu, Y., Iwashita, T., Tomono, K., Kawakami, K., & Kohno, S. (2000). *Microbiol. Immunol.*, 44:849–855.

Aguirre, K., Crowe, J., Haas, A., & Smith, J. (2004). *Med. Mycol.*, 42:15–25.

Aguirre, K. & Miller, S. (2002). *Glia*, 39:184–188.

Akhter, S., McDade, H. C., Gorlach, J. M., Heinrich, G., Cox, G. M., & Perfect, J. R. (2003). *Infect. Immun.*, 71:5794–5802.

Almeida, G. M., Andrade, R. M., & Bento, C. A. (2001). *J. Immunol.*, 167:5845–5851.

Alspaugh, J. A., Cavallo, L. M., Perfect, J. R., & Heitman, J. (2000). *Mol. Microbiol.*, 36:352–365.

Alspaugh, J. A., Perfect, J. R., & Heitman, J. (1997). *Genes Dev.*, 11:3206–3217.

Alspaugh, J. A., Pukkila-Worley, R., Harashima, T., Cavallo, L. M., Funnell, D., Cox, G. M., Perfect, J. R., Kronstad, J. W., & Heitman, J. (2002). *Eukaryot Cell*, 1:75–84.

Alvarez, M. & Casadevall, A. (2006). *Curr. Biol.*, 16:2161–2165.

Apidianakis, Y., Rahme, L. G., Heitman, J., Ausubel, F. M., Calderwood, S. B., & Mylonakis, E. (2004). *Eukaryot Cell*, 3:413–419.

Aratani, Y., Kura, F., Watanabe, H., Akagawa, H., Takano, Y., Ishida-Okawara, A., Suzuki, K., Maeda, N., & Koyama, H. (2006). *J. Med. Microbiol.*, 55:1291–1299.

Bahn, Y. S., Hicks, J. K., Giles, S. S., Cox, G. M., & Heitman, J. (2004). *Eukaryot Cell*, 3:1476–1491.

Bahn, Y. S., Kojima, K., Cox, G. M., & Heitman, J. (2005). *Mol. Biol. Cell*, 16:2285–2300.

Bahn, Y. S., Kojima, K., Cox, G. M., & Heitman, J. (2006). *Mol. Biol. Cell*, 17:3122–3135.

Banerjee, U., Datta, K., Majumdar, T., & Gupta, K. (2001). *Med. Mycol.*, 39:51–67.

Barbosa, F. M., Fonseca, F. L., Holandino, C., Alviano, C. S., Nimrichter, L., & Rodrigues, M. L. (2006). *Microbes Infect.*, 8:493–502.

Barluzzi, R., Brozzetti, A., Mariucci, G., Tantucci, M., Neglia, R. G., Bistoni, F., & Blasi, E. (2000). *J. Neuroimmunol.*, 109:75–86.

Bar-Peled, M., Griffith, C. L., & Doering, T. L. (2001). *Proc. Natl. Acad. Sci. USA*, 98:12003–12008.

Bar-Peled, M., Griffith, C. L., Ory, J. J., & Doering, T. L. (2004). *Biochem. J.*, 381:131–136.

Blackstock, R., Buchanan, K. L., Adesina, A. M., & Murphy, J. W. (1999). *Infect. Immun.*, 67:3601–3609.

Blackstock, R., McCormack, J. M., & Hall, N. K. (1987). *Infect. Immun.*, 55:233–239.

Blackstock, R. & Murphy, J. W. (2004a). *Infect. Immun.*, 72:5175–5180.

Blackstock, R. & Murphy, J. W. (2004b). *Am. J. Respir. Cell. Mol. Biol.*, 30:109–117.

Bose, I., Reese, A. J., Ory, J. J., Janbon, G., & Doering, T. L. (2003). *Eukaryot Cell*, 2:655–663.

Brandt, M. E., Hutwagner, L. C., Klug, L. A., Baughman, W. S., Rimland, D., Graviss, E. A., Hamill, R. J., Thomas, C., Pappas, P. G., Reingold, A. L., & Pinner, R. W. (1996). *J. Clin. Microbiol.*, 34:912–917.

Buchanan, K. L. & Doyle, H. A. (2000). *Infect. Immun.*, 68:456–462.

Bulmer, G. S. & Sans, M. D. (1968). *J. Bacteriol.*, 95:5–8.

Bulmer, G. S., Sans, M. D., & Gunn, C. M. (1967). *J. Bacteriol.*, 94:1475–1479.

Casadevall, A., Cleare, W., Feldmesser, M., Glatman-Freedman, A., Goldman, D. L., Kozel, T. R., Lendvai, N., Mukherjee, J., Pirofski, L. A., Rivera, J., Rosas, A. L., Scharff, M. D., Valadon, P., Westin, K., & Zhong, Z. (1998). *Antimicrob. Agents Chemother.*, 42:1437–1446.

Casadevall, A. & Perfect, J. R. (1998). *Cryptococcus Neoformans*, ASM Press, Washington DC.

Casadevall, A. & Pirofski, L. (2005). *Curr. Mol. Med.*, 5:421–433.

Cenci, E., Bistoni, F., Mencacci, A., Perito, S., Magliani, W., Conti, S., Polonelli, L., & Vecchiarelli, A. (2004). *Cell Microbiol.*, 6:953–961.

Chaka, W., Verheul, A. F., Vaishnav, V. V., Cherniak, R., Scharringa, J., Verhoef, J., Snippe, H., & Hoepelman, I. M. (1997). *Infect. Immun.*, 65:272–278.

Chang, Y. C., Miller, G. F., & Kwon-Chung, K. J. (2003). *Infect. Immun.*, 71:4953–4960.

Chang, Y. C., Penoyer, L. A., & Kwon-Chung, K. J. (2001). *Proc. Natl. Acad. Sci. USA*, 98:3258–3263.

Chang, Z. L., Netski, D., Thorkildson, P., & Kozel, T. R. (2006). *Infect. Immun.*, 74:144–151.

Charlier, C., Chretien, F., Baudrimont, M., Mordelet, E., Lortholary, O., & Dromer, F. (2005). *Am. J. Pathol.*, 166:421–432.

Chaturvedi, V., Wong, B., & Newman, S. L. (1996). *J. Immunol.*, 156:3836–3840.

Chen, S. C., Muller, M., Zhou, J. Z., Wright, L. C., & Sorrell, T. C. (1997). *J. Infect. Dis.*, 175:414–420.

Cherniak, R., Valafar, H., Morris, L. C., & Valafar, F. (1998). *Clin. Diagn. Lab. Immunol.*, 5:146–159.

Chiapello, L. S., Aoki, M. P., Rubinstein, H. R., & Masih, D. T. (2003). *Med. Mycol.*, 41:347–353.

Chiapello, L. S., Baronetti, J. L., Aoki, M. P., Gea, S., Rubinstein, H., & Masih, D. T. (2004). *Immunology*, 113:392–400.

Coenjaerts, F. E., Hoepelman, A. I., Scharringa, J., Aarts, M., Ellerbroek, P. M., Bevaart, L., Van Strijp, J. A., & Janbon, G. (2006). *FEMS Yeast Res.*, 6:652–661.

Coenjaerts, F. E., Walenkamp, A. M., Mwinzi, P. N., Scharringa, J., Dekker, H. A., van Strijp, J. A., Cherniak, R., & Hoepelman, A. I. (2001). *J. Immunol.*, 167:3988–3995.

Cox, G. M., Mukherjee, J., Cole, G. T., Casadevall, A., & Perfect, J. R. (2000). *Infect. Immun.*, 68:443–448.

Cramer, K. L., Gerrald, Q. D., Nichols, C. B., Price, M. S., & Alspaugh, J. A. (2006). *Eukaryot Cell*, 5:1147–1156.

Cruz, M. C., Del Poeta, M., Wang, P., Wenger, R., Zenke, G., Quesniaux, V. F., Movva, N. R., Perfect, J. R., Cardenas, M. E., & Heitman, J. (2000). *Antimicrob. Agents Chemother.*, 44:143–149.

Cruz, M. C., Fox, D. S., & Heitman, J. (2001). *EMBO J.*, 20:1020–1032.

Dadachova, E., Nakouzi, A., Bryan, R. A., & Casadevall, A. (2003). *Proc. Natl. Acad. Sci. USA*, 100:10942–10947.

Davidson, R. C., Nichols, C. B., Cox, G. M., Perfect, J. R., & Heitman, J. (2003). *Mol. Microbiol.*, 49:469–485.

Delfino, D., Cianci, L., Lupis, E., Celeste, A., Petrelli, M. L., Curro, F., Cusumano, V., & Teti, G. (1997). *Infect. Immun.*, 65:2454–2456.

Denning, D. W., Armstrong, R. W., Lewis, B. H., & Stevens, D. A. (1991). *Am. J. Med.*, 91:267–272.

Doering, T. L., Nosanchuk, J. D., Roberts, W. K., & Casadevall, A. (1999). *Med. Mycol.*, 37:175–181.

Dong, Z. M. & Murphy, J. W. (1995a). *Infect. Immun.*, 63:2632–2644.

Dong, Z. M. & Murphy, J. W. (1995b). *Infect. Immun.*, 63:770–778.

Dong, Z. M. & Murphy, J. W. (1996). *J. Clin. Invest.*, 97:689–698.

Dong, Z. M. & Murphy, J. W. (1997). *Infect. Immun.*, 65:557–563.

D'Souza, C. A., Alspaugh, J. A., Yue, C., Harashima, T., Cox, G. M., Perfect, J. R., & Heitman, J. (2001). *Mol. Cell. Biol.*, 21:3179–3191.

D'Souza, C. A. & Heitman, J. (2001). *FEMS Microbiol. Rev.*, 25:349–364.

Edwards, L., Williams, A. E., Krieg, A. M., Rae, A. J., Snelgrove, R. J., & Hussell, T. (2005). *Eur. J. Immunol.*, 35:273–281.

Eisenman, H. C., Nosanchuk, J. D., Webber, J. B., Emerson, R. J., Camesano, T. A., & Casadevall, A. (2005). *Biochemistry*, 44:3683–3693.

Ellerbroek, P. M., Hoepelman, A. I., Wolbers, F., Zwaginga, J. J., & Coenjaerts, F. E. (2002). *Infect. Immun.*, 70:4762–4771.

Ellerbroek, P. M., Ulfman, L. H., Hoepelman, A. I., & Coenjaerts, F. E. (2004). *Cell Microbiol.*, 6:581–592.

Erickson, T., Liu, L., Gueyikian, A., Zhu, X., Gibbons, J., & Williamson, P. R. (2001). *Mol. Microbiol.*, 42:1121–1131.

Feldmesser, M., Kress, Y., Novikoff, P., & Casadevall, A. (2000). *Infect. Immun.*, 68:4225–4237.

Feldmesser, M., Mednick, A., & Casadevall, A. (2002). *Infect. Immun.*, 70:1571–80.

Fleuridor, R., Lees, A., & Pirofski, L. (2001). *J. Immunol.*, 166:1087–1096.

Fox, D. S., Cruz, M. C., Sia, R. A., Ke, H., Cox, G. M., Cardenas, M. E., & Heitman, J. (2001). *Mol. Microbiol.*, 39:835–849.

Fox, D. S. & Heitman, J. (2005). *Eukaryot Cell*, 4:1526–1538.

Fraser, J. A., Giles, S. S., Wenink, E. C., Geunes-Boyer, S. G., Wright, J. R., Diezmann, S., Allen, A., Stajich, J. E., Dietrich, F. S., Perfect, J. R., & Heitman, J. (2005). *Nature*, 437:1360–1364.

Frases, S., Chaskes, S., Dadachova, E., & Casadevall, A. (2006a) *Appl. Environ. Microbiol.*, 72:1542–1550.

Frases, S., Salazar, A., & Casadevall, A. (2006b) *Appl. Environ. Microbiol.*

Fries, B. C. & Casadevall, A. (1998). *J. Infect. Dis.*, 178:1761–1766.

Fries, B. C., Cook, E., Wang, X., & Casadevall, A. (2005). *Antimicrob. Agents Chemother.*, 49:350–357.

Fries, B. C., Goldman, D. L., Cherniak, R., Ju, R., & Casadevall, A. (1999). *Infect. Immun.*, 67:6076–6083.

Fries, B. C., Taborda, C. P., Serfass, E., & Casadevall, A. (2001). *J. Clin. Invest.*, 108:1639–1648.

Garcia-Hermoso, D., Dromer, F., & Janbon, G. (2004). *Infect. Immun.*, 72:3359–3365.

Garcia-Rivera, J., Chang, Y. C., Kwon-Chung, K. J., & Casadevall, A. (2004). *Eukaryot Cell*, 3:385–392.

Garcia-Rivera, J., Eisenman, H. C., Nosanchuk, J. D., Aisen, P., Zaragoza, O., Moadel, T., Dadachova, E., & Casadevall, A. (2005a). *Fungal Genet. Biol.*, 42:989–998.

Garcia-Rivera, J., Tucker, S. C., Feldmesser, M., Williamson, P. R., & Casadevall, A. (2005b). *Infect Immun.*, 73:3124–3127.

Ghosn, E. E., Russo, M., & Almeida, S. R. (2006). *J. Leukoc. Biol.*, 80:36–44.

Giles, S. S., Batinic-Haberle, I., Perfect, J. R., & Cox, G. M. (2005). *Eukaryot Cell*, 4:46–54.

Guerrero, A., Jain, N., Goldman, D. L., & Fries, B. C. (2006). *Microbiology*, 152:3–9.

Gyetko, M. R., Sud, S., Chen, G. H., Fuller, J. A., Chensue, S. W., & Toews, G. B. (2002). *J. Immunol.*, 168:801–809.

He, W., Casadevall, A., Lee, S. C., & Goldman, D. L. (2003). *Infect. Immun.*, 71:930–936.

Heung, L. J., Luberto, C., Plowden, A., Hannun, Y. A., & Del Poeta, M. (2004). *J. Biol. Chem.*, 279:21144–21153.

Hicks, J. K., Bahn, Y. S., & Heitman, J. (2005). *Eukaryot Cell*, 4:1971–1981.

Hill, H. Z. (1992). *Bioessays*, 14:49–56.

Hirano, A., Zimmerman, H. M., & Levine, S. (1964a). *Am. J. Pathol.*, 45:1–19.

Hirano, A., Zimmerman, H. M., & Levine, S. (1964b). *Am. J. Pathol.*, 45:195–207.

Hoang, L. M., Maguire, J. A., Doyle, P., Fyfe, M., & Roscoe, D. L. (2004). *J. Med. Microbiol.*, 53:935–940.

Huang, C., Nong, S. H., Mansour, M. K., Specht, C. A., & Levitz, S. M. (2002). *Infect. Immun.*, 70:5485–5493.

Huffnagle, G. B. & McNeil, L. K. (1999). *J. Neurovirol.*, 5:76–81.

Hull, C. M., Boily, M. J., & Heitman, J. (2005). *Eukaryot Cell*, 4:526–535.

Hull, C. M. & Heitman, J. (2002). *Annu. Rev. Genet.*, 36:557–615.

Husain, S., Wagener, M. M., & Singh, N. (2001). *Emerg. Infect. Dis.*, 7:375–381.

Idnurm, A., Bahn, Y. S., Nielsen, K., Lin, X., Fraser, J. A., & Heitman, J. (2005). *Nat. Rev. Microbiol.*, 3:753–764.

Jacobson, E. S. & Ikeda, R. (2005). *Med. Mycol.*, 43:327–333.

Jain, N., Li, L., McFadden, D. C., Banarjee, U., Wang, X., Cook, E., & Fries, B. C. (2006). *Infect. Immun.*, 74:896–903.

Janbon, G. (2004). *FEMS Yeast Res.*, 4:765–771.

Janbon, G., Himmelreich, U., Moyr&, F., Improvisi, L., & Dromer, F. (2001). *Mol. Microbiol.*, 42:453–467.

Kawakami, K. (2004). *Jpn. J. Infect. Dis.*, 57:137–145.

Kawakami, K., Kinjo, Y., Uezu, K., Yara, S., Miyagi, K., Koguchi, Y., Nakayama, T., Taniguchi, M., & Saito, A. (2001a) *J. Immunol.*, 167:6525–6532.

Kawakami, K., Kinjo, Y., Yara, S., Koguchi, Y., Uezu, K., Nakayama, T., Taniguchi, M., & Saito, A. (2001b) *Infect. Immun.*, 69:213–220.

Kawakami, K., Koguchi, Y., Qureshi, M. H., Kinjo, Y., Yara, S., Miyazato, A., Kurimoto, M., Takeda, K., Akira, S., & Saito, A. (2000). *FEMS Microbiol. Lett.*, 186:121–126.

Kawakami, K., Shibuya, K., Qureshi, M. H., Zhang, T., Koguchi, Y., Tohyama, M., Xie, Q., Naoe, S., & Saito, A. (1999). *FEMS Immunol. Med. Microbiol.*, 25:391–402.

Kelly, R. M., Chen, J., Yauch, L. E., & Levitz, S. M. (2005). *Infect. Immun.*, 73:592–598.

Kingsbury, J. M., Yang, Z., Ganous, T. M., Cox, G. M., & McCusker, J. H. (2004a). *Microbiology*, 150:1547–1558.

Kingsbury, J. M., Yang, Z., Ganous, T. M., Cox, G. M., & McCusker, J. H. (2004b). *Eukaryot Cell*, 3:752–763.

Kozel, T. R., Gulley, W. F., & Cazin, J., Jr. (1977). *Infect. Immun.*, 18:701–707.

Kozel, T. R. & Mastroianni, R. P. (1976). *Infect. Immun.*, 14:62–67.

Kraus, P. R., Fox, D. S., Cox, G. M., & Heitman, J. (2003). *Mol. Microbiol.*, 48:1377–1387.

Kraus, P. R. & Heitman, J. (2003). *Biochem. Biophys. Res. Commun.*, 311:1151–1157.

Kwon-Chung, K. J. & Rhodes, J. C. (1986). *Infect. Immun.*, 51:218–223.

Kwon-Chung, K. J. & Varma, A. (2006). *FEMS Yeast Res.*, 6:574–587.

Larsen, R. A., Pappas, P. G., Perfect, J., Aberg, J. A., Casadevall, A., Cloud, G. A., James, R., Filler, S.. & Dismukes, W. E. (2005). *Antimicrob. Agents Chemother.*, 49:952–958.

Lendvai, N. & Casadevall, A. (1999). *J. Infect. Dis.*, 180:791–801.

Levitz, S. M., Nong, S., Mansour, M. K., Huang, C., & Specht, C. A. (2001). *Proc. Natl. Acad. Sci. USA*, 98:10422–10427.

Levitz, S. M. & Specht, C. A. (2006). *FEMS Yeast Res.*, 6:513–524.

Lindell, D. M., Ballinger, M. N., McDonald, R. A., Toews, G. B., & Huffnagle, G. B. (2006). *Infect. Immun.*, 74:4538–4548.

Lindell, D. M., Moore, T. A., McDonald, R. A., Toews, G. B., & Huffnagle, G. B. (2005). *J. Immunol.*, 174:7920–7928.

Liu, L., Tewari, R. P., & Williamson, P. R. (1999). *Infect. Immun.*, 67:6034–6039.

Loftus, B. J., Fung, E., Roncaglia, P., Rowley, D., Amedeo, P., Bruno, D., Vamathevan, J., (2005). *Science*, 307:1321–1324.

Lortholary, O., Poizat, G., Zeller, V., Neuville, S., Boibieux, A., Alvarez, M., Dellamonica, P., Botterel, F., Dromer, F., & Chene, G. (2006). *Aids*, 20:2183–2191.

Lovchik, J. A., Wilder, J. A., Huffnagle, G. B., Riblet, R., Lyons, C. R., & Lipscomb, M. F. (1999). *J. Immunol.*, 163:3907–3913.

Luberto, C., Martinez-Marino, B., Taraskiewicz, D., Bolanos, B., Chitano, P., Toffaletti, D. L., Cox, G. M., Perfect, J. R., Hannun, Y. A., Balish, E., & Del Poeta, M. (2003). *J. Clin. Invest.*, 112:1080–1094.

Luberto, C., Toffaletti, D. L., Wills, E. A., Tucker, S. C., Casadevall, A., Perfect, J. R., Hannun, Y. A., & Del Poeta, M. (2001). *Genes Dev.*, 15:201–212.

Luo, Y., Cook, E., Fries, B. C., & Casadevall, A. (2006). *Clin. Exp. Immunol.*, 145:380–387.

Maitta, R. W., Datta, K., Lees, A., Belouski, S. S., & Pirofski, L. A. (2004). *Infect. Immun.*, 72:196–208.

Mambula, S. S., Simons, E. R., Hastey, R., Selsted, M. E., & Levitz, S. M. (2000). *Infect. Immun.*, 68:6257–6264.

Mansour, M. K., Latz, E., & Levitz, S. M. (2006). *J. Immunol.*, 176:3053–3061.

Mansour, M. K., Schlesinger, L. S., & Levitz, S. M. (2002). *J. Immunol.*, 168:2872–2879.

Mansour, M. K., Yauch, L. E., Rottman, J. B., & Levitz, S. M. (2004). *Infect. Immun.*, 72:1746–1754.

Mare, L., Iatta, R., Montagna, M. T., Luberto, C., & Del Poeta, M. (2005). *J. Biol. Chem.*, 280:36055–36064.

Marques, S. A., Robles, A. M., Tortorano, A. M., Tuculet, M. A., Negroni, R., & Mendes, R. P. (2000). *Med. Mycol.*, 38(Suppl. 1):269–279.

McClelland, E. E., Adler, F. R., Granger, D. L., & Potts, W. K. (2004b). *Proc. R. Soc. Lond. B Biol. Sci.*, 271:1557–1564.

McClelland, E. E., Granger, D. L., & Potts, W. K. (2003b). *Infect. Immun.*, 71:4815–4817.

McFadden, D. C., De Jesus, M., & Casadevall, A. (2006). *J. Biol. Chem.*, 281:1868–1875.

McGaha, T. & Murphy, J. W. (2000). *Infect. Immun.*, 68:4624–4630.

Mednick, A. J., Feldmesser, M., Rivera, J., & Casadevall, A. (2003). *Eur. J. Immunol.*, 33:1744–1753.

Mirza, S. A., Phelan, M., Rimland, D., Graviss, E., Hamill, R., Brandt, M. E., Gardner, T., Sattah, M., de Leon, G. P., Baughman, W., & Hajjeh, R. A. (2003). *Clin. Infect. Dis.*, 36:789–794.

Missall, T. A. & Lodge, J. K. (2005). *Mol. Microbiol.*, 57:847–858.

Missall, T. A., Moran, J. M., Corbett, J. A., & Lodge, J. K. (2005). *Eukaryot Cell*, 4:202–208.

Missall, T. A., Pusateri, M. E., & Lodge, J. K. (2004). *Mol. Microbiol.*, 51:1447–1458.

Mody, C. H. & Syme, R. M. (1993). *Infect. Immun.*, 61:464–469.

Monari, C., Bistoni, F., Casadevall, A., Pericolini, E., Pietrella, D., Kozel, T. R., & Vecchiarelli, A. (2005a) *J. Infect. Dis.*, 191:127–137.

Monari, C., Kozel, T. R., Bistoni, F., & Vecchiarelli, A. (2002). *Infect. Immun.*, 70:3363–3370.

Monari, C., Kozel, T. R., Casadevall, A., Pietrella, D., Palazzetti, B., & Vecchiarelli, A. (1999). *Immunology*, 98:27–35.

Monari, C., Pericolini, E., Bistoni, G., Casadevall, A., Kozel, T. R., & Vecchiarelli, A. (2005b). *J. Immunol.*, 174:3461–3468.

Monari, C., Pericolini, E., Bistoni, G., Cenci, E., Bistoni, F., & Vecchiarelli, A. (2005c). *J. Infect. Dis.*, 191:307–311.

Monari, C., Retini, C., Casadevall, A., Netski, D., Bistoni, F., Kozel, T. R., & Vecchiarelli, A. (2003). *Eur. J. Immunol.*, 33:1041–1051.

Moyrand, F., Chang, Y. C., Himmelreich, U., Kwon-Chung, K. J., & Janbon, G. (2004). *Eukaryot Cell*, 3:1513–1524.

Moyrand, F. & Janbon, G. (2004). *Eukaryot Cell*, 3:1601–1608.

Moyrand, F., Klaproth, B., Himmelreich, U., Dromer, F., & Janbon, G. (2002). *Mol. Microbiol.*, 45:837–849.

Mozaffarian, N., Casadevall, A., & Berman, J. W. (2000). *J. Immunol.*, 165:1541–1547.

Mukherjee, J., Cleare, W., & Casadevall, A. (1995). *J. Immunol. Methods*, 184:139–143.

Murphy, J. W. & Cozad, G. C. (1972). *Infect. Immun.*, 5:896–901.

Mylonakis, E., Ausubel, F. M., Perfect, J. R., Heitman, J., & Calderwood, S. B. (2002). *Proc. Natl. Acad. Sci. USA*, 99:15675–15680.

Mylonakis, E., Ausubel, F. M., Tang, R. J., & Calderwood, S. B. (2003). *Expert Rev. Anti. Infect. Ther.*, 1:167–173.

Mylonakis, E., Idnurm, A., Moreno, R., El Khoury, J., Rottman, J. B., Ausubel, F. M., Heitman, J., & Calderwood, S. B. (2004). *Mol. Microbiol.*, 54:407–419.

Mylonakis, E., Moreno, R., El Khoury, J. B., Idnurm, A., Heitman, J., Calderwood, S. B., Ausubel, F. M., & Diener, A. (2005). *Infect. Immun.*, 73:3842–3850.

Narasipura, S. D., Ault, J. G., Behr, M. J., Chaturvedi, V., & Chaturvedi, S. (2003). *Mol. Microbiol.*, 47:1681–1694.

Nichols, C. B., Fraser, J. A., & Heitman, J. (2004). *Mol. Biol. Cell*, 15:4476–4489.

Nosanchuk, J. D. & Casadevall, A. (2003a). *Microbiology*, 149:1945–1951.

Nosanchuk, J. D. & Casadevall, A. (2003b) *Cell Microbiol.*, 5:203–223.

Nosanchuk, J. D., Ovalle, R., & Casadevall, A. (2001). *J. Infect. Dis.*, 183:1093–1099.

Nosanchuk, J. D., Rosas, A. L., & Casadevall, A. (1998). *J. Immunol.*, 160:6026–6031.

Nosanchuk, J. D., Rosas, A. L., Lee, S. C., & Casadevall, A. (2000). *Lancet*, 355:2049–2050.

Noverr, M. C., Cox, G. M., Perfect, J. R., & Huffnagle, G. B. (2003). *Infect. Immun.*, 71:1538–1547.

Noverr, M. C., Williamson, P. R., Fajardo, R. S., & Huffnagle, G. B. (2004). *Infect. Immun.*, 72:1693–1699.

Odom, A., Muir, S., Lim, E., Toffaletti, D. L., Perfect, J., & Heitman, J. (1997). *EMBO J.*, 16:2576–2589.

Olszewski, M. A., Noverr, M. C., Chen, G. H., Toews, G. B., Cox, G. M., Perfect, J. R., & Huffnagle, G. B. (2004). *Am. J. Pathol.*, 164:1761–1771.

Palmer, D. A., Thompson, J. K., Li, L., Prat, A., & Wang, P. (2006). *J. Biol. Chem.*, 276:27026–27033.

Panepinto, J., Liu, L., Ramos, J., Zhu, X., Valyi-Nagy, T., Eksi, S., Fu, J., Jaffe, H. A., Wickes, B., & Williamson, P. R. (2005). *J. Clin. Invest.*, 115:632–641.

Parra, C., Gonzalez, J. M., Castaneda, E., & Fiorentino, S. (2005). *Biomedica*, 25:110–119.

Pascon, R. C., Ganous, T. M., Kingsbury, J. M., Cox, G. M., & McCusker, J. H. (2004). *Microbiology*, 150:3013–3023.

Pericolini, E., Cenci, E., Monari, C., Perito, S., Mosci, P., Bistoni, G., & Vecchiarelli, A. (2006). *Med. Mycol.*, 44:119–126.

Pettoello-Mantovani, M., Casadevall, A., Kollmann, T. R., Rubinstein, A., & Goldstein, H. (1992). *Lancet*, 339:21–23.

Pettoello-Mantovani, M., Casadevall, A., Smarnworawong, P., & Goldstein, H. (1994). *AIDS Res. Hum. Retroviruses*, 10:1079–1087.

Pietrella, D., Cherniak, R., Strappini, C., Perito, S., Mosci, P., Bistoni, F., & Vecchiarelli, A. (2001a) *Infect. Immun.*, 69:2808–2814.

Pietrella, D., Lupo, P., Perito, S., Mosci, P., Bistoni, F., & Vecchiarelli, A. (2004). *FEMS Immunol. Med. Microbiol.*, 40:63–70.

Pietrella, D., Mazzolla, R., Lupo, P., Pitzurra, L., Gomez, M. J., Cherniak, R., & Vecchiarelli, A. (2002). *Infect. Immun.*, 70:6621–6627.

Pietrella, D., Perito, S., Bistoni, F., & Vecchiarelli, A. (2001b). *Infect. Immun.*, 69:1508–1514.

Pukkila-Worley, R. & Alspaugh, J. A. (2004). *FEMS Yeast Res.*, 4:361–367.

Pukkila-Worley, R., Gerrald, Q. D., Kraus, P. R., Boily, M. J., Davis, M. J., Giles, S. S., Cox, G. M., Heitman, J., & Alspaugh, J. A. (2005). *Eukaryot Cell*, 4:190–201.

Reed, S. G. (1999). *Microbes Infect.*, 1:1313–1325.

Ren, P., Springer, D. J., Behr, M. J., Samsonoff, W. A., Chaturvedi, S., & Chaturvedi, V. (2006). *Eukaryot Cell*, 5:1065–1080.

Retini, C., Casadevall, A., Pietrella, D., Monari, C., Palazzetti, B., & Vecchiarelli, A. (1999). *J. Immunol.*, 162:1618–1623.

Retini, C., Kozel, T. R., Pietrella, D., Monari, C., Bistoni, F., & Vecchiarelli, A. (2001). *Infect. Immun.*, 69:6064–6073.

Retini, C., Vecchiarelli, A., Monari, C., Bistoni, F., & Kozel, T. R. (1998). *Infect. Immun.*, 66:664–669.

Retini, C., Vecchiarelli, A., Monari, C., Tascini, C., Bistoni, F., & Kozel, T. R. (1996). *Infect. Immun.*, 64:2897–2903.

Rivera, J. & Casadevall, A. (2005). *J. Immunol.*, 174:8017–8026.

Rivera, J., Mukherjee, J., Weiss, L. M., & Casadevall, A. (2002). *J. Immunol.*, 168:3419–3427.

Rivera, J., Zaragoza, O., & Casadevall, A. (2005). *Infect. Immun.*, 73:1141–1150.

Rodrigues, M. L., Nimrichter, L., Oliveira, D. L., Frases, S., Miranda, K., Zaragoza, O., Alvarez, M., Nakouzi, A., Feldmesser, M., & Casadevall, A. (2006). *Eukaryot Cell.*

Rosas, A. L., Nosanchuk, J. D., & Casadevall, A. (2001). *Infect. Immun.*, 69:3410–3412.

Rosas, A. L., Nosanchuk, J. D., Feldmesser, M., Cox, G. M., McDade, H. C., & Casadevall, A. (2000). *Infect. Immun.*, 68:2845–2853.

Santangelo, R., Zoellner, H., Sorrell, T., Wilson, C., Donald, C., Djordjevic, J., Shounan, Y., & Wright, L. (2004). *Infect. Immun.*, 72:2229–2239.

Shea, J. M. & Del Poeta, M. (2006). *Curr. Opin. Microbiol.*, 9:352–358.

Shea, J. M., Kechichian, T. B., Luberto, C., & Del Poeta, M. (2006). *Infect. Immun.*, 74:5977–5988.

Shen, W. C., Davidson, R. C., Cox, G. M., & Heitman, J. (2002). *Eukaryot Cell*, 1:366–377.

Sommer, U., Liu, H., & Doering, T. L. (2003). *J. Biol. Chem.*, 278:47724–47730.

Steenbergen, J. N., Nosanchuk, J. D., Malliaris, S. D., & Casadevall, A. (2003). *Infect. Immun.*, 71:4862–4872.

Steenbergen, J. N., Shuman, H. A., & Casadevall, A. (2001). *Proc. Natl. Acad. Sci. USA*, 98:15245–15250.

Syme, R. M., Bruno, T. F., Kozel, T. R., & Mody, C. H. (1999). *Infect. Immun.*, 67:4620–4627.

Syme, R. M., Spurrell, J. C., Amankwah, E. K., Green, F. H., & Mody, C. H. (2002). *Infect. Immun.*, 70:5972–5981.

Syme, R. M., Wood, C. J., Wong, H., & Mody, C. H. (1997). *Immunology*, 92:194–200.

Taborda, C. P., Rivera, J., Zaragoza, O., & Casadevall, A. (2003). *J. Immunol.*, 170:3621–3630.

Tampieri, M. P. (2006). *Parassitologia*, 48:121–124.

Tang, R. J., Breger, J., Idnurm, A., Gerik, K. J., Lodge, J. K., Heitman, J., Calderwood, S. B., & Mylonakis, E. (2005). *Infect. Immun.*, 73:8219–8225.

Uezu, K., Kawakami, K., Miyagi, K., Kinjo, Y., Kinjo, T., Ishikawa, H., & Saito, A. (2004). *J. Immunol.*, 172:7629–7634.

Uicker, W. C., Doyle, H. A., McCracken, J. P., Langlois, M., & Buchanan, K. L. (2005). *Med. Mycol.*, 43:27–38.

Vaishnav, V. V., Bacon, B. E., O'Neill, M., & Cherniak, R. (1998). *Carbohydr. Res.*, 306:315–330.

Vecchiarelli, A. (2000). *Med. Mycol.*, 38:407–417.

Vecchiarelli, A., Pietrella, D., Lupo, P., Bistoni, F., McFadden, D. C., & Casadevall, A. (2003). *J. Leukoc. Biol.*, 74:370–378.

Vecchiarelli, A., Retini, C., Monari, C., Tascini, C., Bistoni, F., & Kozel, T. R. (1996). *Infect. Immun.*, 64:2846–2849.

Vecchiarelli, A., Retini, C., Pietrella, D., Monari, C., Tascini, C., Beccari, T., & Kozel, T. R. (1995). *Infect. Immun.*, 63:2919–2923.

Vilchez, R. A., Fung, J., & Kusne, S. (2002). *Am. J. Transplant*, 2:575–580.

Walton, F. J., Idnurm, A., & Heitman, J. (2005). *Mol. Microbiol.*, 57:1381–1396.

Wang, P., Cardenas, M. E., Cox, G. M., Perfect, J. R., & Heitman, J. (2001). *EMBO Rep.*, 2:511–518.

Wang, P., Nichols, C. B., Lengeler, K. B., Cardenas, M. E., Cox, G. M., Perfect, J. R., & Heitman, J. (2002). *Eukaryot Cell*, 1:257–272.

Wang, Y., Aisen, P., & Casadevall, A. (1996). *Infect. Immun.*, 64:2420–2424.

Waterman, S. R., Hacham, M., Panepinto, J., Hu, G., Shin, S., & Williamson, P. R. (2006). *Infect. Immun.*

Williamson, P. R. (1994). *J. Bacteriol.*, 176:656–664.

Wills, E. A., Roberts, I. S., Del Poeta, M., Rivera, J., Casadevall, A., Cox, G. M., & Perfect, J. R. (2001). *Mol. Microbiol.*, 40, 610–620.

Wormley, F. L., Jr., Heinrich, G., Miller, J. L., Perfect, J. R., & Cox, G. M. (2005). *Infect. Immun.*, 73:5022–5030.

Wozniak, K. L., Vyas, J. M., & Levitz, S. M. (2006). *Infect. Immun.*, 74:3817–3824.

Wright, L. C., Payne, J., Santangelo, R. T., Simpanya, M. F., Chen, S. C., Widmer, F., & Sorrell, T. C. (2004). *Biochem. J.*, 384:377–384.

Xue, C., Bahn, Y. S., Cox, G. M., & Heitman, J. (2006). *Mol. Biol. Cell*, 17:667–679.

Yang, Z., Pascon, R. C., Alspaugh, A., Cox, G. M., & McCusker, J. H. (2002). *Microbiology*, 148:2617–2625.

Yauch, L. E., Lam, J. S., & Levitz, S. M. (2006). *PLoS Pathog.*, 2.

Yoneda, A. & Doering, T. L. (2006). *Mol. Biol. Cell.*

Yuan, R., Casadevall, A., Spira, G., & Scharff, M. D. (1995). *J. Immunol.*, 154:1810–1816.

Yuan, R. R., Casadevall, A., Oh, J., & Scharff, M. D. (1997). *Proc. Natl. Acad. Sci. USA*, 94:2483–2488.

Yuan, R. R., Spira, G., Oh, J., Paizi, M., Casadevall, A., & Scharff, M. D. (1998). *Infect. Immun.*, 66:1057–1062.

Zaragoza, O., Telzak, A., Bryan, R. A., Dadachova, E., & Casadevall, A. (2006). *Mol. Microbiol.*, 59:67–83.

Zhang, H., Zhong, Z., & Pirofski, L. A. (1997). *Infect. Immun.*, 65:1158–1164.

Zhang, S., Hacham, M., Panepinto, J., Hu, G., Shin, S., Zhu, X., & Williamson, P. R. (2006). *Mol. Microbiol.*, 62:1090–1101.

Zhou, Q., Gault, R. A., Kozel, T. R., & Murphy, W. J. (2006). *Infect. Immun.*, 74:2161–2168.

Zhu, X. & Williamson, P. R. (2003). *Mol. Microbiol.*, 50:1271–1281.

Chapter 7
The Zygomycetes

Eric Dannaoui and Dea Garcia-Hermoso

Introduction

General Description of Zygomycetes and Zygomycosis

The class Zygomycetes is a large group of fungi that comprise two orders of medical interest, the Mucorales and the Entomophthorales. Zygomycoses caused by Entomophthorales are generally chronic infections seen in immunocompetent patients, mostly in tropical areas (Dromer & McGinnis, 2002). General description of Entomophthorales and associated diseases have been reviewed recently (Prabhu & Patel, 2004; Ribes et al., 2000) and as no major advancement has been made recently in the knowledge and management of entomophthoromycosis, these fungi will not be detailed in this chapter. Human pathogens belonging to the order Mucorales are grouped into six families (Table 7.1) and comprise approximately 20 species in approximately 10 genera. The most frequent species responsible for zygomycosis are *Rhizopus* spp., *Mucor* spp., *Rhizomucor* spp., and *Absidia* spp. (Ribes et al., 2000). Other species such as *Cunninghamella bertholletiae*, *Apophysomyces elegans*, and *Saksaenea vasiformis* have been less frequently reported as etiological agents of zygomycosis.

Mucorales are hyaline fungi growing easily on standard culture media. Macroscopically, mycelium is expanding with a woolly appearance. Microscopically, these fungi are characterized by large, rarely septated hyphae. Mucorales can reproduce both by production of asexual sporangiospores or by sexual zygospores. Sexual reproduction is seldom observed in culture as most of the pathogenic mucorales are heterothallic (i.e. production of zygospores need confrontation of two strains of opposite mating type). Identification of the species in culture then mainly relies on the presence of asexual spore-bearing structures. For some species, these structures are sufficiently characteristic for an easy identification. Nevertheless, for the most frequent species (i.e. *Rhizopus* spp., *Mucor* spp., *Absidia* spp., and *Rhizomucor* spp.) the overall morphology is similar and identification is based on detailed observation of presence, location, and morphological characteristics of specific structures (Figure 7.1). For this reason, a precise identification to the species level can only be performed by specialized laboratories. Beside macroscopic

K. Kavanagh (ed.), *New Insights in Medical Mycology.*
© Springer 2007

Table 7.1 Zygomycetes species that have been described as human pathogens

Class	Order	Family	Genus	Species
Zygomycetes	*Mucorales*	*Mucoraceae*	*Absidia*	*A. coerulea*
				A. corymbifera
			Apophysomyces	*A. elegans*
			Chlamydoabsidia	*C. padenii*
			Mucor	*M. amphibiorum*
				M. circinelloides
				M. hiemalis
				M. indicus
				M. racemosus
				M. ramosissimus
			Rhizomucor	*R. miehei*
				R. pusillus
				R. variabilis
			Rhizopus	*R. azygosporus*
				R. microsporus
				R. oryzae
				R. schipperae
				R. stolonifer*
		Thamnidiaceae	*Cokeromyces*	*C. recurvatus*
		Cunninghamellaceae	*Cunninghamella*	*C. bertollethiae*
		Syncephalastraceae	*Syncephalastrum*	*S. racemosum*
		Saksenaeaceae	*Saksenaea*	*S. vasiformis*
	Mortierellales	*Mortierellaceae*	*Mortierella*	*M. polycephala**
				M. wolfii*

* The true pathogenic role *R. stolonifer* and Mortierellales in humans has not been definitely proven.

and microscopic morphology, few other identification criteria could be used such as maximum growth temperature or scanning electronic microscopy (de Hoog & Guarro, 2000; Scholer et al., 1983). Specific physiological tests have been used for discriminative identification between some closely related species (Vágvölgyi et al., 1996; Vastag et al., 1998; Ribes et al., 2000; Scholer et al., 1983), but no extensive studies have been performed on a sufficient number of isolates to date.

Clinical manifestations of zygomycosis can be grouped in five major entities: rhinocerebral, pulmonary, cutaneous, gastrointestinal, and disseminated infections (Dromer & McGinnis, 2002; Kwon-Chung & Bennett, 1992; Spellberg et al., 2005a). Common characteristics of these infections are the rapidity of progression and angioinvasion leading to vascular thrombosis and subsequently to tissue necrosis. Rhino-cerebral zygomycosis is one of the most frequent forms. It initially presents as acute sinusitis with rapid spread of infection to contiguous tissues of the orbit, skin, palate, and to the central nervous system. The main symptoms in patients with pulmonary zygomycosis include fever, cough, and chest pain. There are no specific clinical or radiological signs and this form of zygomycosis can easily be misdiagnosed for another filamentous fungal infection such as pulmonary

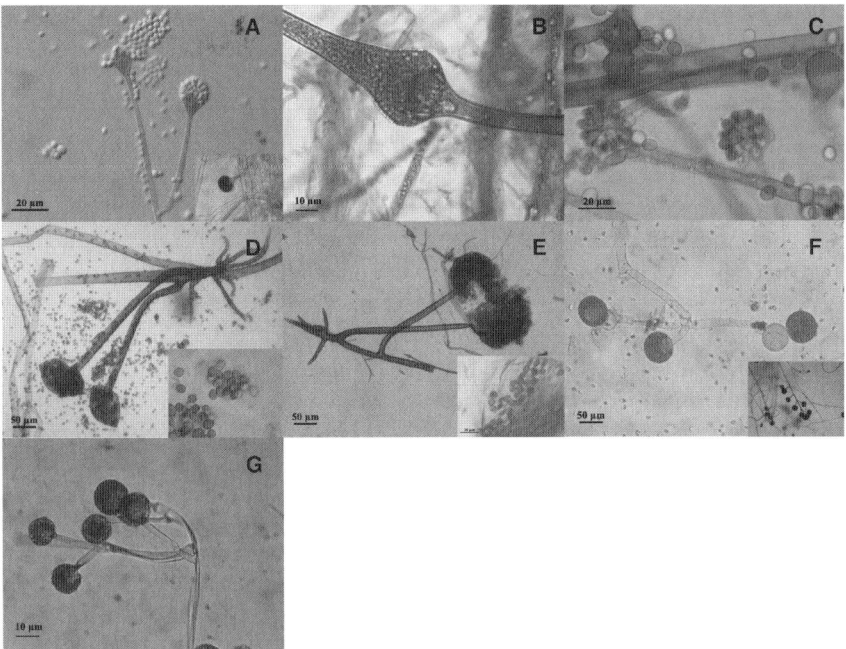

Figure 7.1 Microscopical characteristics of different species of zygomycetes. Some species can be easily identified such as *Absidia corymbifera* (**A**), *Saksenaea vasiformis* (**B**), or *Cunninghamella bertholletiae* (**C**). By contrast, other species share similar morphological traits making the phenotypic identification more complex such as *Rhizopus oryzae* (**D**) and *Rhizopus microsporus* (**E**), or *Mucor circinelloides* (**F**) and *Rhizomucor pusillus* (**G**)

invasive aspergillosis. Cutaneous zygomycosis is mainly acquired in a patient with altered skin barrier. Typical cases are traumatic injuries with contamination by soil but can also occur in patients with extensive burns or following traumatic implantation of the fungi at injection site. Gastrointestinal mucormycosis is a rare infection but generally acute and rapidly fatal. Symptoms are nonspecific and variable depending on the gastrointestinal tract site involved. Disseminated zygomycosis occurs after hematogenous spread of the infection from any primary site and is generally fatal. The brain is the most common site of dissemination but many other organs can be involved.

Ecology and Epidemiology

Zygomycetes are common fungi in the environment and can be recovered from soil, decaying organic matter, or food stuffs. Most of the species have a worldwide distribution. Nevertheless, the two species *Apophysomyces elegans* and *Saksenaea vasiformis* have been mainly reported from tropical and subtropical areas and

Cokeromyces recurvatus seems to be restricted to the USA and some areas of Mexico (Ribes et al., 2000). Some zygomycetes have been recovered in extreme environments such as hot geothermal soil (Redman et al., 1999) or in the Antarctic (Lawley et al., 2004). The route of transmission is mainly airborne, but infection can also occur after ingestion or traumatic implantation. Outbreaks and pseudoepidemics of cutaneous or gastrointestinal zygomycosis have been reported in some instances.

Zygomycetes are mostly opportunistic pathogens responsible for infections in immunocompromised patients or patients with other predisposing factors. Integrity of the skin barrier as well as innate immunity mediated by mononuclear and polymorphonuclear phagocytes prevent development of infection in an immunocompetent individual. Therefore, risk factors for zygomycosis comprise conditions in which these defense mechanisms are altered. The clinical presentation of the disease will vary depending upon the risk factors (Roden et al., 2005). Diabetes mellitus, and particularly in patients with ketoacidosis, is a well-known and common risk factor for zygomycosis. In these patients sinus and rhinocerebral infections are the most common. Iron is an important growth factor for zygomycetes and iron overload has been identified as a risk factor for zygomycosis. Paradoxically, deferoxamine therapy used for treatment of iron and aluminum overload in dialysis patients has also been recognized as an important risk factor for zygomycosis. In fact, it has been shown that deferoxamine, by chelating iron, acts as a siderophore, making iron more easily available to the fungus. A large number of these patients had disseminated zygomycosis. Immunosuppression either in patients with cancer or hematological malignancies or in solid-organ transplant recipients is a major risk factor. In these patients, neutropenia and corticosteroid therapy used for graft-versus-host disease, are the main factors that enhance the risk of infection. In patient with malignancy, bone marrow or solid organ transplantation, pulmonary zygomycosis is the most frequent form. Other underlying conditions predisposing to zygomycosis include traumas and burns, HIV infection, and intravenous drug abuse.

Diagnosis

Diagnosis of zygomycosis remains difficult. There are no specific clinical or radiological signs for zygomycosis and the relative low frequency of the disease explain why the diagnosis is not always suspected. A definite diagnosis could be done by histopathology or direct examination of infected tissues, by demonstration of hyphae characteristic of a zygomycete (Figure 7.2). In tissue sections, hyphae are broad, ribbon-like with irregular diameter, non- or rarely septated, and branched at right angles (Ribes et al., 2000). These morphological characteristics are different from those of hyphomycetes (e.g. *Aspergillus* spp.). Demonstration of fungal elements could be done after routine histopathological staining but are better visualized with fungal-specific staining such as Gomori methenamine-silver (GMS) or periodic acid-Schiff. Direct examination of tissue or other samples such as

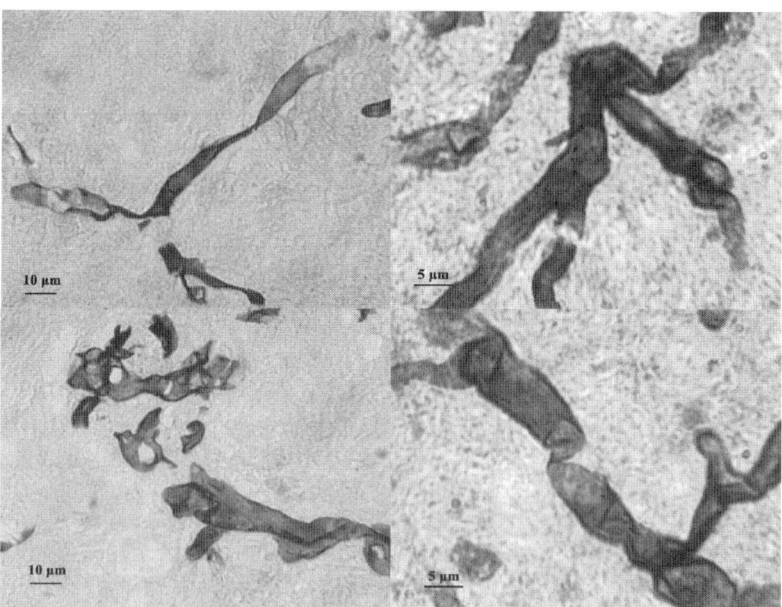

Figure 7.2 Typical morphology of zygomycetes in tissues (Gomori methenamine-silver stain). Characteristics of hyphae are identical for all species. In tissue sections, hyphae are broad, ribbon-like with irregular diameter, non- or rarely septated, and branched at right angles

bronchoalveolar lavage, preferably after specific fungal stain (e.g. GMS, Calcofluor white), should also be performed in the microbiology laboratory before culture. Demonstration of typical hyphae is of prime importance for at least two reasons. First, because mucorales are common laboratory contaminants and the signification of a positive culture from a non-sterile site is difficult to interpret when histology or direct examination is negative or has not been performed. Second, culture of infected samples is often negative. Probably due to their coenocytic nature, the hyphae seem to lose viability very easily. In this respect, it is not recommended to grind tissues before culture (Dromer & McGinnis, 2002), but alternatively to place small pieces of intact tissues directly on the culture medium. As zygomycetes are sensitive to cycloheximide, the culture medium should not contain this antimicrobial agent. Although, zygomycetes have the capability to invade vessels and to disseminate, blood cultures are usually negative, even in case of disseminated disease. There are no serological methods or antigen detection tests currently available. PCR and other molecular-based diagnostic methods have been recently evaluated but are not yet standardized and are not commercially available.

Treatment

Zygomycosis is regularly lethal in absence of treatment. Zygomycetes are resistant to most of the currently available antifungal drugs. Amphotericin B is the only drug that exhibit potent *in vitro* activity and *in vivo* efficacy and it represents the first-line treatment of all forms of zygomycosis. Nevertheless, the clinical efficacy of amphotericin B remains suboptimal with an overall mortality rate of approximately 40% (Roden et al., 2005). This mortality rate has been stable during the last 50 years. This poor prognosis could also be related to the rapid progression of the disease, the difficulty of an early diagnosis, and the poor penetration of antifungal drugs at the site of infection related to thrombosis and tissue necrosis. Adjunctive therapy is of importance, particularly surgery and management of underlying risk factors. Combination of amphotericin B with surgery significantly increases survival rate. Surgery has been shown to improve survival in cutaneous zygomycosis, but also in the rhinocerebral and pulmonary forms of the disease (Spellberg et al., 2005a).

Recent Developments

Taxonomy and Phylogeny

Historically, members of the kingdom Fungi (Eumycota) are recognized in four phyla: Ascomycota, Basidiomycota, Chytridiomycota, and Zygomycota. In contrast to Ascomycota and Basidiomycota, which form regularly septate mycelia, most members of the phylum Zygomycota are characterized by possessing coenocytic mycelia (lacking cell septa). The sexual reproduction is by formation of highly resistant zygotes after fusion of isogamic sex organs and the asexual reproduction is by nonmotile spores (de Hoog & Guarro, 2000).

The phylum Zygomycota consists of two classes, the Trichomycetes (associated within living arthropods) and Zygomycetes (common in nature and surviving on decaying vegetation). Ten orders of Zygomycetes with diverse morphology and ecology have been described in the recent edition of the Dictionary of the Fungi (Kirk et al., 2001). These include saprobes obligate parasite of insects or microorganisms, endocommensals in crustacean guts and ecto- or endo-mycorrhizae. The ten orders of Zygomycetes are composed of 32 families, 124 genera, and 870 species. Organisms of clinical significance are mostly included in three orders, the group of *Entomophtorales* (having spores forcibly discharged), *Mucorales* and *Mortierellales* (spores liberated after cleavage of sporangial plasma). Therefore, species implicated in pathology are included in the Mucorales within the *Mucoraceae* (18 species covering genera such as *Absidia, Apophysomyces, Chlamydoabsidia, Mucor, Rhizomucor*, and *Rhizopus*), the *Thamnidiaceae* (*Cokeromyces recurvatus*), the *Cunninghamellaceae* (*Cunninghamella bertholletiae*), the *Syncephalastraceae* (*Syncephalastrum racemosum*), and the *Saksenaeaceae* (*Saksenaea vasiformis*); and in the *Mortierellales* within the family *Mortierellaceae* (2 species) (de Hoog & Guarro, 2000).

Although the taxonomy of the Zygomycetes is still mainly based on morphology, molecular phylogenetic data have contributed to the understanding of the evolution and taxonomy of the group. Over the past years, phylogenetic interactions among members of the order of Mucorales and Entomophtorales have been studied by sequencing ribosomal DNA (rDNA) genes as well as actin elongation factor of translation EF-1 genes.

An 18S and 28S rDNA phylogeny of clinically relevant species of 42 Zygomycetes was provided by Voigt ct al. (1999). The analysis of the 18S data indicated that *Absidia corymbifera* was misplaced taxonomically. The genus *Absidia* harbors species producing small, apophysate sporangia with deliquescent walls, the production of stolons and rhizoids and appendage zygospores suspensors. Instead, *A. corymbifera* possess non-appendaged suspensors. Results of the molecular phylogeny also supported the transfer of *Rhizomucor variabilis* to *Mucor*. This species seemed to be more closely related phylogenetically to *Mucor mucedo* and *Mucor hiemalis*. Species of *Rhizomucor* are thermophilic and form simple or weakly branched rhizoids from hyphae, stolons, and sporangia while *R. variabilis* fails to do this.

For a better resolution of closely related species of some genera the same group (Voigt & Wostemeyer, 2001) used intron-containing nuclear genes (actin and translation elongation factor EF-1) to evaluate the phylogenetic positions of the Mucorales and Mortierellales. Results of this study indicated that a taxonomic revision of these orders is necessary.

A large-scale phylogenetic analysis of the Mucorales was carried out by O'Donnell and collaborators (2001). The analysis was based on partial nucleotide sequences from 18S and 28S rDNA genes together with elongation factor-1 exons and morphological data. Relationships within 13 morphologically defined families (including 54 genera and 63 species) were investigated. The results suggested that several of the largest families (Thamnidiaceae, Mucoraceae, and Chaetocladiaceae) as well as the genera *Absidia* and *Mucor* were polyphyletic.

Rhizopus species are classically divided by phenotypical criteria into three groups: the *microsporus-, oryzae-,* and *stolonifer*-group. A recent study investigated the molecular phylogeny of this genus using rDNA sequencing (18S, ITS, and 28S regions). The results indicated that phylogenetic relationships were similar to the morphological grouping, except for *R. schipperae*, a newly described human pathogen belonging to the *microsporus* group and *R. stolonifer* var. *lyococcos*, a nonpathogenic species. The *microsporus-, oryzae-,* and *stolonifer*-groups were supported by the analysis of 18S, ITS, and 28S sequences. Nevertheless, these molecular data placed *R. schipperae* distantly from the other members of this group. Similarly, *R. stolonifer* var. *lyococcos* clustered independently from *R. stolonifer* var. *stolonifer* in all trees suggesting an eventual reclassification as an independent species. These genetic data are not in accordance with the phenotypic taxonomy because these two varieties differ only by the morphology of the sporiangophore and are sexually compatible indicating that they should belong to the same species (Abe et al., 2006).

The evolutionary history of Zygomycetes and the interrelations among the different orders continue to be investigated by various groups which will contribute to a better knowledge of the organisms.

Pathophysiology and Biology

Pathophysiology

The main host defense mechanisms against zygomycetes are known and explain the risk factors for zygomycosis. Nevertheless, the pathophysiology of the disease remains largely unknown and there have been almost no studies performed during the last 10 years to uncover potential virulence factors or to evaluate the cellular and molecular interactions leading to infection. Angioinvasion with subsequent thrombosis and tissue necrosis is usually seen during zygomycosis. For this reason, one recent study focused on the interactions between zygomycetes and endothelial cells (Ibrahim et al., 2005b). It was shown that spores and germ tubes of *R. oryzae* adhere to endothelial cells, but only spores adhere to subendothelial matrix. Furthemore, damages of endothelial cells was demonstrated and required phagocytosis of the fungus. Interestingly, viability of the fungus was not required as dead *R. oryzae* was able to damage endothelial cells (Ibrahim et al., 2005b).

Biology

More advances have been made in the field of the biology of zygomycosis. In particular, very interesting data have been obtained concerning the production of toxins by zygomycetes (Partida-Martinez et al., 2006; Partida-Martinez & Hertweck, 2005). *R. microsporus* is a human pathogen but is also known to cause rice seedling blight, an economically important agricultural disease. This disease is mediated by rhizoxin, a macrocyclic polyketide metabolite. This toxin binds to beta-tubulin of the plant cells resulting in cell cycle arrest. Interestingly, rhizoxin is also known as a strong antimitotic agent against various human cancer cell lines. A very elegant study has recently shown that rhizoxin was not synthesized by the fungus itself, but by endosymbiont bacteria (Partida-Martinez & Hertweck, 2005). Several lines of evidence demonstrated that endosymbiotic bacteria belonging to the genus *Burkholderia* were living in the cytoplasm of rhizoxin producing strains of *R. microsporus* and were responsible for the synthesis of the toxin. Curing the fungus by antibiotic treatment resulted in symbiont-free strains that were unable to produce rhizoxin and toxin production was recovered after reintroduction of the bacterium in the symbiont-free fungus host. Moreover, pure culture of the *Burkholderia* sp. strain was able to produce rhizoxin as well as some other rhizoxin derivatives (Partida-Martinez & Hertweck, 2005). Fungus–bacteria associations are very rare (Lumini et al., 2006) and *R. microsporus* is the first human pathogenic fungus known to harbor endosymbiotic bacteria. This discovery may have medical implications. First, rhizoxin derivatives produced by the endosymbiont have been shown to be potent antimitotic agents (Scherlach et al., 2006). Second, as recently highlighted (Chamilos et al., 2006), it could be possible that the pathogenicity of some zygomycetes species may be enhanced by the presence of endosymbiotic toxin-producing bacteria. It has also been hypothesized that the recent emergence of zygomycosis in cancer patients could be related to the higher frequency of multi-drug-resistant *Pseudomonas* spp. induced by the extensive use of broad-spectrum antibiotics in these patients (Chamilos et al., 2006). These hypotheses need to be tested by experimental studies in animals and to be supported by clinical and micro-biological data obtained in patients.

Some *R. microsporus* strains are also known to produce mycotoxins such as rhizonin A and B, two cyclic heptapeptides that are highly hepatotoxic. Since closely related *Rhizopus* species are used in many part of the world for the production of fermented foods, detection of toxinogenic strains is of prime importance. Once again, it has been demonstrated that rhizonin was not synthesized by the fungus itself but by endosymbiont *Burkholderia* sp. living in the fungal cytosol (Partida-Martinez et al., 2006).

Advances have also been made in the knowledge of zygomycetes genome and in the technical field of genetic manipulation of these fungi. Firstly, a genome-sequencing project of *R. oryzae* is currently under way (http://www.broad.mit.edu/annotation/genome/rhizopus_oryzae/Info.html). The available data have already been used, for example for identification and study of the 14-alpha-demethylase of *R. oryzae* (Chau et al., 2006). Secondly, new tools for genetic manipulation of zygomycetes are now available (Ibrahim & Skory, 2006).

Epidemiology of Zygomycosis

Because zygomycosis is a rare infection, difficult to diagnose, and occurring in different patient populations with varied predisposing factors, epidemiology of this disease remains poorly understood. The only evaluation of incidence of zygomycosis among the general population has been published about 10 years ago after a population-based laboratory active surveillance performed in three counties of California, USA, during 1992–1993. Ten cases of zygomycosis have been reported for a total population of 2.94 million, which correspond to an incidence of 1.7 cases per million per year (Rees et al., 1998). Nevertheless, advances have been made during the past years, particularly in selected populations such as patients with malignancies although there have been no large-scale multicenter surveillance studies. Overall, these studies have reported an increase of the number of zygomycosis over time. In particular, analysis of the records of 5,589 patients who underwent hematopoietic stem cell transplantation from 1985 to 1999 at one center showed that the number of zygomycosis increased from seven cases in the period 1985–1989 to 15 cases in the period 1995–1999 (Marr et al., 2002). Similarly, in a large cancer center, 24 cases of zygomycosis have been reported over a 10-year period, with an increase of incidence from eight cases for 100,000 admissions during 1989–1993 to 20 cases for 100,000 admissions during 1994–1998 (Kontoyiannis et al., 2000). Interestingly all cases were diagnoses in patients with hematologic malignancy and none in patients with solid cancer. Of importance, since 2004 several publications have reported cases of breakthrough zygomycosis in patients receiving voriconazole as prophylaxis or empiric therapy (Marty et al., 2004; Siwek et al., 2004; Vigouroux et al., 2005; Oren, 2005; Kontoyiannis et al., 2005; Imhof et al., 2004). Voriconazole is known to have no activity against zygomycetes and this could explain the breakthrough infections. In a prospective case-control study in a cancer center it has been shown that voriconazole prophylaxis was an independent factor for zygomycosis (Kontoyiannis et al., 2005). Nevertheless, as recently highlighted (Kauffman, 2004), it is not clear if the use of voriconazole is directly responsible for the increased incidence of zygomycosis. It has to be noticed that the increase in the incidence of zygomycosis has started in the 1990s before the availability of voriconazole (Kontoyiannis et al., 2000; Marr et al., 2002), and that practices for management of patients with hematologic malignancy have changed with an increase frequency of graft-versus-host disease that requires aggressive immunosuppressive therapy (Kauffman, 2004).

Zygomycosis is acquired from environmental sources and there is no person to person transmission. Nevertheless, one should be aware that nosocomial outbreaks may occur and that grouped cases of zygomycosis should prompted investigations. Several outbreaks and pseudoepidemics have been reported in the past, and other reports have been published recently. An outbreak of gastric mucormycosis involving five patients in an intensive care unit, with an attributable mortality of 40%, was linked to the use of wooden tongue depressors contaminated with *R. microsporus* (Maravi-Poma et al., 2004). Outbreaks of cutaneous zygomycosis caused by *R. oryzae* and *A. corymbifera* have also been reported recently (LeMaile-Williams et al., 2006; Christiaens et al., 2005).

Identification of Zygomycetes and Diagnosis of Zygomycosis

Identification of zygomycetes and diagnosis of zygomycosis is notoriously difficult. Three main issues are of importance: the identification to the species level of a strain isolated in culture, the identification of a zygomycete as the etiologic agent of infection in tissues (either from fresh biopsy samples or from formalin-fixed paraffin-embedded material), and the identification of a particular strain (i.e. typing of isolates within a given species). Recent advances have been made in these three areas, in particular with the availability and evaluation of new molecular approaches.

Identification of Strains in Culture

By using several molecular targets (Table 7.2) and by increasing the number of available DNA sequences in international databases (e.g. GenBank), several studies have shown that accurate molecular identification to the species level of the zygomycetes pathogenic for humans is feasible (Machouart et al., 2006; Schwarz

Table 7.2 Molecular methods used for identification of zygomycetes strains in culture

Species	Target rDNA [a]region	Method	References
Several species	28S	PCR (specific primers for 13 taxons)	Voigt et al.(1999)
Several species	ITS	PCR + Sequencing	Schwarz et al. (2006)
Several species	18S	PCR + RFLP	Machouart et al. (2006)
Several species	ITS	PCR + Sequencing	
Several species[b]	28S	MicroSeq D2 large-subunit ribosomal DNA sequencing kit	Hall et al. (2004)
Rhizopus species	ITS	Multiplex PCR	Nagao et al. (2005)
Apophysomyces elegans	ITS	PCR + RFLP	Chakrabarti et al. (2003)
Rhizopus microsporus (Case report)	ITS	PCR + hybridization with specific probes on DNA microarray	Monecke et al. (2006)
Rhizomucor pusillus (Case report)	ITS	PCR + Sequencing	Iwen et al. (2005)
Rhizopus microsporus (Case report)	18S	PCR + RFLP	Larche et al. (2005)

[a] ribosomal DNA.
[b] Inaccurate identification in almost 50% of the cases.

et al., 2006; Voigt et al., 1999). In a large study, Voigt et al. (1999) have PCR amplified and sequenced the nuclear small subunit (18S) ribosomal DNA and the variable domains of the nuclear large subunit (28S) ribosomal DNA from 42 zygomycetes isolates including all the species known to be pathogenic for human and animals. They have shown that the 18S sequences are highly conserved but that sequences of the D1 and D2 domains of the 28S rDNA were more variable. This variability was used to design specific primers for 13 different species that could be used for identification. Analysis of intra and interspecies variability of ITS sequences obtained from 54 zygomycetes isolates belonging to 16 different species has also been evaluated recently (Schwarz et al., 2006). The whole ITS1-5.8S-ITS2 region was PCR amplified with universal fungal primers, and sequenced. Overall, sequences comparison showed that for a given species, the variability between isolates was very low. In contrast, sequences were very different between species (Table 7.3). These results showed that sequencing of ITS region is a reliable method for an accurate identification of zygomycetes to the species level. A different approach combining PCR and RFLP has been used in another study (Machouart et al., 2006). By amplification of a part of the 18S r DNA with a mix of 4 specific sense primers and a degenerated antisense primer followed by restriction enzyme digestion, it has been shown that specific patterns were obtained, allowing for precise species identification. Both ITS sequencing and PCR amplification of 18S combined with RFLP have been successfully employed for strain identification in

Table 7.3 Intraspecies and interspecies variability of ITS1-5.8S-ITS2 sequences for five zygomycetes pathogenic for humans

	Percentage of similarity	
Species	Within species	Between species
Rhizopus oryzae	99–100	36–61
Rhizopus microsporus	99–100	33–61
Rhizomucor pusillus	100	33–42
Absidia corymbifera	99–100	34–37
Mucor circinelloides	99	37–48

individual case reports (Larche et al., 2005; Iwen et al., 2005). Alternative techniques have also been used. In one case report of *R. microporus* peritonitis, identification of the causative organism was achieved by hybridization of the panfungal PCR product against a set of specific probes on a DNA microarray (Monecke et al., 2006). More specific studies (Table 7.2) have also demonstrated the usefulness of molecular-based technique for identification of *Apophysomyces elegans* (Chakrabarti et al., 2003) and *Rhizopus* species (Nagao et al., 2005).

The ITS sequence-based method has been used in a recent prospective surveillance study of 27 patients with zygomycosis in a large cancer center (Kontoyiannis et al., 2005). Interestingly, molecular identification of strains showed that morphological-based identification was erroneous in >20% of the cases. These results highlighted the difficulty of identification of zygomycetes to the species level by mycological standard procedures for a nonspecialized laboratory. Although these different recent studies (Kontoyiannis et al., 2005; Schwarz et al., 2006; Voigt et al., 1999; Machouart et al., 2006) demonstrated that molecular approaches are useful for species identification, it should be noted that these techniques need reliable methodologies and accurate as well as comprehensive sequence databases for comparison purposes. Indeed, as previously discussed (Greenberg et al., 2004) the use of the MicroSeq D2 large-subunit DNA sequencing kit, a commercial kit for molecular identification of filamentous fungi, led to misidentification of the zygomycetes isolates (including the most frequently pathogenic species) in almost 50% of the cases (Hall et al., 2004). These misidentifications could be related, at least in part, to the incomplete sequences library associated to the kit.

Identification in Tissues

Cultures from infected tissue are often negative (Ribes et al., 2000) and the different zygomycetes shared similar morphological characteristics by histopathology. Furthermore, in some instances, differentiation of a zygomycetes with another hyalohyphomycete could be difficult by histopathology. Then, alternative methods for diagnosis of zygomycosis and for species identification directly from tissues are needed. For this purpose, molecular methods have been recently evaluated both on unfixed fresh/frozen material (Kobayashi et al., 2004; Iwen et al., 2005; Machouart et al., 2006; Lau et al., 2006; Schwarz et al., 2006) and on formalin-fixed paraffin-embedded biopsies (Lau et al., 2006; Bialek et al., 2005; Nagao et al., 2005; Rickerts et al., 2006b; Hayden et al., 2002; Kobayashi et al., 2004).

In an experimental model of zygomycosis in mice, it has been shown that molecular identification of the most common zygomycetes species could be done from frozen tissue samples by PCR amplification with panfungal primers followed by direct sequencing of the ITS region (Schwarz et al., 2006). Very recently, a similar approach has been successfully used in 7 tissue samples from three patients with proven zygomycosis (Lau et al., 2006). Molecular identification has also been used in selected case reports to identify zygomycetes such as *Cunninghamella bertholletiae* (Kobayashi et al., 2004), *Rhizomucor pusillus* (Iwen et al., 2005), *Rhizopus microsporus* (Machouart et al., 2006), and *Absidia corymbifera* (Machouart et al., 2006) from various unfixed tissue samples.

Molecular detection of fungi from formalin-fixed paraffin-embedded tissues could be more difficult than from fresh specimens due to the degradation of DNA during processing of the samples. Nevertheless, paraffin-embedded tissues are often the only available samples, because zygomycosis has not been suspected initially. Several molecular techniques have been evaluated recently for identification of zygomycetes in such samples. *In situ* hybridization with panfungal and zygomycetes-specific oligonucleotide probes directed against 18S rDNA has been recently assessed in 13 samples from culture-proven cases of zygomycosis (Hayden et al., 2002). Although, good sensitivity and specificity was obtained, low staining intensity of the zygomycetes hyphae and sometimes high background staining limited the interpretation in some cases. A panfungal PCR assay targeting the ITS region has been recently evaluated on a large number of formalin-fixed paraffin-embedded tissue samples from patients with histologically proven fungal infection (Lau et al., 2006). Among the nine samples diagnosed as zygomycosis by histopathology, PCR amplification was negative in four samples, indicating that this molecular technique was less sensitive than the standard histological diagnosis. Nevertheless, in case of positive PCR, a molecular identification of the species was achieved while it is not possible by histology. In two other studies (Bialek et al., 2005; Rickerts et al., 2006b), an 18S-targeted semi-nested PCR specific for zygomycetes has been evaluated and showed promising results. Nevertheless, sensitivity was not optimal, as shown in one of the studies, with negative PCR for 9 out of 23 samples tested (Bialek et al., 2005). In some other clinical cases, molecular techniques have been successfully used for species identification (Kobayashi et al., 2004; Nagao et al., 2005). Overall, these results are very encouraging and warrant further studies to improve sensitivity.

Typing

Typing of specific species of zygomycetes could be of importance in case of epidemics of human infections, but could also be of interest for epidemiological studies and in the field of food biotechnology or pharmaceutical industry for discrimination of isolates with particular enzymatic activity. Until now very few typing methods have been evaluated and none have been proven useful or sufficiently evaluated in the field of medical mycology. Intraspecies variability has been tested for clinical isolates of *A. elegans* by PCR-RFLP of ITS regions and by microsatellite markers (Chakrabarti et al., 2003) but PCR-RFLP failed to discriminate between isolates and microsatellite typing was not discriminatory enough. Recently, typing of 20 zygomycete isolates, in a single institution, by a PCR technique based on the amplification of repetitive genomic sequences showed different typing patterns suggesting the lack of clonality and then the absence of a common source of contamination (Kontoyiannis et al., 2005). In one case report of a patient with a hepatic zygomycosis after ingestion of naturopathic medicines (Oliver et al., 1996), the same species, *M. indicus*, was isolated from the patient's liver abscess and from the pills of the naturopathic medicine. Arbitrary-primed PCR analysis suggested that both isolates were genetically identical.

More extensive data have been obtained for *R. oryzae* isolates for biotechnology purposes. Sequencing of the ITS regions of 64 strains of *R. oryzae* showed that some polymorphism was present (Abe et al., 2003). This polymorphism was not important enough for typing purposes but allowed to cluster the isolates in two distinct groups. Interestingly, there was a correlation between ITS type and production of lactic acid by the strain. These results allowed the design of specific primers for molecular detection of strains that produce lactic acid (Abe et al., 2003). The low variability of ITS sequences of *R. oryzae* has also been reported recently for clinical isolates (Schwarz et al., 2006) suggesting that ITS sequencing is not discriminatory enough for epidemiological purposes. More genetic variability has been observed by sequencing the lactate dehydrogenase genes from a large panel of *R. oryzae* isolates (Saito et al., 2004) but this approach has not been tested in clinical isolates.

Antifungal Susceptibility

Zygomycetes have been considered for a long time as a homogenous group of fungi in respect to their antifungal susceptibility. Recent studies both *in vitro* and in animal models of zygomycosis have shown that the different species exhibited different susceptibility to antifungal drugs. Moreover, some of the new drugs recently marketed or still in development have shown potent activity against zygomycetes and some studies have explored the potential of antifungal combinations. Standardized techniques for *in vitro* susceptibility testing of filamentous fungi are now available and allow comparison between studies. Nevertheless, due to the limited number of available studies both *in vitro* and *in vivo*, the clinical significance of some of these preclinical data remain unclear and should be further evaluated.

In vitro Data

The good *in vitro* activity of polyenes, particularly amphotericin B is known for many years (Otcenasek & Buchta, 1994) and has been regularly confirmed in several studies using standardized methodologies (Sun et al., 2002a; Minassian et al., 2003;

Gomez-Lopez et al., 2001; Dannaoui et al., 2002b, 2003b). Some differences could be noted between species (Table 7.4), *Apophysomyces elegans* and *Cunninghamella bertholletiae* being less susceptible to amphotericin B than other species (Dannaoui et al., 2002b, 2003b). The relative resistance of *C. bertholletiae* has been confirmed in an animal model of pulmonary zygomycosis in severely immunosuppressed mice (Honda et al., 1998) and is in accordance with the poor clinical response in patients (Ribes et al., 2000). For *A. elegans* an animal model of infection in non-neutropenic mice has also been setup and showed that this fungus is considerably more pathogenic than other zygomycetes (Dannaoui et al., 2002c, 2003c). Nevertheless, in this model, despite high MICs of amphototericin B, the drug was fully active *in vivo* (Dannaoui et al., 2002c).

Azoles are a widely used antifungal group of compounds that comprise marketed drugs with activity against filamentous fungi, such as itraconazole, voriconazole, and posaconazole. Other azoles are still in development. All azole drugs mainly act as inhibitors of ergosterol biosynthesis by inhibition of 14-alpha-demethylase (Cyp51). It was generally assumed that azole antifungal agents had no activity against zygomycetes (Sheehan et al., 1999; Kwon-Chung & Bennett, 1992). The recent *in vitro* data have shown that voriconazole had indeed no activity against all tested species of zygomycetes (Dannaoui et al., 2003b; Espinel-Ingroff, 1998b; Gomez-Lopez et al., 2001; McGinnis et al., 1998; Sabatelli et al., 2006; Sun et al., 2002a). In contrast, itraconazole exhibited varied *in vitro* activity depending on the species. Low MICs has been observed for *Absidia* spp. and *Rhizomucor* spp. while higher MICs, with some variability between isolates, were noted for *Rhizopus* spp. and *Mucor* spp. (Sabatelli et al., 2006; Sun et al., 2002a; Pfaller et al., 1998; Johnson et al., 1998; Dannaoui et al., 2003c). More interestingly, posaconazole, a new triazole with broad-spectrum activity, showed good *in vitro* activity against various species of zygomycetes (Table 7.4). Again, some differences were noted between species, with significantly lower MICs for *Absidia* spp. and *Rhizomucor* spp. compared to *Rhizopus* spp. and *Mucor* spp. (Dannaoui et al., 2003b; Sun et al., 2002a; Sabatelli et al., 2006). Very recently, the comparative activity of posaconazole, itraconazole, voriconazole, and fluconazole has been evaluated *in vitro* against *Rhizopus oryzae* and *Absidia corymbifera* (Chau et al., 2006). It has been shown that

Table 7.4 *In vitro* antifungal susceptibility of different species to amphotericin B (AMB), voriconazole (VRZ), and posaconazole (PSZ)

Species (no. of strains)	MIC* (µg/mL) of		
	AMB	VRZ	PSZ
Rhizopus spp. (20)	0.06–1	4–64	\–1
Mucor spp. (10)	0.03–0.25	16–64	0.5–2
Rhizomucor spp. (10)	0.06	2–4	0.06–0.5
Absidia corymbifera (15)	0.06–0.5	2–16	0.06–0.25
Cunninghamella bertholletiae (2)	2–8	8–32	0.06–0.5
Apophysomyces elegans (1)	2	16	0.5

* MIC: Minimum Inhibitory Concentration.

posaconazole and itraconazole were more active than fluconazole and voriconazole for inhibition of ergosterol synthesis in cell extract and that posaconazole was the most active azole for inhibition of ergosterol synthesis in whole cells. Moreover, expression of Cyp51 from *R. oryzae* in an azole-susceptible *Saccharomyces cerevisiae* decreased susceptibility to voriconazole by >32-fold, but only by eight- and twofold for itraconazole and posaconazole, respectively (Chau et al., 2006).

Among other azole drugs that are still in development, some have shown potential activity against zygomycetes. For example, ravuconazole, which was considered initially as poorly active against zygomycetes (Fung-Tomc et al., 1998), has been recently reevaluated *in vitro*. Modal MICs of 0.5–2 μg/mL were found for most zygomycetes tested except for *Mucor* spp. that were less susceptible (Minassian et al., 2003).

Echinocandins such as caspofungin that are glucan-synthase inhibitors, are not active *in vitro* against zygomycetes with MIC or MEC (Minimum Effective Concentration) values usually >8 μg/mL (Espinel-Ingroff, 1998a; Pfaller et al., 1998; Singh et al., 2005; Gil-Lamaignere et al., 2005). Nevertheless, a recent study showed the presence in *R. oryzae* of both a FKS gene and a membrane-associated glucan-synthase activity, demonstrating that the drug target is present in this fungus (Ibrahim et al., 2005a). Although this glucan-synthase activity was inhibited by caspofungin, a 1,000-fold higher concentration was needed to achieve similar inhibition than those observed against other fungi such as *Candida albicans* or *Aspergillus fumigatus*.

Among other antifungal drugs, terbinafine has shown *in vitro* activity against some zygomycetes species (Jessup et al., 2000; Dannaoui et al., 2003b). Interestingly, a sharp difference was noted between *R. oryzae* (geometric mean MIC of terbinafine of 64 μg/mL) and *R. microsporus* (geometric mean MIC of terbinafine of 0.15 μg/mL) (Dannaoui et al., 2003b) and this could be used as one of the identification criteria, along with others, to distinguish these two morphologically close species.

Other compounds such as statins, which are not used as antifungal drugs, have been shown to have inhibitory activity against a variety of fungi. Recently, the *in vitro* activity of lovasatin has been evaluated against some species of zygomycetes (Chamilos et al., 2006). It has been shown that lovastatin exhibited fungicidal activity by a microdilution broth technique with MIC50s and MFC50s of 48 and 56 μg/mL, respectively. The fungicidal effect against hyphae was further confirmed by fluorescence microscopy after staining with DiBAC, a dye that specifically stain damaged hyphae. The effect of lovastatin has also been recently evaluated against two *Rhizomucor* species (Lukacs et al., 2004). There was a dramatic difference of susceptibility between the species, *R. pusillus* being inhibited by 1–2 μg/mL of lovastatin while concentration of >8 μg/mL was necessary for growth inhibition of the *R. meihei* isolates. This difference of lovastatin susceptibility has been proposed as a simple test for differentiating these two closely related species (Lukacs et al., 2004).

Drug Efficacy in Animal Models of Zygomycosis

Animal models of zygomycosis are useful for the study of the pathophysiology of the disease or to develop new diagnostic tools, but could also be used for the *in vivo* evaluation of antifungal activity (Dannaoui, 2006). Many different models have been developed in different animal species, both in immunocompetent and immucompromized animals (Kamei, 2001). Testing antifugal treatments in animal models that take into account host factors and the pharmacokinctic properties of the drugs remains a critical bridge between the *in vitro* evaluation of a drug and its potential use in patients. Indeed, for zygomycetes, the correlation between *in vitro* activity of an antifungal and its *in vivo* efficacy has not been always very good (Dannaoui et al., 2002c; Odds et al., 1998; Sun et al., 2002b; Ibrahim et al., 2005a).

Evaluation of antifungal efficacy in animal models of zygomycosis in recent years has addressed several issues. First, the good activity of amphotericin B has been confirmed in most of the studies. In nonimmunosupressed mice infected intravenously by *R. oryzae*, amphotericin B at doses of 0.63 to 2.5 mg/kg/day prolonged survival, although it did not sterilize target organs (Odds et al., 1998). Similarly, in a model of *R. oryzae* disseminated infection in immunocompetent mice, treatment of animals with amphotericin B at 1 mg/kg/day resulted in 100% survival (Dannaoui et al., 2002c). Even in more severe models of *R. oryzae* infection such as pulmonary infection in cortisone-treated mice, amphotericin B at 1 mg/kg/day led to a 90–100% survival rate compared to 0% in untreated animals (Sugar & Liu, 2000). Amphotericin B at 0.5 to 1 mg/kg/day has also shown a good *in vivo* efficacy against *A. corymbifera* (Dannaoui et al., 2002c; Mosquera et al., 2001), *R. microsporus* (Dannaoui et al., 2002c), *Apophysomyces elegans* (Dannaoui et al., 2002c), and *Mucor circinelloides* (Sun et al., 2002b). Few studies have compared the efficacy of conventional deoxycholate-amphotericin B to the newer lipid formulations of the drug. In one study, liposomal amphotericin B has been shown to be more effective than deoxycholate amphotericin B in a model of *R. oryzae* infection in diabetic mice (Ibrahim et al., 2003). While only 40% of untreated animal survived, survival rates for mice treated with amphotericin B at 1 mg/kg/day or liposomal amphotericin B at 15 mg/kg/day were 67%, and 90%, respectively.

In vitro studies have documented low MICs for terbinafine against several species of zygomycetes. For this reason, the *in vivo* efficacy of the drug has been evaluated *in vivo* against one strain of *R. microsporus* (terfinafine MIC of 0.25 µg/mL) and one strain of *A. corymbifera* (MIC of 0.12 µg/mL). Despite documented absorption of terbinafine with tissues concentration of 25 and 10 µg/g in kidney and brain, respectively, treatment was ineffective either in mice infected with *R. micro sporus* or *A. corymbifera* (Dannaoui et al., 2002c). Reasons for these discrepancies between the *in vitro* activity of terbinafine and its *in vivo* efficacy remain unknown, but similar findings have been observed for other filamentous fungi.

Among the azoles, itraconazole, and posaconazole have been tested in animal models of zygomycosis. Itraconazole showed a certain degree of efficacy against *A. corymbifera* (Mosquera et al., 2001; Dannaoui et al., 2002c, 2003a) but was less effective than amphotericin B. Similarly, itraconazole treatment reduced mortality in mice infected with *A. elegans* (Dannaoui et al., 2002c) but was ineffective against *R. oryzae* (Dannaoui et al., 2003a; Odds et al., 1998), *R. microsporus* (Dannaoui et al., 2002c, 2003a) and two species of *Mucor* (Sun et al., 2002b). Posaconazole has been tested in animal model of zygomycosis in three recent studies. Against *M. ramosissimus* and *M. circinelloides*, posaconazole at 30 and 60 mg/kg/ day significantly prolonged survival in neutropenic mice and reduced fungal burden in kidneys compared to untreated controls but was less effective than amphotericin B at 1 mg/kg/day (Sun et al., 2002b). In nonimmunocompromised mice with disseminated infection, posaconazole tested at 5, 25, and 40 mg/kg/day was not effective against *R. oryzae*, showed a clear dose–response effect against *R. microsporus*, and was partially active against *A. corymbifera* (Dannaoui et al., 2003a). In the third study, the prophylactic efficacy of posaconazole, given three consecutive days before infection with either *R. oryzae* or *A. corymbifera*, was assessed and compared with prophylaxis with amphotericin B (Barchiesi et al., 2007). Overall, it was shown that posaconazole was more effective against *A. corymbifera* than against *R. oryzae* and that amphotericin B was more effective than posaconazole in prolonging survival of mice infected with *R. oryzae*.

Although caspofungin MICs against zygomycetes are consistently high, this drug has been tested in a model of *R. oryzae* infection in diabetic mice. Overall, the efficacy of caspofungin treatment was limited (Ibrahim et al., 2005a). Interestingly, low dose (1 mg/kg/day) caspofungin improved survival but high doses (5 or 10 mg/ kg/day) did not. Moreover, this effect was observed in animal infected with a low inoculum but not a large inoculum.

Beside antifungal drugs, other therapeutic strategies have been evaluated in animal. In particular, in one study, the efficacy of hyperbaric oxygen as an adjunct of amphotericin B treatment has been tested in mice infected with *R. oryzae* (Barratt et al., 2001). In this model the addition of hyperbaric oxygen to amphotericin B but did not improve survival compared to amphotericin B alone.

Another study evaluated the effect of deferiprone, an iron chelator, in a mouse model of *R. oryzae* infection (Ibrahim et al., 2006). Treatment with deferiprone increased survival rate and reduced fungal burden in brain compared to untreated controls with an efficacy similar to that of liposomal amphotericin B given at 15 mg/kg/day. Administration of free iron was shown to suppress the effect of deferiprone therapy demonstrating that the mechanism of protection was chelation of iron (Ibrahim et al., 2006).

Antifungal Combinations

Because of the poor *in vitro* activity of most of the antifungal drugs against zygomycetes and the limited *in vivo* efficacy of monotherapies in zygomycosis, combination of two or more drugs could be of interest. Several techniques are available to study antifungal combination *in vitro* (Vitale et al., 2005). Few studies either *in vitro* or in experimental animal models have been conducted so far. Evaluation of several combinations against 35 isolates of zygomycetes belonging to different species showed that amphotericin B interacted synergistically with rifampicin or with terbinafine against 69, and 20% of the isolates, respectively (Dannaoui et al., 2002a). Combination of terbinafine with voriconazole was also synergistic in 44% of the cases. In contrast combination of amphotericin B with flucytosine was indifferent. Antagonism was never observed. Two *in vitro* studies have evaluated combination of nikkomycin Z (a chitin synthase inhibitor) with an echinocandin (glucan synthase inhibitors) against a very limited number of *R. oryzae* isolates (Chiou et al., 2001; Stevens, 2000) and different results have been obtained. Anidulafungin in combination with nikkomycin Z showed synergy, although both drug were inactive when tested alone (Stevens, 2000). In contrast, micafungin combined with nikkomycin Z showed indifferent interaction for four strains of *R. oryzae* that were susceptible to nikkomycin Z alone (Chiou et al., 2001). Few studies have been carried out in animal models but have shown interesting results. In a model of pulmonary zygomycosis due *R. oryzae* it has been shown that fluconazole combined with a quinolone improved survival of mice, while none of these drugs were active alone (Sugar & Liu, 2000). In another study, synergy was observed between a lipid formulation of amphotericin B and caspofungin in a model of *R. oryzae* infection in diabetic mice (Spellberg et al., 2005b). By using an original *Drosophila* model of zygomycosis, it has been demonstrated that the combination of voriconazole with lovastatin was synergistic against *R. oryzae* and *M. circinelloides* (Chamilos et al., 2006).

New Treatment Strategies in Patients

High doses of deoxycholate amphotericin B associated with surgery was considered until recently the treatment of choice for zygomycosis (Dromer & McGinnis, 2002) but was associated with a high rate of nephrotoxicity. Development of the lipid formulations of amphotericin B, which cause less nephrotoxicity, allowed for administration of higher dosages. Although there have been no prospective trials to compare efficacy of deoxycholate amphotericin B with lipid formulations, there is a consensus that a lipid formulation of amphotericin B represents the best treatment for zygomycosis (Spellberg et al., 2005a; Chayakulkeeree et al., 2006). Case series and retrospective studies have shown favorable outcomes in patients treated with either amphotericin B colloidal dispersion (Herbrecht et al., 2001), amphotericin B lipid complex (Walsh et al., 1998), or liposomal amphotericin B (Gleissner et al., 2004). The relative efficacy of the different lipid formulations remains unknown. Nevertheless, based on pharmacokinetic data and results obtained in experimental models, liposomal amphotericin B at 10 mg/kg/day, in combination with surgery, has been proposed as the first-line therapy of zygomycosis (Spellberg et al., 2005a).

Among other antifungals, posaconazole is the only drug that has been used in a reasonable number of patients. The first case of successful treatment of zygomycosis with posaconazole has been reported in 2003 (Tobon et al., 2003) and several other individual case reports have been published afterward. The analysis of outcome in the first 24 patients who received posaconazole as salvage therapy for zygomycosis (as a second-line therapy after failure of or intolerance to amphotericin B) showed promising results with an overall surviving rate of 79% (Greenberg et al., 2006). These results have been confirmed in the retrospective analysis of 91 patients treated with posaconazole for zygomycosis refractory to or intolerant of prior antifungal therapy (van Burik et al., 2006). Nevertheless, posaconazole is not currently approved for treatment of zygomycosis and its role in the management of zygomycosis remains to be further evaluated in prospective studies.

Combination treatment has been rarely used in patients with zygomycosis until now (Voitl et al., 2002; Vazquez et al., 2005; Nivoix et al., 2006; Rickerts et al., 2006a). It is therefore difficult to draw any conclusion. In one case of rhinocerebral zygomycosis in a neutropenic patient (Vazquez et al., 2005) a combination of liposomal amphotericin B with caspofungin was successfully used. In contrast, the same combination was not effective in two other reported cases (Nivoix et al., 2006; Voitl et al., 2002). Combination of liposomal amphotericin B with posaconazole has been reported to be effective in treating one patient (Rickerts et al., 2006a).

Despite antifungal therapy, mortality rate in zygomycosis remains very high and there is a need for new therapeutic strategies. Iron chelation could be one of these strategies. Very recently, deferasirox, an iron chelator approved for treatment of iron overload has been successfully used as salvage therapy in a patient with rhinocerebral zygomycosis (Reed et al., 2006).

Conclusions

Zygomycetes are known as human pathogens for more than 100 years but have remained largely understudied partly due to the low frequency of the disease. The recent increase of incidence of zygomycosis in immunocompromised patients and the high mortality rate of the disease despite therapy have prompted many recent studies. Although very important advances have been made during the last 5 years, a better understanding of the biology of the pathogens and the epidemiology of the disease will be of particular importance in the coming years. New diagnostic tools and improved therapeutic strategies are also urgently needed.

Zygomycetes comprise very diverse genera and species. Recent molecular studies have allowed revising the taxonomy and phylogenetic relationships of these pathogens. Some species are clearly misplaced taxonomically and will probably be reassigned to different genera.

Pathophysiology of zygomycosis remains largely unknown and has been poorly explored recently. The few recent studies have focused on the role of iron in pathogenesis and on the interactions between the fungus and the endothelial cells with interesting results. A better knowledge of the biology of zygomycetes would greatly help the understanding of the pathogenesis of the disease. Very unexpected data have shown that some isolates of *Rhizopus* spp. harbored endosymbiont bacteria responsible for toxin production. The potential role of these bacteria and of the produced toxin as a virulence factor for humans needs to be explored in experimental models. It is sure that the recent availability of the complete genome of *R. oryzae*, as well as the recent development of new techniques for genetic manipulations of these fungi will considerably facilitate the analysis of putative virulence factors.

Incidence of zygomycetes infections has risen in the last two decades but reasons for emergence of the disease is still not completely understood. The role of new antifungal drugs has been suspected but increase in the incidence could also be linked to recent changes in immunosuppression regimens given to the patients. To answer these clinically relevant issues, surveillance studies on a national and international basis are needed.

In a patient with a suspected invasive filamentous fungi infection, diagnosis of zygomycosis remains very difficult. Unfortunately no major advances have been made for improving diagnosis. In particular, there is neither serological nor antigen-detection tests currently available and PCR techniques are not yet fully standardized. In contrast, very important advances have been made for identification of the strains in culture. Evaluation of intra- and interspecies variability of several DNA regions and increase of sequence data in public databases have made accurate identification to the species level possible. These new and reliable identification methods will allow to precise the respective role of each species and probably to identify new species responsible for human infections.

Zygomycosis remains a life-threatening infection and treatment efficacy is still suboptimal. Recent data obtained in several *in vitro* studies demonstrated that the different species exhibited different antifungal susceptibilities. Nevertheless, the clinical relevance of these differences has to be confirmed. New antifungal drugs such as posaconazole have also shown potential clinical efficacy on a limited number of patients. The exact role of these new antifungals, either alone or in combination, for the management of zygomycosis remains to be further evaluated.

References

Abe, A., Oda, Y., Asano, K., & Sone, T. (2006). *Biosci. Biotechnol. Biochem.*, 70:2387–2393.

Abe, A., Sone, T., Sujaya, I. N., Saito, K., Oda, Y., Asano, K., & Tomita, F. (2003). *Biosci. Biotechnol. Biochem.*, 67:1725–1731.

Barchiesi, F., Spreghini, E., Santinelli, A., Fothergill, A. W., Pisa, E., Giannini, D., Rinaldi, M. G., & Scalise, G. (2007). *Antimicrob. Agents Chemother.*, 51:73–77.

Barratt, D. M., Van Meter, K., Asmar, P., Nolan, T., Trahan, C., Garcia-Covarrubias, L., & Metzinger, S. E. (2001). *Antimicrob. Agents Chemother.*, 45:3601–3602.

Bialek, R., Konrad, F., Kern, J., Aepinus, C., Cecenas, L., Gonzalez, G. M., Just-Nubling, G., Willinger, B., Presterl, E., Lass-Florl, C., & Rickerts, V. (2005). *J. Clin. Pathol.*, 58:1180–1184.

Chakrabarti, A., Ghosh, A., Prasad, G. S., David, J. K., Gupta, S., Das, A., Sakhuja, V., Panda, N. K., Singh, S. K., Das, S., & Chakrabarti, T. (2003). *J. Clin. Microbiol.*, 41:783–788.

Chamilos, G., Lewis, R. E., & Kontoyiannis, D. P. *Fungal Genet. Biol.* Published online 2006 Sep 19: doi: 10.1016/j.fgb.2006.1007.1011.

Chamilos, G., Lewis, R. E., & Kontoyiannis, D. P. (2006). *Antimicrob. Agents Chemother.*, 50:96–103.

Chau, A. S., Chen, G., McNicholas, P. M., & Mann, P. A. (2006). *Antimicrob. Agents Chemother.*, 50:3917–3919.

Chayakulkeeree, M., Ghannoum, M. A., & Perfect, J. R. (2006). *Eur. J. Clin. Microbiol., Infect. Dis.*, 25:215–229.

Chiou, C. C., Mavrogiorgos, N., Tillem, E., Hector, R., & Walsh, T. J. (2001). *Antimicrob. Agents Chemother.*, 45:3310–3321.

Christiaens, G., Hayette, M. P., Jacquemin, D., Melin, P., Mutsers, J., & De Mol, P. (2005). *J. Hosp. Infect.*, 61:88.

Dannaoui, E. (2006) Animal models for evaluation of antifungal efficacy against filamentous fungi. In: *Medical Mycology: Cellular and Molecular Techniques*. Kavanagh, K., Ed., Wiley, Bognor Regis, UK, pp. 114–135.

Dannaoui, E., Afeltra, J., Meis, J. F., & Verweij, P. E. (2002a). *Antimicrob. Agents Chemother.*, 46:2708–2711.

Dannaoui, E., Meis, J. F., Mouton, J. W., & Verweij, P. E. (2002b). *J. Antimicrob. Chemother.*, 49:741–744.

Dannaoui, E., Mouton, J. W., Meis, J. F., & Verweij, P. E. (2002c). *Antimicrob. Agents Chemother.*, 46:1953–1959.

Dannaoui, E., Meis, J. F., Loebenberg, D., & Verweij, P. E. (2003a). *Antimicrob. Agents Chemother.*, 47:3647–3650.

Dannaoui, E., Meletiadis, J., Mouton, J. W., Meis, J. F., & Verweij, P. E. (2003b). *J. Antimicrob. Chemother.*, 51:45–52.

Dannaoui, E., Rijs, A. J., & Verweij, P. E. (2003c). *Mayo Clin. Proc.* 78:252–253.

de Hoog, G. S. & Guarro, J. (Eds.) (2000). *Atlas of Clinical Fungi*, 2nd edn., Centraalbureau voor Schimmelcultures, Baarn, The Netherlands.

Dromer, F. & McGinnis, M. R. (2002). Zygomycosis. In: *Clinical Mycology*, Anaissie, E., McGinnis, M. R., & Pfaller, M. A., Eds., Churchill Livingstone, New York, pp. 297–308.

Espinel-Ingroff, A. (1998a). *J. Clin. Microbiol.*, 36:2950–2956.

Espinel-Ingroff, A. (1998b). *J. Clin. Microbiol.*, 36:198–202.

Fung-Tomc, J. C., Huczko, E., Minassian, B., & Bonner, D. P. (1998). *Antimicrob. Agents Chemother.*, 42:313–318.

Gil-Lamaignere, C., Hess, R., Salvenmoser, S., Heyn, K., Kappe, R., & Muller, F. M. (2005). *J. Antimicrob. Chemother.*, 55:1016–1019.

Gleissner, B., Schilling, A., Anagnostopolous, I., Siehl, I., & Thiel, E. (2004). *Leuk. Lymphoma*, 45:1351–1360.

Gomez-Lopez, A., Cuenca-Estrella, M., Monzon, A., & Rodriguez-Tudela, J. L. (2001). *J. Antimicrob. Chemother.*, 48:919–921.

Greenberg, R. N., Mullane, K., van Burik, J. A., Raad, I., Abzug, M. J., Anstead, G., Herbrecht, R., Langston, A., Marr, K. A., Schiller, G., Schuster, M., Wingard, J. R., Gonzalez, C. E., Revankar, S. G., Corcoran, G., Kryscio, R. J., & Hare, R. (2006). *Antimicrob. Agents Chemother.*, 50:126–133.

Greenberg, R. N., Scott, L. J., Vaughn, H. H., & Ribes, J. A. (2004). *Curr. Opin. Infect. Dis.*, 17:517–525.

Hall, L., Wohlfiel, S., & Roberts, G. D. (2004). *J. Clin. Microbiol.*, 42:622–626.

Hayden, R. T., Qian, X., Procop, G. W., Roberts, G. D., & Lloyd, R. V. (2002). *Diagn. Mol. Pathol.*, 11:119–126.

Herbrecht, R., Letscher-Bru, V., Bowden, R. A., Kusne, S., Anaissie, E. J., Graybill, J. R., Noskin, G. A., Oppenheim, Andres, E., & Pietrelli, L. A. (2001). *Eur. J. Clin. Microbiol. Infect. Dis.*, 20:460–466.

Honda, A., Kamei, K., Unno, H., Hiroshima, K., Kuriyama, T., & Miyaji, M. (1998). *Mycopathologia*, 144:141–146.

Ibrahim, A. S., Avanessian, V., Spellberg, B., & Edwards, J. E., Jr. (2003). *Antimicrob. Agents Chemother.*, 47:3343–3344.

Ibrahim, A. S., Bowman, J. C., Avanessian, V., Brown, K., Spellberg, B., Edwards, J. E. Jr., & Douglas, C. M. (2005a). *Antimicrob. Agents Chemother.*, 49:721–727.

Ibrahim, A. S., Edwards, J. E., Jr., Fu, Y., & Spellberg, B. (2006). *J. Antimicrob. Chemother.*, 58:1070–1073.

Ibrahim, A. S. & Skory, C. D. (2006). Genetic manipulation of zygomycetes. In: *Medical Mycology: Cellular & Molecular Techniques*, Kavanagh, K., Ed., Wiley, Bognor Regis, UK, pp. 305–326.

Ibrahim, A. S., Spellberg, B., Avanessian, V., Fu, Y., & Edwards, J. E., Jr. (2005b). *Infect. Immun.*, 73:778–783.

Imhof, A., Balajee, S. A., Fredricks, D. N., Englund, J. A., & Marr, K. A. (2004). *Clin. Infect. Dis.*, 39:743–746.

Iwen, P. C., Freifeld, A. G., Sigler, L., & Tarantolo, S. R. (2005). *J. Clin. Microbiol.*, 43:5819–5821.

Jessup, C. J., Ryder, N. S., & Ghannoum, M. A. (2000). *Med. Mycol.*, 38:155–159.

Johnson, E. M., Szekely, A., & Warnock, D. W. (1998). *J. Antimicrob. Chemother.*, 42:741–745.

Kamei, K. (2001). *Mycopathologia*, 152:5–13.

Kauffman, C. A. (2004). *Clin. Infect. Dis.*, 39:588–590.

Kirk, P. M., Cannon, P. F., David, J. C., & Stalpers, J. (2001). *Ainsworth and Bisby's Dictionary of the Fungi*, CABI Publishing, Wallingford, UK.

Kobayashi, M., Togitani, K., Machida, H., Uemura, Y., Ohtsuki, Y., & Taguchi, H. (2004). *Respirology*, 9:397–401.

Kontoyiannis, D. P., Lionakis, M. S., Lewis, R. E., Chamilos, G., Healy, M., Perego, C., Safdar, A., Kantarjian, H., Champlin, R., Walsh, T. J., & Raad, II (2005). *J. Infect. Dis.*, 191:1350–1360.

Kontoyiannis, D. P., Wessel, V. C., Bodey, G. P., & Rolston, K. V. (2000). *Clin. Infect. Dis.*, 30:851–856.

Kwon-Chung, K. J. & Bennett, J. E. (1992). *Medical Mycology*, Lea & Febiger, Philadelphia, Pa.

Larche, J., Machouart, M., Burton, K., Collomb, J., Biava, M. F., Gerard, A., & Fortier, B. (2005). *Clin. Infect. Dis.*, 41:1362–1365.

Lau, A., Chen, S., Sorrell, T., Carter, D., Malik, R., Martin, P., & Halliday, C. (2006). *J. Clin. Microbiol.*

Lawley, B., Ripley, S., Bridge, P., & Convey, P. (2004). *Appl. Environ. Microbiol.*, 70:5963–5972.

LeMaile-Williams, M., Burwell, L. A., Salisbury, D., Noble-Wang, J., Arduino, M., Lott, T., Brandt, M. E., Iiames, S., Srinivasan, A., & Fridkin, S. K. (2006). *Clin. Infect. Dis.*, 43:e83–88.

Lukacs, G., Papp, T., Nyilasi, I., Nagy, E., & Vagvolgyi, C. (2004). *J. Clin. Microbiol.*, 42:5400–5402.

Lumini, E., Ghignone, S., Bianciotto, V., & Bonfante, P. (2006). *New Phytol.*, 170:205–208.

Machouart, M., Larche, J., Burton, K., Collomb, J., Maurer, P., Cintrat, A., Biava, M. F., Greciano, S., Kuijpers, A. F., Contet-Audonneau, N., de Hoog, G. S., Gerard, A., & Fortier, B. (2006). *J. Clin. Microbiol.*, 44:805–810.

Maravi-Poma, E., Rodriguez-Tudela, J. L., de Jalon, J. G., Manrique-Larralde, A., Torroba, L., Urtasun, J., Salvador, B., Montes, M., Mellado, E., Rodriguez-Albarran, F., & Pueyo-Royo, A. (2004). *Intensive Care Med.*, 30:724–728.

Marr, K. A., Carter, R. A., Crippa, F., Wald, A., & Corey, L. (2002). *Clin. Infect. Dis.*, 34:909–917.

Marty, F. M., Cosimi, L. A., & Baden, L. R. (2004). *N. Engl. J. Med.*, 350:950–952.

McGinnis, M. R., Pasarell, L., Sutton, D. A., Fothergill, A. W., Cooper, C. R., Jr., & Rinaldi, M. G. (1998). *Med. Mycol.*, 36:239–242.

Minassian, B., Huczko, E., Washo, T., Bonner, D., & Fung-Tomc, J. (2003). *Clin. Microbiol. Infect.*, 9:1250–1252.

Monecke, S., Hochauf, K., Gottschlich, B., & Ehricht, R. (2006). *Mycoses*, 49:139–142.

Mosquera, J., Warn, P. A., Rodriguez-Tudela, J. L., & Denning, D. W. (2001). *J. Antimicrob. Chemother.*, 48:583–586.

Nagao, K., Ota, T., Tanikawa, A., Takae, Y., Mori, T., Udagawa, S., & Nishikawa, T. (2005). *J. Dermatol. Sci.*, 39: 23–31.

Nivoix, Y., Zamfir, A., Lutun, P., Kara, F., Remy, V., Lioure, B., Rigolot, J. C., Entz-Werle, N., Letscher-Bru, V., Waller, J., Leveque, D., Koffel, J. C., Beretz, L., & Herbrecht, R. (2006). *J. Infect.*, 52:67–74.

O'Donnell, K., Lutzoni, F., Ward, T. J., & Benny, G. L. (2001). *Mycologia*, 93:286–296.

Odds, F. C., Van Gerven, F., Espinel-Ingroff, A., Bartlett, M. S., Ghannoum, M. A., Lancaster, M. V., Pfaller, M. A., Rex, J. H., Rinaldi, M. G., & Walsh, T. J. (1998). *Antimicrob. Agents Chemother.*, 42:282–288.

Oliver, M. R., Van Voorhis, W. C., Boeckh, M., Mattson, D., & Bowden, R. A. (1996). *Clin. Infect. Dis.*, 22:521–524.

Oren, I. (2005). *Clin. Infect. Dis.*, 40:770–771.

Otcenasek, M. & Buchta, V. (1994). *Mycopathologia*, 128:135–137.

Partida-Martinez, L. P., de Looss, C. F., Ishida, K., Ishida, M., Roth, M., Buder, K., & Hertweck, C. *Appl. Environ. Microbiol.* Published online 2006 Nov 22: doi:10.1128/AEM.01784 –01706.

Partida-Martinez, L. P. & Hertweck, C. (2005). *Nature*, 437:884–888.

Pfaller, M. A., Marco, F., Messer, S. A., & Jones, R. N. (1998). *Diagn. Microbiol. Infect. Dis.*, 30:251–255.

Prabhu, R. M. & Patel, R. (2004). *Clin. Microbiol. Infect.*, 10(Suppl. 1):31–47.

Redman, R. S., Litvintseva, A., Sheehan, K. B., Henson, J. M., & Rodriguez, R. (1999). *Appl. Environ. Microbiol.*, 65:5193–5197.

Reed, C., Ibrahim, A., Edwards, J. E., Jr., Walot, I., & Spellberg, B. (2006). *Antimicrob. Agents Chemother.*, 50:3968–3969.

Rees, J. R., Pinner, R. W., Hajjeh, R. A., Brandt, M. E., & Reingold, A. L. (1998). *Clin. Infect. Dis.*, 27:1138–1147.

Ribes, J. A., Vanover-Sams, C. L., & Baker, D. J. (2000). *Clin. Microbiol. Rev.*, 13:236–301.

Rickerts, V., Atta, J., Herrmann, S., Jacobi, V., Lambrecht, E., Bialek, R., & Just-Nubling, G. (2006a). *Mycoses*, 49(Suppl. 1):27–30.

Rickerts, V., Just-Nubling, G., Konrad, F., Kern, J., Lambrecht, E., Bohme, A., Jacobi, V., & Bialek, R. (2006b). *Eur. J. Clin. Microbiol. Infect. Dis.*, 25:8–13.

Roden, M. M., Zaoutis, T. E., Buchanan, W. L., Knudsen, T. A., Sarkisova, T. A., Schaufele, R. L., Sein, M., Sein, T., Chiou, C. C., Chu, J. H., Kontoyiannis, D. P., & Walsh, T. J. (2005). *Clin. Infect. Dis.*, 41:634–653.

Sabatelli, F., Patel, R., Mann, P. A., Mendrick, C. A., Norris, C. C., Hare, R., Loebenberg, D., Black, T. A., & McNicholas, P. M. (2006). *Antimicrob. Agents Chemother.*, 50:2009–2015.

Saito, K., Saito, A., Ohnishi, M., & Oda, Y. (2004). *Arch. Microbiol.*, 182:30–36.

Scherlach, K., Partida-Martinez, L. P., Dahse, H. M., & Hertweck, C. (2006). *J. Am. Chem. Soc.*, 128:11529–11536.

Scholer, H. J., Vanover-Sams, C. L., & Baker, D. J. (1983). Mucorales. In: *Fungi Pathogenic for Humans and Animals, Part A. Biology*, Howard, D. H., Ed., Marcel Dekker, NY.

Schwarz, P., Bretagne, S., Gantier, J. C., Garcia-Hermoso, D., Lortholary, O., Dromer, F., & Dannaoui, E. (2006). *J. Clin. Microbiol.*, 44:340–349.

Sheehan, D. J., Hitchcock, C. A., & Sibley, C. M. (1999). *Clin. Microbiol. Rev.*, 12:40–79.

Singh, J., Rimek, D., & Kappe, R. (2005). *Mycoses*, 48:246–250.

Siwek, G. T., Dodgson, K. J., De Magalhaes-Silverman, M., Bartelt, L. A., Kilborn, S. B., Hoth, P. L., Diekema, D. J., & Pfaller, M. A. (2004). *Clin. Infect. Dis.*, 39:584–587.

Spellberg, B., Edwards, J., Jr., & Ibrahim, A. (2005a). *Clin. Microbiol. Rev.*, 18:556–569.

Spellberg, B., Fu, Y., Edwards, J. E., Jr., & Ibrahim, A. S. (2005b). *Antimicrob. Agents Chemother.*, 49:830–832.

Stevens, D. A. (2000). *Antimicrob. Agents Chemother.*, 44:2547–2548.

Sugar, A. M. & Liu, X. P. (2000). *Antimicrob. Agents Chemother.*, 44:2004–2006.

Sun, Q. N., Fothergill, A. W., McCarthy, D. I., Rinaldi, M. G., & Graybill, J. R. (2002a). *Antimicrob. Agents Chemother.*, 46:1581–1582.

Sun, Q. N., Najvar, L. K., Bocanegra, R., Loebenberg, D., & Graybill, J. R. (2002b). *Antimicrob. Agents Chemother.*, 46:2310–2312.

Tobon, A. M., Arango, M., Fernandez, D., & Restrepo, A. (2003). *Clin. Infect. Dis.*, 36:1488–1491.

Vágvölgyi, C., Magyar, K., Papp, T., Palágyi, Z., Ferenczy, L. & Nagy, Á. (1996) *Can. J. Microbiol.*, 42:613–615.

Van Burik, J. A., Hare, R. S., Solomon, H. F., Corrado, M. L., & Kontoyiannis, D. P. (2006). *Clin. Infect. Dis.*, 42:e61–65.

Vastag, M., Papp, T., Kasza, Z., & Vágvölgyi, C. (1998). *J. Clin. Microbiol.*, 36:2153–2156.

Vazquez, L., Mateos, J. J., Sanz-Rodriguez, C., Perez, E., Caballero, D., & San Miguel, J. F. (2005). *Haematologica*, 90: ECR39.

Vigouroux, S., Morin, O., Moreau, P., Mechinaud, F., Morineau, N., Mahe, B., Chevallier, P., Guillaume, T., Dubruille, V., Harousseau, J. L., & Milpied, N. (2005). *Clin. Infect. Dis.*, 40: e35–37.

Vitale, R. G., Afeltra, J., & Dannaoui, E. (2005). *Methods Mol. Med.*, 118:143–152.

Voigt, K., Cigelnik, E., & O'Donnell, K. (1999). *J. Clin. Microbiol.*, 37:3957–3964.

Voigt, K. & Wostemeyer, J. (2001). *Gene*, 270:113–120.

Voitl, P., Scheibenpflug, C., Weber, T., Janata, O., & Rokitansky, A. M. (2002). *Eur. J. Clin. Microbiol. Infect. Dis.*, 21:632–634.

Walsh, T. J., Hiemenz, J. W., Seibel, N. L., Perfect, J. R., Horwith, G., Lee, L., Silber, J. L., DiNubile, M. J., Reboli, A., Bow, E., Lister, J., & Anaissie, E. J. (1998). *Clin. Infect. Dis.*, 26:1383–1396.

Chapter 8
The Virulence of *Aspergillus fumigatus*

Nir Osherov

Introduction

The aim of this chapter is to analyze *Aspergillus fumigatus* virulence in light of recent developments in genomics and our growing understanding of the complexity of the host–pathogen interaction. Readers interested in comprehensive summaries describing putative virulence determinants of *A. fumigatus* are directed to several excellent recent reviews (Latge, 1999; Rementeria et al., 2005; Brakhage, 2005).

Aspergillus fumigatus

The Versatile Saprophyte

A. fumigatus is a saprophytic fungus with a worldwide prevalence. It is one of the most common species of airborne spore-producing fungi in the world (Shelton et al., 2002). Its buoyant airborne spores (conidia), ability to utilize varied carbon sources and thermotolerance enable it to survive in areas with widely different climates and environments. Conidia of this fungus can be isolated nearly everywhere, from the winds of the Sahara to the snows of Antarctica (Braude, 1986). It sporulates profusely, with each conidiophore producing thousands of tiny (2–3 micron diameter), bluish-green hydrophobic conidia that are easily dispersed into the atmosphere. Environmental surveys indicate that in areas of human habitation, approximately 1–100 *A. fumigatus* conidia are typically found per cubic meter of air, and thus all humans inhale at least several hundred conidia per day (Latge, 1999).

The species exhibits little variation, either within geographic regions or on a global scale, suggesting that the population is in continual gene flow across continents (Pringle, 2005; Rydholm, 2006). *A. fumigatus* is abundant in moist soil rich in organic materials and feeds mainly upon decaying vegetation undergoing aerobic decomposition (Thom & Raper, 1945; Mullins, 1976). It is especially plentiful in man-made environments and in disturbed soils (Rydholm, 2006). The thermophylic ability of *A. fumigatus* to grow at temperatures of 55°C and

K. Kavanagh (ed.), *New Insights in Medical Mycology.*
© Springer 2007

survive at up to 70°C enables it to successfully dominate the compost fungal microflora (Beffa et al., 1998; Ryckeboer et al., 2003).

Recent genomic analysis strongly suggest that *A. fumigatus* is primarily a herbivorous leaf eater: a survey of its genome has shown that it contains the full armamentarium of enzymes necessary to degrade plants, but lacks ligninases necessary to digest wood (Tekaia & Latge, 2005).

Although considered to lack a sexual life cycle, the presence of mating type and sex-related genes in its genome support the assertion that *A. fumigatus* exhibits the potential to reproduce by sexual means (Dyer & Paoletti, 2005; Paoletti et al., 2005). In contrast, the closely related *Neosartorya fischeri* (94% average protein identity with *A. fumigatus*) exhibits a functional sexual life cycle. Interestingly *N. fischeri* rarely infects humans despite its ubiquitous distribution, raising the possibility that a genomic comparison between it and *A. fumigatus* may identify specific virulence determinants in the latter (Wortman et al., 2006). Recent typing analysis of clinical *A. fumigatus* isolates has revealed that a substantial number were misidentified and are in fact closely related species. They include the potentially drug resistant *Aspergillus lentulus* and *Aspergillus udagawae* (Balajee et al., 2006).

A. fumigatus causes a spectrum of diseases depending on the immune status of the host: *A. fumigatus* is exceptional among microorganisms in being both an opportunistic pathogen and a major allergen, capable of causing a wide range of clinical manifestations depending on the immune status of the host (Denning, 1998; Casadevall & Pirofski, 2003; Denning et al., 2006). It causes damage primarily at the extremes of both weak and strong immune responses (Figure 8.1). In immunocompetent hypersensitive individuals, conidial inhalation can result in various allergic responses such as allergic bronchopulmonary aspergillosis (ABPA), extrinsic allergic alveolitis (EAA), allergic aspergillus sinusitis, and asthma with fungal sensitization (Moss, 2005; Denning et al., 2006) (Figure 8.1B). These reactions are defined as an immune hypersensitivity reaction to specific *Aspergillus* antigens, with tissue damage being mostly self-inflicted by infiltrating neutrophils and eosinophils (Tillie-Leblond & Tonnel, 2005). They can be effectively control-led during their early stages by administration of corticosteroids to temper the immune response and in the case of ABPA, antifungals such as itraconazole or voriconazole to inhibit fungal growth (TillieLeblond & Tonnel, 2005).

In immunocompetent persons, *A. fumigatus* infection may very rarely lead to chronic conditions such as aspergilloma or fungus ball of the lung and sinus. Hyphal growth is restricted to the aspergilloma, with no invasion of the surrounding tissue. Treatment ranges from observation without intervention to antifungal administration and surgical resection in extreme cases.

In immunocompromised patients, inhaled *A. fumigatus* conidia germinate and invade the sinus, airway, or lung tissues, causing acute or subacute invasive infection depending on the severity of neutropenia (Figure 8. 1C). In severely compromised patients (<200 neutrophils/mm³) hyphae invade and penetrate surrounding tissue and blood vessels, causing acute invasive aspergillosis (IA) a severe and generally fatal infection (30–90% mortality when treated) (Denning, 1998; Kontoyiannis & Bodey, 2002). Damage is caused primarily by the toxic

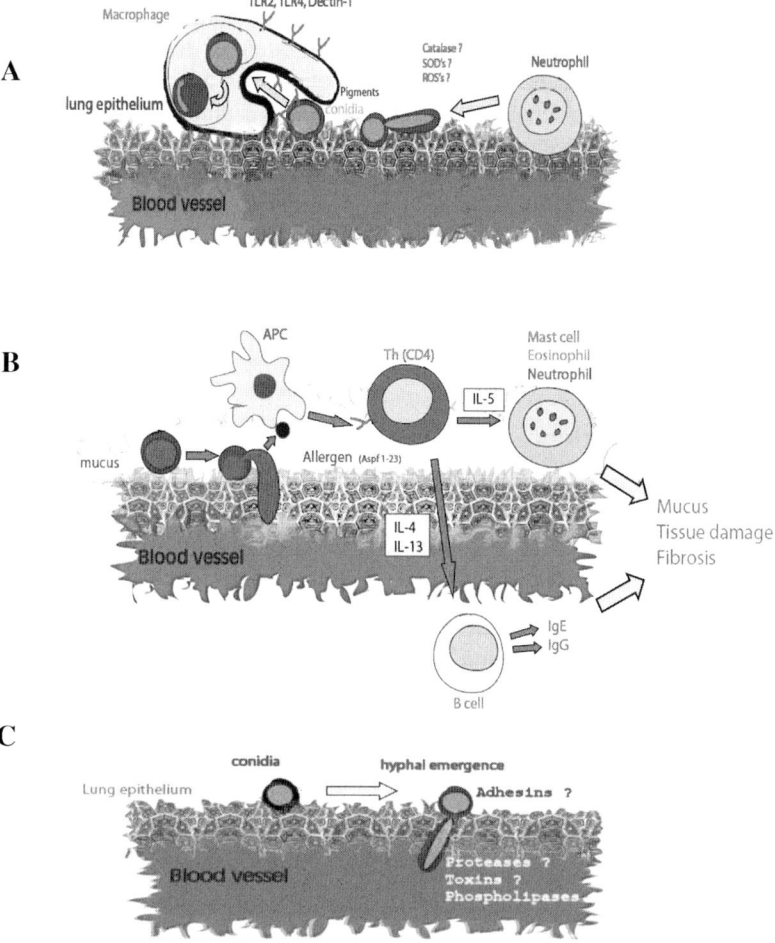

Figure 8.1 The outcome of *A. fumigatus* lung infection depends on the immune status of the host. (**A**) In the immunocompetent host, conidia are rapidly destroyed by alveolar macrophages and neutrophils. (**B**) In the hypersensitive host, conidia initiate an allergic response leading to self-inflicted tissue damage. (**C**) In the immunodeficient host conidia germinate, invading the lung tissue and blood vessels

soup of enzymes and secondary metabolites secreted by the fungus during its vigorous growth in the host organs. This growth occurs because the host lacks effective lines of defense, primarily composed of resident alveolar macrophages and recruited neutrophils. The high mortality rate following treatment can only partially be attributed to a lack of truly effective fungicidal drugs; additional factors complicating an effective cure are the weak general condition of the patients and the difficulty in diagnosing the disease (Kontoyiannis & Bodey,

2002). Because of the increase in the number of immunosuppressed patients following aggressive modern chemotherapy and immunosuppressive regimens, the incidence of IA has increased fourfold during the last 15 years (Steinbach & Stevens, 2003). Ten percent of all deaths in patients who undergo allogeneic bone marrow transplants are attributed to IA, which has a mortality rate of approximately 90% in that setting. IA is found at autopsy in about 20% (95/484) of patients with hematologic malignancies. Furthermore, the financial burden of IA-associated hospitalization is enormous: US data from 1996 estimated the total cost of IA treatment to be US $633 million, with an average cost per case of US $65,000 (Dasbach, 2000). Because of its importance, most research involving *A. fumigatus* virulence factors has centered on identifying those factors involved in IA, and they will be the primary focus of this review.

A. fumigatus Virulence Factors

Overview

There are two opposing views of *A. fumigatus* virulence in IA. The 'fungal-centric' approach argues that the virulence of *A. fumigatus* is a result of specific fungal virulence factors. It advocates identifying a lengthy list of *A. fumigatus* virulence factors that contribute to its ability to invade the host and cause disease. According to this approach, only by carefully dissecting these factors, alone and in combination, can we begin to understand the molecular basis for the various infections caused by *A. fumigatus*. In contrast, the 'immuno-centric' approach argues that *A. fumigatus* virulence results from the immunosuppression or genetic deficiency of the host rather than from specific and unique fungal determinants. According to this approach the fungus infects the host not because it has developed unique systems but because the colonized host exhibits very weak defense immunity. It is a pathogen only because of very simple biological reasons: it is a hardy and ubiquitous generalist, and all humans come into daily contact with it. The approach taken in this chapter is that these two 'world-views' are complementary: in *Aspergillus* infections there is a complex interplay between pathogen virulence factors and the host immune system. These two sides of the coin will be described and integrated in the following sections.

   Virulence factors: a definition. Virulence factors are defined as pathogen determinants that cause damage to the infected host (Casadevall, 2005). Damage caused by the pathogen is either direct or indirect. Direct damage can be offensive (i.e. by secreted enzymes, toxins, adhesins), or defensive (thermotolerance, melanin, hydrophobins, ROS detoxifying enzymes, efflux pumps, etc.). Indirect damage is caused by the host response (necrosis, inflammation) to specific pathogenic determinants (i.e. antige ns). This definition naturally excludes fungal genes encoding key biosynthetic proteins such as PABA synthase (Brown et al., 2000) and orotidine-5'-phosphate decarboxylase (D'enfert, 1996). An additional constraint is that deletion of the gene encoding a virulence factor reduces virulence without affecting normal *in vitro* growth.

***A. fumigatus* virulence factors were developed for life in decaying vegetation**. A central assertion of this review is that the virulence factors of *A. fumigatus* were not developed to infect humans. Rather, they are abilities that enable it to survive and compete in the soil environment. When the fungus accidentally infects humans, it fortuitously uses some of these natural capabilities to survive. For example, to poison its competitors in the soil *A. fumigatus* has become adept at chemical warfare: it contains 26 clusters of genes controlling the production of secondary metabolites, primarily toxic molecules. To facilitate efflux of toxic compounds secreted by competing soil microorganisms, its genome contains over 40 ABC cassette transporters and 100 major facilitator transporters, far more than in yeast (Nierman et al., 2005). These genes may enable it to poison and weaken its human host, and resist the action of drugs and chemicals. Additional virulence factors, such as the presence of a tough melanized conidial cell wall highly resistant to degradative enzymes and oxidative stress, may have developed to enable *A. fumigatus* to survive being consumed by organisms such as soil amoeba and nematodes. Such factors may coincidentally protect it from macrophages that use similar mechanisms to dispose of pathogens (Fuchs & Mylonakis, 2005).

***A. fumigatus* virulence is multifactorial**. In *A. fumigatus* multiple survival traits contribute towards virulence. For example, *A. fumigatus* is neither unique among the filamentous fungi in possessing a large number of efflux pumps or genes for producing secondary metabolites, nor is it more commonly found in human surroundings (Hospenthal et al., 1998; Shelton et al., 2002). Yet it is the most common human mold pathogen. One possible explanation is that *A. fumigatus* possesses a unique combination of different traits that together make it the primary mold pathogen in the world. For example, none of the other, less pathogenic Aspergilli or other common airborne fungi such as *Alternaria, Penicillium*, and *Cladosporum* species may have the multifactorial combination of small conidial size, resistance to oxidants and thermotolerance in one package.

Different virulence factors for different diseases. Because *A. fumigatus* is capable of causing a wide range of diseases depending on the immune system of the host, it is quite obvious that those *A. fumigatus* virulence factors causing allergic aspergillosis will differ considerably from those involved in invasive aspergillosis. In the former, we can expect to find such factors as immunogenic surface and secreted proteins and the ability to withstand a strong immune response, whereas in the latter, crucial factors may be the ability to withstand acidic necrotic surroundings, and overcome the lack of specific vital nutrients and elevated body temperatures found in the infected immunocompromised host.

***A. fumigatus* virulence factors are found in large, often redundant gene families**. Complicating matters further, many of the suspected *A. fumigatus* virulence factors are found in large, often redundant gene families. Their analysis presents a formidable technical challenge. For example, the *A. fumigatus* genome encodes for at least 38 different secreted proteases, five catalases, and six lysophospholipases (see below) and all are suspected virulence factors. To conclusively establish the importance of these gene families in *A. fumigatus* virulence, it will be necessary to delete all the known members in a single mutant strain. This is technically problematic: *A. fumigatus* lacks a sexual life cycle and therefore repeated rounds of transformation are required to delete each gene in turn, using a limited available number of dominant selectable markers. Because the process of transformation is inherently mutagenic, repeated rounds of transformation greatly increases the risk of introducing secondary nonspecific mutations in the final strain. Possible technical solutions to this problem may include the use of RNAi designed to hybridize with a conserved family sequence (although gene repression is never total and the RNAi plasmid is unstable over time) (Mouyna et al., 2004; Mouyna et al., 2006, in press) or the deletion of family-specific transcription factors (Bok & Keller, 2004).

Offensive Virulence Factors

Offensive virulence factors are molecules that enable the fungus to interact directly with the host, causing it damage. They include secreted enzymes such as proteases, peptidases and phospholipases, secreted toxins, and adhesins.

Proteases

Proteases (also known as peptidases) play a major role in the virulence of several fungal pathogens by digesting host proteins and thereby enhancing tissue penetration (Monod et al., 2002). During the previous decade, considerable efforts were made to identify *A. fumigatus* proteases and assess their role in infection and virulence (Tomee & Kauffmann, 2000; Monod et al., 2002). The major findings were that secreted *A. fumigatus* proteases induce pro-inflammatory cytokine release in infected macrophages and epithelial cells, thereby alerting the immune system (Kauffmann et al., 2000). Infected lung epithelial cells also undergo major changes to the actin cytoskeleton, leading to cell peeling and death (Tomee et al., 1997; Kogan et al., 2004). However, deletion analyses of selected proteases (*Alp1, MEP, PEP1*), have failed to conclusively demonstrate a significant role in virulence in animal models (Monod et al., 2002). This is probably a result of the large number of proteases secreted by *A. fumigatus* and the functional redundancy between them (Robson et al., 2005). Its genome encodes for at least 38 proteases (Table 8.1). Of these, 15 are predicted secreted endoproteases (cleaving peptide bonds internally, within a polypeptide), including eight aspartic proteases, four serine/subtilisin proteases and three metalloproteases. Only five were studied to date (Table 8.1, underlined). The *A. fumigatus* genome also contains at least 23 predicted secreted exopeptidases (cleaving peptide bonds only at the N- or C-terminal of the polypeptide chain) (Table 8.1). Of these only two (DppIV and DppV) were cloned, characterized enzymatically but not deleted (Beauvais et al., 1997a, b). DppV is

Table 8.1 *A. fumigatus* putative proteases, lysophospholipases, and non-ribosomal peptide synthase (NRPS) genes

Classification‡	Genes
Endopeptidases	
Aspartic endopeptidase	*PEP1*/aspergillopepsin A/I/ F/*asp f10**, *PEP2*, OpsB,CtsD, Afu6g03260/AP1, Afu4g09400/AP3, Afu3g01220, Afu2g15950
Serine endopeptidase	*Alp1*/subtilisin N, *Alp2/asp f13*, AorO, KexB/Afu4g12970,
Metalloproteases	*MEP/ Asp f5*, Afu6g14420 Afu4g13750
Exopeptidases	CpdS/Afu6g00310, Afu8g04120, Afu5g01200,
Carboxypeptidases	Afu5g01200, Afu2g04370, Afu1g00420, Afu6g13540, Afu4g07270, Afu1g08940, Afu5g14610, Afu5g07330, Afu3g12210, Afu2g0879
Dipeptidyl peptidases	DppIV, DppV, Afu2g0903
Tripeptidyl peptidases	Afu4g03490, Afu4g14000, Afu3g0893
Serine exodopeptidases	Afu4g03790, Afu2g01250, Afu2g17330
Aminopeptidase	Afu3g00650
Lysophospholipases	
Phospholipase A	Afu2g10760
Phospholipase B	AfPLB1, AfPLB3
Phospholipase C	Afu7g04910, Afu1g17590
Phospholipase D	Afu2g16520
Non-ribosomal peptide *synthases (NRPS)*	gliP, sidC, pes1, Afu6g08560, Afu6g09610, Afu6g12050, Afu6g12080, Afu4g14440, Afu5g12730, Afu5g10120, Afu15270, Afu3g13730, Afu3g12920

‡ Classification based on *A. fumigatus* genome annotation (Nierman 2005) and available at http://www.genome.jp/kegg-bin/show_organism?menu_type = gene_catalogs & org = afm. All genes except the non-ribosomal peptide synthases contain a putative signal peptide sequence as analyzed by Signal P 3.0 (http://www.cbs.dtu.dk/services/SignalP/).
* Underlined: Genes that have been studied to date.

strongly antigenic and is useful for the diagnosis of aspergilloma and ABPA (Sarfati et al., 2006).

Phospholipases

Phospholipases are a heterogeneous group of enzymes which hydrolyze one or more ester bonds in phosphoglycerides. They have an important role in the virulence of bacteria and several pathogenic fungi including *Candida* species and *Cryptococcus neoformans* (Ghannoum, 2000). Spectrometric analysis of *A. fumigatus* filtrates indicated phospholipase A, B, C, and D activities (Birch et al., 1996). The *A. fumigatus* genome contains six predicted secreted phospholipases (Table 8.1). Of these, two phospholipase B (PLB) genes, *AfPLB1* and *AfPLB3*, were cloned and studied. These proteins contain a signal peptide sequence and are secreted enzymes (Shen et al., 2004). None of the *A. fumigatus* phospholipases were deleted to date and their contribution to virulence remains unknown.

Toxins

A. fumigatus produces a rich cocktail of toxins and apparently uses them to fight off competition and predators in its natural environment. They include the low molecular weight toxic secondary metabolites (gliotoxin, helvolic acid/fumigacin, fumagillin, fumigaclavine A-C, festuclavine, aurasperone C, aflatoxin B1, and G1 and verruculogen), and small proteins inhibiting ribosome function (Restrictocin/ mitogillin/As f1) or inducing erythrocyte hemolysis (Asp-hemolysin) (Kamei & Watanabe, 2005; Rementeria et al., 2005; Yu & Keller, 2005). Gliotoxin and restrictocin are the only two toxins produced at toxic levels in infected animals and patients by most clinical strains of *A. fumigatus* (Kamei & Watanabe, 2005; Lewis et al., 2005). Deletion of the gene encoding restrictocin did not affect fungal virulence in mice (Smith et al., 1993).

Until recently, the involvement of the toxic secondary metabolites in virulence had only been demonstrated *in vitro*, through their immunosuppressive action on macrophages, and inhibitory effects on epithelial cell function. Identifying the genes responsible for their synthesis has proven difficult. The sequenced *A. fumigatus* genome indicates the potential for the biosynthesis of up to 40 novel metabolites from 26 clusters of genes encoding predicted prenyltransferases, oxidoreductases, methyltransferases, transporters, 13 nonribosomal peptide synthetases (NRPSs) (Table 8.1), polyketide synthases (PKSs), and transcription factors. Thirteen of these clusters are unique to *A. fumigatus*. (Nierman et al., 2005). The gene clusters involved in ergot alkaloid biosynthesis and gliotoxin biosynthesis were identified (Gardiner & Howlett, 2005; Coyle & Panaccione, 2005).

dmaW and Ergot Alkaloid Biosynthesis

Deletion of *dmaW*, encoding dimethylallyltryptophan synthase, a key enzyme in ergot alkaloid biosynthesis, eliminates the biosynthesis of all known ergot alkaloids (fumigaclavine A-C, festuclavine) from *A. fumigatus*. However, the virulence of this strain has yet to be characterized (Coyle & Panaccione, 2005).

gliP and Gliotoxin Biosynthesis

Deletion of the NRPS *gliP*, the first enzyme in the pathway for gliotoxin biosynthesis, results in an *A. fumigatus* strain which fails to produce gliotoxin. *In vitro*, the culture supernatant of the *gliP*-deficient strains has a reduced cytotoxic effect on both macrophage-like cells and T cell lines. Surprisingly, the *gliP* null mutant is normally virulent in immunosuppressed mice, indicating that gliotoxin alone is not a virulence factor in *A. fumigatus* (Cramer et al., 2006; Kupfahl et al., 2006; May GS submitted) A possible explanation for this result is that *in vivo*, gliotoxin acts in combination with additional secreted toxins. This is supported by the recent work of Bok et al. (2005): Deletion of *laeA*, a transcription factor and global regulator of secondary metabolite biosynthesis, blocked production of all major types of secondary metabolites including gliotoxin and resulted in decreased virulence in immunosuppressed mice.

pes1

Deletion of the NRPS *pes1* results in significantly reduced virulence in the *Galleria* insect larva model system and increased sensitivity to oxidative stress in culture and during neutrophil-mediated phagocytosis. The mutant also exhibits smoother conidial surface morphology and increased hydrophobicity (Reeves et al., 2006).

In the coming years, deletion of additional gene clusters and transcription factors involved in secondary metabolite biosynthesis will greatly enhance our understanding of their role in virulence.

Adhesins

Fungal adhesion to host tissues is mediated primarily by adhesins (Verstrepen & Klis, 2006). Adhesins have been cloned in several fungi, including *S. cerevisiae* and *C. albicans*. All fungal adhesins share a common three-domain structure which includes a glycophosphatidylinositol-anchored (GPI) C-terminal part, a large middle domain characterized by the presence of multiple serine- and threonine-rich repeats and an N-terminal part containing a carbohydrate or peptide-binding domain. The repeats undergo frequent recombination dependent expansion or contraction resulting in isolates containing different length versions of this region. Adhesins with more repeats generally show greater adherence, whereas shorter adhesins with less repeats generally exhibit decreased adhesion, possibly because the N-terminal domain remains buried in the cell wall. Several putative *A. fumigatus* adhesins have been characterized biochemically but none were cloned (reviewed in Latge, 1999). BLAST homology searches with sequences of known fungal adhesins (e.g. *S. cerevisiae FLO* genes, *C. albicans ALS* or *Hwp1* genes among others) have failed to identify significant homologs in *Aspergillus* spp. However, the genome of *A. fumigatus* contains at least 20 genes with a putative C-terminal GPI anchor and multiple tandem repeats. Of these, four genes (*Afu3g08990*, *Afu2g05150 (MP-2)*, *Afu4g09600*, and *Afu6g14090*) were shown to vary in the number of repeats in different *A. fumigatus* isolates (Levdansky et al., submitted). Deletion of theses genes will help clarify whether they are involved in adhesion or host interactions.

Defensive Virulence Factors

Defensive virulence factors are traits that enable the fungus to proliferate in the hostile environment of the human body. They include the ability to grow at human body temperature (thermotolerance), the protective capacity of the fungal cell wall and its components (polysaccharides, proteins, pigments) and the ability to degrade reactive oxygen species produced by the host immune system.

Thermotolerance

A. fumigatus thermotolerance is a trait critical for survival in decomposing organic matter. However, genetic evidence suggests that whereas the ability of *A. fumigatus* to grow at 37°C is important for virulence, the ability to grow at the high temperatures in which this fungus specializes (i.e. above 48°C), is not. Seven *A. fumigatus* chemical mutants that grow normally up to 42°C, but fail to grow at 48°C displayed wild-type virulence in mice (Chang et al., 2004). Also, a mutant disrupted for *CgrA*, a nucleolar gene involved in ribosome biogenesis, which grew normally at 25°C but poorly at 37°C exhibited reduced virulence in mice (Bhabhra et al., 2004).

Recent microarray analysis following a shift of growth to elevated temperatures (37°C and 48°C), identified many upregulated *A. fumigatus* heat-shock genes which may have a role in thermotolerance (Nierman et al., 2005). Interestingly, the heat-responsive genes activated in *A. fumigatus* differ greatly from those in the non-pathogenic yeast *S. cerevisiae*. Recently, proteomic analysis of *A. fumigatus* has been developed, and this should yield further insights into the response to heat at the protein level (Kniemeyer et al., 2006).

In the near future, a comparison of the genome-wide heat-shock response of *A. fumigatus* to related, non thermophylic Aspergilli, such as *A. nidulans*, and other fungal pathogens, may identify genes or pathways that enable *A. fumigatus* to withstand high temperatures and affect pathogenicity.

The Fungal Cell Wall

The cell wall of *A. fumigatus* can be viewed as a multifactorial protective shield. It is composed of a polysaccharide skeleton of glucan, chitin, and galactomannan polymers interlaced and coated with cell wall-associated proteins (CWAPs) and embedded pigments (Latge et al., 2005). Past studies have identified pigments such as melanin and CWAPs such as the hydrophobins as important protective cell wall-virulence factors and these are described below. However, the recent analysis of the *A. fumigatus* genome has identified over 80 putative plasma membrane or cell wall-associated GPI-anchored proteins of which almost half are of unknown function, providing a rich starting point for the discovery of additional protective virulence factors (Table 8.2) (Nierman et al., 2005).

Cell wall mutants: Recently, two *A. fumigatus* cell wall mutants (Δ *ecm33* and Δ *ags3*) which are hypervirulent in infected immunosuppressed mice have been described (Romano et al., 2006; Maubon et al., 2006). The reason for the hypervirulence of these mutants is not entirely clear. Hypervirulence in the Δ *ags3* mutant could be a result of its faster germination (thereby evading macrophage killing) or higher concentrations of conidial melanin (leading to increased resistance to ROS).

Table 8.2 *A. fumigatus* genes encoding for putative GPI-anchored proteins

Classification[‡]	Genes
Glycosyl hydrolase	*Afu2g00680, Afu4g00620, Afu3g00700*
Glucanosyl transferase	*Afu1g16190 (Crf1[†])[e], Afu2g01170 (GEL1[†]), Afu2g03120 (Utr2p-like[†]), Afu2g14360S/T, Afu3g00270 (Bgl2p-like) S/T, Afu6g08510 (Crf1-like) S/T, Afu6g11390 (GEL2), Afu8g06030 (MutA-like), Afu3g03080, Afu6g12410 (GEL7), Afu8g02130 S/T+C(GEL5), Afu2g12850 (GEL3), Afu3g13200(GEL6), Afu2g05340(GEL4)*
Spore coat hydrophobin	*Afu5g09580 (HYP1), Afu8g07060*
Aspartic protease	*Afu4g07040 (CtsD), Afu6g05350 (OpsB-like),*
Chitin-related proteins	*Afu5g03760S/T+C (ChiA1), Afu6g10430*(chitin deacetylase), Afu6g00500 (chitosanase)*
CFEM[a]-domain protein	*Afu6g10580 S/T (proline-rich antigen 2-like)[f], Afu6g14090 [†]S/T Afu6g06690*
ecm33-like	*Afu4g06820 (SPS-2 like; AfuEcm33;spot10[†]) S/T*
Immunoreactive protein	*Afu2g05150S/T+C (MP-2), Afu6g02800 (spot 19[†]) S/T, Afu4g03240 (MP-1) S/T*
SUN[b] domain protein	*Afu1g13940(Adg3p-like) S/T+C*
Lysophospholipase	*Afu3g14680 (Plb3), Afu4g08720,*
Alpha-amylase Agglutinin/Ecm	*Afu2g03230 (AmyA), Afu2g13460 (AmyA-like), Afu3g00900 (AmyA) Afu4g09600 (agglutinin-like) f, Afu8g05410* (ecm-like) f*
Amino acid permease	*Afu6g00410*
Acid phosphatase	*Afu1g03570 (phoA)*
Sexual development	*Afu5g10400 (LsdA-like)*
WSC domain protein	*Afu3g07050 S/T,*
Acetyltransferase	*Afu8g04000*
Conserved hypothetical[c]	*Afu1g03630 S/T, Afu1g05790 S/T, Afu1g09510 S/T, Afu1g09590, Afu1g11220 S/T, Afu2g01140 S/T, Afu3g00880 S/T, Afu3g08990 S/T, Afu3g13110 S/T+C, Afu3g14210 S/T, Afu4g12370, Afu5g08800, Afu7g00450 S/T, Afu7g00580 S/T, Afu7g00970 S/T+C, Afu7g03970 S/T+C, Afu8g02450 S/T, Afu8g04370 S/T, Afu5g07800, Afu5g09960* S/T+C, Afu6g13710, Afu7g02440 S/T, Afu8g04860S/T, Afu1g09650 S/T+C, Afu1g11680, Afu1g13760, Afu2g02440 S/T, Afu3g01150 S/T, Afu3g13640 S/T, Afu4g03500, Afu5g01920, Afu8g00830 S/T+,*
Hypothetical[d]	*Afu5g10010 S/T Afu4g06370 S/T, Afu8g01770 S/T+C, Afu6g14010 S/T, Afu2g07800 S/T*

[‡] Classification was based on functional, structural, or immunological criteria.
Underlined – Genes that have been studied to date.
S/T – serine/threonine-rich protein.
S/T+C – serine/threonine and cysteine-rich protein.
[a] –An eight-cysteine-containing domain unique to a group of fungal membrane and cell wall proteins (Kulkarni et al., 2003).
[b] – Similar to *S. cerevisiae* proteins of the SUN family (Sim1p, Uth1p, Nca3p, Sun4p) that may participate in DNA replication, autophagy and cell death.
[c] – ORFs with significant homology (BlastP value $<10^{-10}$) to other uncharacterized proteins.
[d] – ORFs with no significant homology to any other proteins.
[e] – Annotation provided by CADRE unless otherwise specified (http://www.cadre.man.ac.uk/).
[f] – Annotation performed by BlastP search (BlastP value $<10^{-10}$).
[†] – Previously identified by Bruneau et al. (2001).

Hypervirulence in the Δ *ecm33* mutant could also occur because of its faster germination. Other factors may include its tendency towards conidial clumping (leading to occlusion of blood vessels) or increased immunogenicity (because of changes in the cell wall structure and exposure of immunogenic antigens). Increased immunogenicity and the induction of septic shock in infected animals, was previously shown in hypervirulent *S. cerevisiae* and *C. glabrata* mutants (Wheeler et al., 2003; Kamran et al., 2004).

Pigments: Pigments such as melanin and its derivatives are important protective virulence factors in fungal pathogens of both plants and humans. They provide mechanical strength, UV protection and the scavenging of free oxygen radicals (Gomez, 2003). Melanins are dark-brown or black pigments formed by the oxidative polymerization of phenolic compounds. In *A. fumigatus* melanin is produced by a cluster of six genes via the DHN-melanin pathway (reviewed in Brakhage and Leibmann, 2005). Homologs of this cluster are found in most filamentous fungi. Melanin is found as discrete black particles in the cell wall of dormant conidia. In culture, it is rapidly degraded during germination, disappearing by the time of germ-tube emergence (Youngchim et al., 2004). However, melanin or its intermediates might also be produced during invasive growth: *PksP*, the enzyme involved in the first step of melanin biosynthesis, which is active in vitro only during conidiogenesis, continues to be expressed during hyphal growth in the lungs of infected mice (Langfelder et al., 2001). This activation may occur via the PKA-dependent cAMP pathway as reduced *PksP* transcription is seen in mutants of this pathway (Liebmann et al., 2003).

A mutant producing pigmentless white conidia was generated by deletion of *PksP* (*Alb1*) (Tsai et al., 1999). This mutant was more sensitive to oxygen radicals and macrophages, showed increased ingestion by neutrophils and reduced virulence in a mouse model for IA (Jahn et al., 1997; Tsai et al., 1998). Deletion of the five additional genes participating in the DHN-melanin pathway *Abr1/Abr2*, *Arp1/Arp2* and *Ayg1* results in brown, pink and yellowish-green conidia, respectively, but the virulence of these strains was not reported (Tsai et al., 1999, 2001).

It is important to note that many non-pathogenic fungi also contain melanin; therefore it is unlikely that melanin alone is responsible for the virulence of *A. fumigatus*.

Hydrophobins: Fungal hydrophobins are small cysteine-rich proteins involved in generating a thin layer of proteinaceous rod-like structures on the surface of aerial hyphae and conidia (Linder et al., 2005). This hydrophobic layer, by retarding water, improves conidial dispersion in the atmosphere. *A. fumigatus* contains six hydrophobin genes of which two, *rodA* and *rodB* were studied. They are conidial-specific proteins, disappearing during germination and hyphal growth. Deletion of *rodA* results in a loss of conidial rodlet structure, increased wettability, reduced conidial adhesion to some substrates, increased sensitivity to killing by alveolar macrophages but normal virulence in mice (Thau et al., 1994). This suggests that *rodA* forms a protective conidial-specific retardant barrier against macrophage oxidants. Deletion of *rodB* does not result in any observable phenotype despite its similarity to *rodA* (Paris et al., 2003).

ROS protection (catalases, SODs, antioxidants): During infection, *A. fumigatus* is attacked by reactive oxygen species (superoxide O_2^-, hydrogen peroxide H_2O_2, and hydroxyl radicals OH) produced by neutrophils and macrophages. It is therefore logical to assume that detoxification of H_2O_2 and O_2^- by *A. fumigatus* catalases, superoxide dismutases (SODs), and antioxidants such as glutathione might function as defensive virulence factors. The *A. fumigatus* genome contains five catalase and two SOD-encoding genes. The catalases include the closely related *catA, catB/cat1, catC*, and *catE* genes and the more distantly related *cat2* gene encoding a bifunctional catalase/peroxidase. Deletion analysis of *catA, catB/cat1*, and *cat2* indicated that only the last two together are involved in virulence, exhibiting delayed infection in a rat model of aspergillosis (Calera et al., 1997; Paris et al., 2003b). This limited effect might be due to functional redundancy or compensatory mechanisms.

The SODs include the immunogenic-secreted Cu/Zn *SOD1* and intracellular Mn-SOD *SodA/Asp f 6* (Holdom et al., 2000; Fluckiger et al., 2002). Neither *SOD1* nor *SodA* have yet been deleted, so their role in virulence remains unknown.

Additional protection against oxidants is afforded by the antioxidant glutathione (GSH). GSH is a tripeptide (Glu-Cys-Gly) generated by the glutaredoxin and thioredoxin protein systems (reviewed in Grant, 2001). Mutants in these pathways have not been constructed in *A. fumigatus* and their role in virulence remains unknown (Chauhan et al., 2006).

Efflux pumps: *A. fumigatus* efflux pumps are involved in antifungal resistance and might also be used to detoxify components of the immune system. The *A. fumigatus* genome contains at least 327 genes encoding multidrug resistance (MDR) efflux transporters, at least threefold more than in the yeast *S. cerevisiae*, but roughly equal to that of *Aspergillus nidulans* and *Aspergillus oryzae*. These include 49 genes of the ATP-binding cassette (ABC) and 278 genes of the major facilitator superfamily (MFS) classes (Nierman et al., 2005; Ferreira et al., 2005). None have yet been deleted. Transcriptional analysis indicates that four ABC family members (*MDR1, MDR2, atrF, abcA*, and *MDR4*) (Tobin et al., 1997; Slaven et al., 2002; Langfelder et al., 2002; Nascimento et al., 2003) and one MFS member (*MDR3*)(Nascimento et al., 2003) are upregulated in azole-resistant isolates of *A. fumigatus* exposed to the drug. Five additional ABC transporters (*abcA-E*) and three MFS members (*mfsA-E*) are upregulated in wild-type *A. fumigatus* in response to voriconazole (da Silva, 2006).

Large-scale inactivation experiments for *A. fumigatus* transporter-encoding genes and the identification and deletion of their transcriptional repressors and activators could help define which of these genes are involved in resistance to drugs and toxic molecules in this species.

Aspergillus fumigatus biofilms: Fungal biofilms are three-dimensional structures composed of cells embedded in an extracellular matrix. Biofilms of the pathogenic yeast *C. albicans* are intrinsicaly resistant to almost all antifungals in clinical use (reviewed in d'Enfert, 2006). *A. fumigatus* in culture produces an amorphous extracellular matrix which can be visualized by cryo-scanning electron microscopy and confocal microscopy (Beauvais et al., *Cell Microbiol.*, accepted for publication).

Further studies to reveal the components of the matrix and its role in defending the fungus against the host defenses and antifungal drugs are in progress.

Other Genes Affecting Virulence

Additional genes affecting virulence that do not fall neatly into the categories described above are described in this section. They can be loosely grouped as genes that enable the pathogen to survive in and adapt to the host environment. For example, the host environment is surprisingly poor in several essential elements such as iron and phosphate, and the fungus must activate pathways and genes to overcome these limitations.

Iron uptake: The ability to acquire iron *in vivo* is essential for most microbial pathogens (reviewed in Haas, 2003). The host contains very little free iron as most of it is bound to carrier proteins such as transferrin. *A. fumigatus* employs three iron-uptake mechanisms: ferrous iron uptake, reductive iron assimilation, and siderophore-mediated iron uptake, with only the last showing involvement in virulence (Schrettl et al., 2004). Siderophores are small cyclic peptides that function as ferric iron-specific chelators. They are excreted during iron starvation and recovered together with the bound iron by specific transporters. *A. fumigatus* synthesizes at least five different siderophores, the most common being intracellular ferricrocin and secreted triacetylfusarinine C (Hissen et al., 2004). The pathway involved in their biosynthesis contains at least five genes, including *sidA* (L-ornithine-N^5-monooxygenase) which catalyzes the first committed step of hydroxamate-type siderophore biosynthesis and *sidC* a nonribosomal peptide synthase (NRPS) (Eisendle, 2004) (Table 8.1). Siderophore-mediated iron uptake was found to be essential for *A. fumigatus* virulence in a murine model of invasive aspergillosis. Deletion of *sidA* results in severely attenuated virulence in immunosuppressed mice (Schrettl et al., 2004; Hissen et al., 2005). Mutant conidia fail to germinate in the infected lungs suggesting that this environment contains very little free iron. Mammals do not contain a *sidA* ortholog or a siderophore system in general. Therefore, it represents a promising new target for the development of antifungal therapies.

Magnesium uptake: The ability to acquire magnesium *in vivo* is essential for bacterial pathogens such as *Mycobacterium tuberculosis* and *Salmonella typhimurium* because it enables them to survive digestion in the phagolysosome (Maguire, 2006). Magnesium is essential for the growth of *A. fumigatus*. It contains four genes with putative magnesium acquisition function including *Afu4g00930*, an ortholog of the prokaryotic *CorA* and yeast *ALR1* magnesium plasma membrane transporters and *Afu7g05060*, an ortholog of the bacterial *MgtC* transporters (Table 8.3). Interestingly, *A. fumigatus* is the only sequenced fungus that contains an *MgtC* prokaryotic-related transporter. *Afu7g05060/MgtC* is currently being studied in *A. fumigatus* (Gastebois and Mouyna, personal communication, 2005).

Phosphate acquisition: *A. fumigatus* needs about 10 mM free phosphate to grow in culture, whereas serum contains only about 1 mM. Its genome contains four homologs of the *S. cerevisiae* low-affinity and high-affinity inorganic phosphate

Table 8.3 *A. fumigatus* putative essential element transporters

Classification[‡]	Genes
Iron uptake	
Siderophore biosynthesis	*sidA (Afu2g07680)*, *sidC (Afu1g17200)*, Afu3g03400, Afu1g04450
Magnesium uptake	
Magnesium transporters	*Afu4g00930 (CorA-like), Afu7g05060 (MgtC-like) Afu5g05830, Afu6g02550*
Phosphate uptake	
Phosphate transporters	*Afu7g06350, Afu1g04290, Afu4g09210, Afu2g10690*
Acid phosphatases	*PhoA (Afu1g03570), Afu4g01070, Afu4g03660, Afu5g01330, Afu6g1330, and Afu3g14570*
Zinc uptake	
Zinc transporters	*zrfA (Afu1g01550), zrfB (Afu2g03860), Afu6g00470, Afu8g04010, Afu2g01460, Afu6g14170, Afu4g09560*

[‡] Classification based on A. fumigatus genome annotation (Nierman et al., 2005) and available at http://www.genome.jp/kegg-bin/show_organism?menu_type=gene_catalogs&org=afm.
[*] Underlined: genes that have been studied to date.

transporters, and at least six secreted acid phosphatases including *PhoA* (Table 8.3) and six phospholipases (Table 8.1) that release phosphate from organic sources. Only one of these 14 genes, *PhoA*, has been studied. It is a GPI-anchored cell wall protein with phosphate-repressible acid phosphatase activity (Bernard et al., 2002). The deletion of *phoA* has not been described.

Zinc uptake: Zinc is an essential micronutrient that *A. fumigatus* must obtain from the environment in order to grow. The *A. fumigatus* genome contains seven putative zinc transporters including *ZrfA* and *ZrfB* (Table 8.3). Deletion of both *ZrfA* and *ZrfB* dramatically reduces growth under acid, zinc-limiting conditions (Vicentefranqueira et al., 2005). Their role in virulence has not been described. The other five transporters have not been studied to date.

Signaling pathways: Signaling pathways are key elements in sensing and transmitting the response of cells to environmental conditions. The major known fungal-signaling pathways are described below, with an emphasis on their role in the growth and virulence of *A. fumigatus*.

Histidine kinases: Histidine kinases have an important role in virulence of both plant and human pathogens. They act primarily as sensors that respond to changes in osmolarity, resistance to fungicides and cell wall assembly (Nemecek et al., 2006). The *A. fumigatus* genome contains approximately 15 putative histidine kinase genes of which only two, *fos-1* and *tcsB* were studied (Table 8.4). Deletion of *fos-1* results in no obvious defect during growth on agar, but the mutant is attenuated in virulence in infected mice (Pott et al., 2000; Clemons et al., 2002). Deletion of *tcsB* results in no obvious phenotype suggesting there is some redundancy in this family (Du et al., 2006). Mammalian cells do not have any two-component or phosphorelay systems, suggesting they might be useful antifungal drug targets for treatment of human mycoses.

G-proteins The fungal G-protein superfamily is divided into the small G-protein and the heterotrimeric receptor coupled G-protein families.

Table 8.4 *A. fumigatus* putative signaling pathway genes

Classification[‡]	Genes
Histidine kinases	*fos-1 (Afu6g10240)*, *tcsB (Afu2g0066)*, *Afu4g01020*, *Afu4g00660*, *Afu4g00320*, *Afu4g02900*, *Afu8g06140*, *Afu7g08550*, *Afu2g03560*, *Afu6g09260*, *Afu4g07400*, *Afu5g10020*, *Afu3g12550*, *Afu3g12530*, *Afu3g07130*
Small G-proteins	
Ras-like	*rasA (Afu5g11230)*, *rasB (Afu2g07770)*, *rsr1 (Afu5g08950)*, *rhbA (Afu5g05480)*, *Afu3g05770*, *Afu7g02540*, *Afu1g02190*, *Afu1g07680*, *Afu2g15570*
Rho-like	*Afu4g04810*
Rac-like	*Rho1 (Afu6g06900)*, *rho2 (Afu3g10340)*, *rho3 (Afu3g06690)*, *rho4 (Afu5g14060)*, *Afu2g05740 Afu3g06300*
MAP-kinase pathway	
MAPKKK	*Afu1g10940*, *Afu5g06420*, *Afu3g11080*
MAPKK	*Afu1g05800*, *Afu1g15950*, *Afu3g05900*
MAPK	*sakA (Afu1g12940)*, *mpkA (Afu4g13720) mpkB (Afu6g12820)*, *mpkC (Afu5g09100)*
cAMP signaling	
PKA catalytic subunits	*pkaC (Afu2g12200)*, *Afu1g06400*, *Afu5g08570*
PKA regulatory subunits	*pkaR (Afu3g10000)*
Adenylate cyclase	*acyA (Afu6g08520)*

[‡] Classification based on *A. fumigatus* genome annotation. (Nierman et al., 2005.) Available at http://www.genome.jp/kegg-bin/show_organism?menu_type=gene_catalogs &org=afm.
* Underlined: genes that have been studied to date.

Small G-proteins such as *ras, rho, rac*, and *cdc42* have a major role in fungal nutrient sensing, polarity and growth (Wendland, 2001). The *A. fumigatus* genome contains at least 16 genes encoding small G-proteins, of which only three *rasA, rasB, rhbA* were studied (Table 8.4). The genes *rasA* and *rasB* appear to have different, but overlapping roles in the germination, vegetative growth and asexual development of *A. fumigatus* (Fortwendel et al., 2004). The virulence of these mutants was not reported. *RhbA* is a small G-protein that is upregulated when *A. fumigatus* is grown under nitrogen starvation and in culture with human endothelial cells. Deletion of *rhbA* results in slow growth on medium containing a poor nitrogen source and reduced virulence in mice (Panepinto et al., 2003). This might be because of their reduced ability to utilize nitrogen sources in infected tissue.

Heterotrimeric receptor-coupled G-proteins which interact with G-protein coupled transmembrane receptors (GPCRs) are involved in growth, development, mating, biosynthesis of secondary metabolites and virulence (Yu & Keller, 2005; Lafon et al., 2006). The *A. fumigatus* genome contains 15 putative GPCRs, three G-alpha subunit, one G-beta, and one G-gamma encoding genes. They also contain five homologs of RGS (regulator of G-protein signaling) predicted to participate in attenuation of G-protein activity (Lafon et al., 2006). Of these, only *gpaA (Afu1g13140)* and *gpaB (Afu1g12930)* encoding G-alpha subunits were studied in *A. fumigatus*. Deletion of *gpaB* resulted in decreased conidiation (but normal growth rates), increased sensitivity to macrophage killing. Most importantly, the *gpaB* null mutant showed greatly reduced virulence in a low-dose inhalation mouse-infection model, despite the fact that the mutant conidia germinated and colonized the infected lungs normally (Leibman et al., 2003, 2005).

MAPK pathways: MAP-kinase (MAPK) cascades have a pivotal role in fungal nutrient sensing, osmolarity response, cell wall integrity and pheromone response pathways. These kinase cascades contain three protein kinases (MAPKKK, MAPKK, and MAPK) that act in series by phosphorylation (May et al., 2005). The *A. fumigatus* genome contains three putative MAPKKK, three MAPKK, and four MAPK genes (Table 8.4). Of these 10 genes, only two, the Hog-family *sakA*/ MAPK and *mpkC* have been studied. Deletion of *sakA* results in an inability to grow under hypotonic stress, sensitivity to oxidative stress, and impaired nitrogen sensing (Xue et al., 2004, Du et al., 2006), whereas deletion of *mpkC* resulted in impaired conidial germination and growth in the presence of sorbitol and mannitol as sole carbon sources (Reyes et al., 2006, in press). The virulence of these strains was not reported.

cAMP signaling: The cyclic AMP-dependent protein kinase (PKA) regulates morphology, stress response and virulence in a number of fungal pathogens of humans and plants. The *A. fumigatus* genome contains three PKA catalytic subunits, one regulatory subunit and one gene encoding adenylate cyclase (Table 8.4). Of these five genes, three, *acyA* adenylate cyclase, *pkaC* catalytic subunit and *pkaR* regulatory subunit, were studied. Deletion of *acyA* results in decreased growth and conidiation and increased sensitivity to macrophages (Liebmann et al., 2004).

Deletion of *A. fumigatus pkaC* and *pkaR* results in decreased growth and reduced tolerance to oxidative stress. Reduced virulence in mice was also seen, although this is most likely a general outcome of impaired growth (Liebmann et al., 2004; Zhao et al., 2006).

Allergens

A. fumigatus produces a considerable number of allergenic molecules, including at least 23 identified proteins (Asp f 1-Asp f 23) and polysaccharides (Kurup, 2005). Among them are several secreted enzymes (*Asp f 10/PEP, Asp f 13/Alp2, Asp f 5/MEP*) which are associated with host invasion. Interestingly, most of the other allergens are not predicted cell wall or secreted molecules (i.e. *Asp f 8* and *Asp f 23* ribosomal proteins, *Asp f 12* hsp90 family, *Asp f 22* enolase). One possibility is that these abundant cytosolic proteins are released from the fungus following damage by the immune system. Another is that these proteins may be naturally present at the cell surface and have some unknown role (e.g. energetic role, antigenicity, surface receptors) (Lopez-Ribot et al., 2004).

Allergies inflicted by *A. fumigatus* include asthma, ABPA, allergic fungal sinusitis, and hypersensitivity pneumitis. They occur in hypersensitive individuals as a result of

excessive IgE-mediated mast cell degranulation and eosinophilic infiltration and tissue damage in response to inhaled *A. fumigatus* allergens (Figure 8.1B). Both
humoral (elevated antigen specific IgG, IgE, and IgA) and cell-mediated immunity
(Th-2 cell cytokine secretion) are activated. Some of the *A. fumigatus* allergens identified could be useful for diagnosis of ABPA. Allergen-specific immunotherapy should
be developed to induce tolerance in susceptible hypersensitive individuals.

The Host Immune Response

Recently, there have been important advances in our understanding of the host
response to *A. fumigatus* infection (Figure 8.1A). (i) Several of the receptors
(Toll-like receptors TLR-2, TLR-4, and dectin-1 receptor) and pathways (ERK,
NFkB) used by the cells of the innate immune system to recognize and respond to
A. fumigatus infection have been identified; (ii) macrophage-mediated phagocytosis
of conidia has been delineated in greater detail; (iii) the role of dendritic cells (DCs)
in activating the T cell response through antigen presentation has also been
examined. These exciting and evolving fields will be discussed. Readers interested
in comprehensive summaries describing the host immune response to *A. fumigatus*
infection are directed to several excellent recent reviews (Roeder et al., 2004; Walsh
et al., 2005; Brown 2006; Rivera et al., 2006).

Receptors

Toll-like Receptors

Toll-like receptors (TLRs) have a major role in mediating cellular immunity. The
family comprises 13 members. They are expressed by macrophages, neutrophils,
DCs and epithelial cells. Two members, TLR2 and TLR4 are implicated in fungal
recognition, apparently in response to fungal surface polysaccharides (Uematsu &
Akira, 2006). Signaling is mediated via two pathways:

(a) TLR2/TLR4 signaling dependent on the adaptor protein MyD88, leading to the
    activation of the nuclear transcription factor NFK-B and transcription of inflam-
    matory Th1-response cytokines including TNF-α, IL-6, IL-12, and IL1-β.
(b) TLR-4 signaling dependent on the adaptor proteins TRAM and TRIF, leading
    to the activation of the nuclear transcription factor IRF3 and transcription of
    IFN-β and interferon inducible genes.

There is some disparity in the data regarding the relative roles of TLR2
and TLR4 in the recognition of *A. fumigatus* by macrophages during dormancy and
germination. Different groups show a major role for either TLR2 alone (Mambula
et al., 2002), TLR4 alone (Wang et al., 2001; Netea et al., 2003), or both receptors

together (Meier et al., 2003; Gersuk et al., 2006). In contrast, Dubordeau et al. (2006) recently demonstrated that TLR2 and TLR4 are not involved in the cytokine response to *A. fumigatus* infection, nor do they contribute to resistance against infection in knockout mice. According to these authors, the main pathway involved in the cytokine response of alveolar macrophages to *A. fumigatus* infection is via ERK signaling.

Dectin-1 Receptor

Dectin-1 is a transmembrane C-type lectin receptor involved in the innate immune response to fungal pathogens. It is widely expressed and present on monocytes, macrophages, DCs, neutrophils, eosinophils, and a subset of T cells. Dectin-1 binds fungal cell wall β-glucans, activating *src*-family tyrosine kinase signaling to unknown downstream effectors. Dectin-1 activation mediates various cellular functions, including fungal binding, uptake, and oxidative killing and induction of cytokines and chemokine production (Brown, 2006). Recent evidence suggests that upon germination, *A. fumigatus* conidia shed their coating of hydrophobins to reveal underlying layers of β-glucan. The macrophage dectin-1 receptor binds to the exposed conidial β-glucan, enabling their subsequent phagocytocis and elimination (Hohl et al., 2005; Steele et al., 2005; Gersuk et al., 2006). This provides a mechanism by which pulmonary macrophages differentiate between nonactivating resting fungal spores, which are constantly inhaled, and activating potentially invasive (maturing) forms of the organism.

Recent studies suggest that there is cross talk between the dectin-1 and TLR pathways: Dectin-1 cooperates with TLR2 in mounting a strong phagocytic and inflammatory cytokine response to germinating *A. fumigatus* conidia (Luther et al., 2006; Gersuk et al., 2006).

Phagocytosis

It is surprising how little is known about the cellular mechanisms of internalization and killing of *A. fumigatus* by alveolar macrophages. Recent work indicates that following recognition by the macrophage dectin-1 receptor (see above) *A. fumigatus* conidia undergo F-actin-dependent phagosomal internalization and subsequent fusion with endosomes to form an acidic phagolysosome (Ibrahim-Granet et al., 2003). During the first few hours, conidia undergo swelling, an early stage in germination. Swelling activates a complex phagolysosomal-killing mechanism involving acidic proteases and NADPH-oxidase bleaching. This gradually (24–48h) kills and disintegrates the ingested spore, deforming it into a sickle-shaped ghost. Importantly, alveolar macrophages collected from corticosteroid immune-suppressed mice, while showing no reduction in number, were significantly inhibited in their ability to produce reactive oxygen intermediates and to kill conidia (Philippe et al., 2003).

Dendritic Cells (DCs) and the Th1/Th2 Response

DCs are key members of the innate immune system. They function to link the innate immune and adaptive immune responses. Immature DCs act as sentinels that take up antigens such as fragments of invading fungi. Thus activated, the DCs mature and migrate to regional lymph nodes. There they activate naïve antigen-specific T cells, setting into motion the adaptive immune response, which can be a predominantly Th1 (cell-mediated) or Th2 (antibody mediated) (Buentke & Scheynius, 2003). It is generally accepted that Th1 responses confer protective immunity against infections with *Aspergillus* conidia, whereas a switch towards a Th2 response contributes to the development of IA (Bozza et al., 2002).

DCs recognize and react differently to *A. fumigatus* conidia and hyphae: phagocytosed conidia are bound to lectin-like receptors (murine MR, DEC-205, and human DC-SIGN receptors) on the surface of DCs and initiate an IL-12 Th1 mediated response (Bozza et al., 2002; Serrano-Gomez et al., 2005; Gafa et al., 2006). In contrast, hyphae bind CR3/FcγR II and III DC receptors, and initiate an IL-4/IL-10 mediated Th2 response (Bozza et al., 2002). The strong Th1 response initiated by conidial-activated DCs could be useful as a form of immunotherapy against IA (Bellochio et al., 2005).

Future Research

Many important questions remain unanswered regarding the host immune response to *A. fumigatus* infection. Are there additional host receptors that recognize *A. fumigatus* during infection? Are they found only on cells of the innate immune system? What are the main signaling pathways activated? What is the degree of cross talk between them? What are the molecular details of antigen presentation to T cells and how do they react? What are the relative contributions of the Th1 and Th2 responses to combating infection? Research at the interface between immunology and pathogen biology will no doubt help answer these questions in the coming years.

Identification of Novel Virulence Factors

The study of *A. fumigatus* virulence has recently passed the threshold of genomics. Based on our new knowledge of its genome, what approaches should be used in the coming years to identify novel virulence factors?

In silico analysis

With the availability of the *A. fumigatus* genome, researchers can now compile comprehensive lists containing their particular 'genes of interest' categorized by sequence motifs, conserved domains etc (Tables 8.1–8.4). Genes can be grouped by putative function, localization or pathway. Comparisons between the sequenced genomes of *A. fumigatus, A. nidulans, A. oryzae,* and *A. niger* can be used to identify conserved hypothetical genes and genes unique to *A. fumigatus* (Nierman et al., 2005; Galagan et al., 2005; Lafon et al., 2006). The 'molecular toolbox' for work in *A. fumigatus* has improved sustantially over the last few years (Brakhage & Langfelder 2002; Archer & Dyer 2004; Krappmann, 2006). Deletion mutants can be generated fairly rapidly with novel transposon and PCR-based methods to rapidly construct knockout vectors (Jadoun et al., 2004; Yu et al., 2004) and with the new KU70 *A. fumigatus* mutant strains in which integration occurs overwhelmingly at the target site (Krappmann et al., 2006).

Examples for such 'category-wide' gene deletions in Aspergillus include the in silico identification and deletion of six of nine G-protein coupled receptors in *A. nidulans* (Han et al., 2004).

Microarrays

A. fumigatus microarrays have now become both reliable and widely available: (http://pfgrc.tigr.org/slide_html/array_descriptions/A_fumigatus_2.shtml).

To date they have been used to analyze the transcriptional response of *A. fumigatus* to the following perturbations: the antifungal drug voriconazole (da Silva et al., 2006), elevated temperatures (Nierman et al., 2005), H_2O_2, neutrophil and macrophage contact, low pH or salt concentrations, anaerobic conditions and during germination and development (G. S. May, personal communication, 2005). Future transcriptome studies could include comparisons between the response of wild-type and mutant *A. fumigatus* strains, as has been described for the *A. fumigatus stuA* developmental mutant (Sheppard et al., 2005).

Proteomic Analysis

Conditions for the reliable proteomic analysis of *A. fumigatus* were recently described (Kniemeyer et al., 2006). Proteome analysis has identified a large set of membrane bound *A. fumigatus* GPI-anchored proteins (Bruneau et al., 2001) over 50 hyphal cytosolic proteins (Carberry et al., 2006) and over 30 specific proteins expressed during growth on glucose and ethanol (Kniemeyer et al., 2006). Studies are underway to analyze the *A. fumigatus* proteome following temperature shift, H_2O_2 addition and neutrophil contact (A. A. Brakhage, personal communication, 2005). Future proteome studies should include the identification of all *A. fumigatus* cell-wall proteins in the presence and absence of cell wall destabilizing agents, secreted proteins ('Secretome'), conidial and hyphal-specific proteins, etc.

Protein Localization and Protein/protein Interactions

Genome-wide protein tagging has been a very effective tool in the yeast *S. cerevisiae*. It has been used to group genes by the localization of their gene product and to determining protein/protein interactions (Gavin et al., 2002). A rapid method to tag *A. nidulans* proteins with red, green and blue fluorescent markers or haemagglutinin epitopes was recently described and should be transferable to *A. fumigatus* (Toews et al., 2004).

Random Mutagenesis, Mutant Libraries, and High-Throughput Screening of Mutants

Large genome-saturating mutant *A. fumigatus* libraries can be extremely effective tools to identify virulence genes, especially when coupled to high-throughput screens. Remarkably, to date, only one collection of randomly tagged *A. fumigatus* mutants has been described (Brown et al., 2000). There are several novel methods to construct such a mutant library, including *in vivo* transposon mutagenesis (Firon et al., 2003) and *Agrobacterium tumefaciens*-mediated transformation (Sugui et al., 2005).

A. fumigatus mutant libraries and collections of deletion mutants can be tested in high throughput screens to identify genes involved in specific processes. For example, screens that identify sensitivity or resistance towards antifungal drugs, oxidants, cell wall damaging agents and even phagocytosis by soil amoebae will enhance our knowledge of *A. fumigatus* pathogenesis and drug resistance (Krappmann, 2006). Alternative, faster *in vivo* models, using infected moth larvae and immunodeficient fruit flies have also been developed recently and could be amenable to high-throughput screening (Renwick et al., 2006; Lionakis et al., 2005).

Data Collection and Analysis

A centralized and standardized database containing the phenotypic, transcriptional, proteomic, and virulence data should be constructed and made publicly available (Garwood et al., 2004). An elegant example is the PACLIMS data management software used to collect and integrate the *Magnaporthe grisea* data (Weld et al., 2006). Computerized integration of the entire data sets can reveal unexpected connections between genes and pathways (Tanay et al., 2005).

As the number of known *A. fumigatus* virulence factors increases, it will be necessary to perform statistical analyses to estimate the relative contribution of each factor to the overall virulence phenotype. Such an analysis, using multivariate linear regression was recently used to estimate the relative contributions of individual *Cryptococcus neoformans* and *Bacillus anthracis* virulence factors (McLelland et al., 2005).

Virulence Factors and Novel Therapies?

Current antifungals target a relatively small number of essential fungal-specific gene functions, such as inhibition of ergosterol biosynthesis, cell wall biosynthesis, etc. Developments have been largely limited to improving antifungals through chemical modifications and developing synergistic drug combinations (reviewed in Kontoyiannis & Lewis, 2004).

Will the identification of novel *A. fumigatus* virulence factors and a more comprehensive understanding of the disease process ultimately lead to the development of novel and more effective therapies? Thus far, drugs targeting fungal virulence factors have not been developed. Neither have genomics-based strategies for discovery of novel antifungal drug targets produced any agents in clinical development (Odds, 2005). However, this does not mean that such novel drugs will not emerge in the future. There are several recent examples for small molecule inhibitors of bacterial virulence factors (Lee et al., 2003): high-throughput screening identified virstatin, a small molecule that inhibits a transcriptional regulator in *Vibrio cholerae*, prevents the expression of cholera toxin and blocks virulence (Hung et al., 2005). Rational drug design has been used to develop peptide 'pilicides' that inhibit the formation of pili in uropathogenic *Escherichia coli* (Svensson et al., 2001). However, no 'anti-virulence' compounds have entered clinical use to date.

Perhaps new compounds should be targeted at unique and essential fungal pathways rather than against virulence factors? Elitra Pharmaceuticals (San Diego, CA.) recently performed a genomewide screen to identify essential genes conserved both in *S. cerevisiae* and *C. albicans* (Haselbeck et al., 2002). A similar but partial screen was also performed in *A. fumigatus*, leading to the identification of 20 previously uncharacterized essential genes (Firon et al., 2003). Essential fungal-specific genes can be developed as novel antifungal targets in the future.

What Should Be the Goals of *A. fumigatus* Research in the Next 5–10 Years?

The aims of *A. fumigatus* research in the coming 'post-genomic' years, relating to its ability to cause disease, are to use new high-throughput methodologies to identify novel virulence factors and gain a better understanding of the host–pathogen interaction leading to the development of novel therapies for the treatment of IA and ABPA. As described in this review, many of the tools and methods to carry this out already exist, or have been developed for other fungi such as *S. cerevisiae* and *C. albicans*. The major operational goals for research laboratories specializing in *A. fumigatus* include the transcriptomic and proteomic analysis of *A. fumigatus* and the construction of knockout strains for most or all the ~9,900 genes in the *A. fumigatus* genome. The primary phenotype of the resulting mutant strains should be characterized and select mutants tested for virulence in several *in vivo* models of infection. An online database to house the accumulating data should be constructed. Finally, avirulent and hypervirulent mutants should be channeled into a host–pathogen and drug-discovery pipeline that could eventually yield novel treatments. Together, these approaches will allow our understanding of the pathogenesis of *A. fumigatus* to advance much more rapidly and yield substantial benefits in the near future.

Acknowledgements I thank the following members in my lab for their help in reading, editing, and preparing the figure for this manuscript: E. Levdansky, J. Romano, C. Sharon, and Y. Shadchkan. I thank D..S. Askew and J. C. Rhodes for critical reading of the section describing thermotolerance. I thank G. S. May, P. Dyer, I. Mouyna, and A. Beauvais for carefully reading this manuscript and adding appropriate comments and suggestions.

References

Archer, D.B., & Dyer, P.S., (2004). *Curr. Opin. Microbiol.*, 7(5):499–504.
Balajee, S.A., Nickle, D., Varga, J., & Marr, K.A. (2006). *Eukaryot Cell*, 5(10):1705–1712.
Beauvais, A., Monod, M., Wyniger, J., Debeaupuis, J. P., et al. (1997a). *Infect. Immun.*, 65(8):3042–3047.
Beauvais, A., Monod, M., Debeaupuis, J. P., Diaquin, M., et al. (1997b). *J. Biol. Chem.*, 272(10):6238–6244.
Beffa, T., Staib, F., Lott Fischer, J., Lyon, P. F., et al. (1998). *Med. Mycol.*, 36(1):137–145.
Bellocchio, S., Bozza, S., Montagnoli, C., Perruccio, K., et al. (2005). *Med. Mycol.*, 43(1): S181–188.
Bernard, M., Mouyna, I., Dubreucq, G., Debeaupuis, J. P., et al. (2002). *Microbiology*, 148(9): 2819–2829.
Bhabhra, R., Miley, M.D., Mylonakis, E., Boettner, D., et al. (2004). *Infect. Immun.*, 72(8): 4731–4740.
Birch, M., Robson, G., Law, D., & Denning, D. W. (1996). *Infect. Immun.*, 64(3):751–755.
Bok, J. W. & Keller, N. P. (2004). *Eukaryot Cell*, 3(2):527–535.
Bok, J. W., Balajee, S. A., Marr, K. A., Andes, D., et al. (2005). *Eukaryot Cell*, 4(9):1574–1582.
Bozza, S., Gaziano, R., Spreca, A., Bacci, A., et al. (2002). *J. Immunol.*, 168(3):1362–1371.
Brakhage, A. A. & Langfelder, K. (2002). *Annu. Rev. Microbiol.*, 56:433–455.
Brakhage, A. A. & Liebmann, B. (2005). *Med. Mycol.*, 43(1):S75–82.
Brakhage, A. A. (2005). *Curr. Drug Targets*, 6(8):875–886.
Braude, A. I. (1986). *Infectious Diseases and Medical Microbiology*, in: Braude, A. I., Davis, C. E., Fierer, J. Eds., 2nd edn., WB Saunders Company, Philadelphia, pp. 592–597.
Brown, J. S., Aufauvre-Brown, A., Brown, J., Jennings, J. M., et al. (2000). *Mol. Microbiol.*, 36(6):1371–1380.
Brown, G. D. (2006). *Nat. Rev. Immunol.*, 6(1):33–43.
Bruneau, J. M., Magnin, T., Tagat, E., Legrand, R., et al. (2001). *Electrophoresis*, 22(13): 2812–2823.
Buentke, E. & Scheynius, A. (2003). *APMIS*, 111(7–8):789–796.
Calera, J. A., Paris, S., Monod, M., Hamilton, A.J., et al. (1997). *Infect Immun.*, 65(11):4718–4724.
Carberry, S., Neville, C. M., Kavanagh, K. A., Doyle, S., et al. (2006). *Biochem. Biophys. Res. Commun.*, 341(4):1096–1104.
Casadevall, A. & Pirofski, L. A. (2003). *Nat. Rev. Microbiol.*, 1(1):17–24.
Casadevall, A. (2005). *Fungal Genet. Biol.*, 42(2):98–106.
Chang, Y. C, Tsai, H. F, Karos, M., & Kwon-Chung, K.J. (2004). *Fungal Genet. Biol.*, 41(9):888–896.
Chauhan, N., Latge, J. P., & Calderone, R. (2006). *Nat. Rev. Microbiol.*, 4(6):435–444.

Clemons, K. V., Miller, T. K., Selitrennikoff, C. P., & Stevens, D. A. (2002). *Med. Mycol.*, 40(3):259–262.

Coyle, C. M. & Panaccione, D. G. (2005). *Appl. Environ. Microbiol.*, 71(6):3112–3118.

Cramer, R. A., Gamcsik, M. P., Brooking, R. M., Najvar, L. K., et al. (2006). *Eukaryot Cell*, 5(6):972–980.

Dasbach, E. J. (2000). *Clin. Infect. Dis.*, 31:1524–1528.

da Silva Ferreira, M. E., Malavazi, I., Savoldi, M., Brakhage, A. A., et al.(2006). *Curr. Genet.*, 50(1):32–44.

D'Enfert, C., Diaquin, M., Delit, A., Wuscher, N., et al. (1996). *Infect. Immun.* 64(10): 4401–4405.

D'Enfert, C. (2006). *Curr. Drug Targets*, 7(4):465–470.

Denning, D. W. (1998). *Clin. Infect. Dis.*, 26(4):781–803.

Denning, D. W., O'Driscoll, B. R., Hogaboam, C. M, Bowyer, P., et al. (2006). *Eur. Respir. J.* 27(3):615–626.

Du, C., Sarfati, J., Latge, J. P., & Calderone, R. (2006). *Med. Mycol.*, 44(3):211–218.

Dubourdeau, M., Athman, R., Balloy, V., Huerre, M., et al. (2006). *J. Immunol.*, 177(6):3994–4001.

Dyer, P. S. & Paoletti, M. (2005). *Med. Mycol.*, 43(Suppl. 1):S7–S14.

Eisendle, M., Oberegger, H., Zadra, I., & Haas, H. (2003). *Mol. Microbiol.*, 49:359–375.

Ferreira, M. E., Colombo, A. L., Paulsen, I., Ren, Q., et al. (2005). *Med. Mycol.*, 43(Suppl. 1): S313–S319.

Firon, A., Villalba, F., Beffa, R., & D'Enfert, C. (2003). *Eukaryot Cell*, 2(2):247–255.

Fluckiger, S., Mittl, P. R., Scapozza, L., & Fijten, H. (2002). *J. Immunol.*, 168(3):1267–1272.

Fortwendel, J. R., Panepinto, J. C., Seitz, A. E., Askew, D. S., et al. (2004). *Fungal Genet. Biol.*, 41(2):129–139.

Fuchs, B. B. & Mylonakis, E. (2006). *Curr. Opin. Microbiol.*, 9(4):346–351.

Gafa, V., Lande, R., Gagliardi, M. C., & Severa, M. (2006). *Infect. Immun.*, 74(3):1480–1489.

Galagan, J. E., Calvo, S. E., Cuomo, C., Ma, L. J., et al. (2005). *Nature*, 438(7071):1105–1115.

Gardiner, D. M. & Howlett, B. J. (2005). *FEMS Microbiol. Lett.*, 248(2):241–248.

Ghannoum, M. A. (2000). *Clin. Microbiol. Rev.*, 13(1):122–143.

Garwood, K., McLaughlin, T., Garwood, C., Joens, S., et al. (2004). *BMC Genomics*, 5(1):68.

Gavin, A. C., Bosche, M., Krause, R., Grandi, P., et al. (2002). *Nature*, 415(6868):141–147.

Gersuk, G. M., Underhill, D. M., Zhu, L., & Marr, K. A. (2006). *J. Immunol.*, 176(6): 3717–3724.

Gomez, B. L. & Nosanchuk, J. D. (2003). *Curr. Opin. Infect. Dis.*, 16(2):91–96.

Grant, C. M. (2001). *Mol Microbiol.*, 39(3):533–541.

Haas, H. (2003). *Appl. Microbiol. Biotechnol.*, 62:316–330

Han, K. H., Seo, J. A., & Yu, J. H. (2004). *Mol. Microbiol.*, 51(5):1333–1345.

Haselbeck, R., Wall, D., Jiang, B., & Ketela, T. (2002). *Curr. Pharm. Des.*, 8:1155–1172.

Hissen, A. H., Chow, J. M., Pinto, L. J., & Moore, M. M. (2004). *Infect. Immun.*, 72:1402–1408.

Hissen, A. H., Wan, A. N., Warwas, M. L., Pinto, L. J., et al. (2005). *Infect Immun.*, 73(9): 5493–5503.

Hohl, T. M., Van Epps, H. L., Rivera, A., Morgan, L. A., et al. (2005). *PLoS Pathog.*, 1(3):e30.

Holdom, M. D., Lechenne, B., Hay, R. J., Hamilton, A. J., et al. (2000). *J. Clin. Microbiol.*, 38(2):558–562.

Hospenthal, D. R., Kwon-Chung, K. J., & Bennett, J. E. (1998). *Med. Mycol.*, 36:165–168

Hung, D. T., Shakhnovich, E. A., Pierson, E., & Mekalanos, J. J. (2005). *Science*, 310(5748):670–674.

Ibrahim-Granet, O., Philippe, B., Boleti, H., Boisvieux-Ulrich, E., et al. (2003). *Infect. Immun.*, 71(2):891–903.

Jadoun, J., Shadkchan, Y., & Osherov, N. (2004). *Curr. Genet.*, 45(4):235–241.

Jahn, B., Koch, A., Schmidt, A., Wanner, G., et al. (1997). *Infect. Immun.*, 65(12):5110–5117.

Kamei, K. & Watanabe, A. (2005). *Med. Mycol.*, 43(Suppl. 1):S95–S99.

Kamran, M., Calcagno, A. M., Findon, H., Bignell, E., et al. (2004). *Eukaryot Cell*, 3(2):546–552.

Kauffman, H. F., Tomee, J. F., van de Riet, M. A., Timmerman, A. J. et al. (2000). *J. Allergy Clin. Immunol.*, 105(6) Pt 1:1185–1193.

Kniemeyer, O., Lessing, F., Scheibner, O., Hertweck, C., et al. (2006). *Curr. Genet.*, 49(3):178–189.

Kogan, T. V., Jadoun, J., Mittelman, L., Hirschberg, K., & Osherov, N. (2004). *J. Infect. Dis.*, 189(11):1965–1973.

Kontoyiannis, D. P. & Bodey, G. P. (2002). *Eur. J. Clin. Microbiol. Infect. Dis.*, 21(3):161–172.

Kontoyiannis, D. P. & Lewis, R. E. (2004). *Br. J. Haematol.*, 126(2):165–175.

Krappmann, S. (2006). *Trends Microbiol.*, 14(8):356–364.

Krappmann, S., Sasse, C., & Braus, G. H. (2006). *Eukaryot Cell*, 5(1):212–215.

Kupfahl, C., Heinekamp, T., Geginat, G., Ruppert, T., et al. (2006). *Mol. Microbiol.*, 62(1):292–302

Kurup, V. P. (2005). *Med. Mycol.*, 43(Suppl. 1):S189–S196.

Lafon, A., Han, K. H., Seo, J. A., Yu, J. H., et al. (2006). *Fungal Genet. Biol.*, 43(7):490–502.

Langfelder, K., Philippe, B., Jahn, B., Latge, J. P., et al. (2001). *Infect. Immun.*, 69(10):6411–6418.

Langfelder, K., Gattung, S., & Brakhage, A. A. (2002). *Curr. Genet.*, 41:268–274

Latge, J. P. (1999). *Clin. Microbiol. Rev.*, 12(2):310–350.

Latge, J. P., Mouyna, I., Tekaia, F., Beauvais, A., et al. (2005). *Med. Mycol.*, 43(1):S15–22.

Lee, Y. M., Almqvist, F., & Hultgren, S. J. (2003). *Curr. Opin. Pharmacol.* 3(5): 513–519.

Lewis, R. E., Wiederhold, N. P., Chi, J., Han, X.Y., et al. (2005). *Infect. Immun.*, 73(1):635–637.

Liebmann, B., Gattung, S., Jahn, B., & Brakhage, A. A. (2003). *Mol. Genet. Genomics*, 269(3): 420–435.

Liebmann, B., Muller, M., Braun, A., & Brakhage, A. A. (2004). *Infect. Immun.*, 72(9): 5193–5203.

Linder, M. B., Szilvay, G. R., Nakari-Setala, T., & Penttila, M. E. (2005). *FEMS Microbiol. Rev.*, 29(5):877–896.

Lionakis, M. S., Lewis, R. E., May, G. S., Wiederhold, N. P., et al. (2005). *J. Infect. Dis.*, 191(7): 1188–1195.

Lopez-Ribot, J. L., Casanova, M., Murgui, A., & Martinez, J. P. (2004). *FEMS Immunol. Med. Microbiol.*, 41(3):187–196.

Luther, K., Torosantucci, A., Brakhage, A. A, Heesemann, J., et al. (2006). *Cell. Microbiol.* 2006 Aug 31; [Epub ahead of print]

Maguire, M. E. (2006). *Front. Biosci.*, 11:3149–3163.

Mambula, S. S., Sau, K., Henneke, P., Golenbock, D. T., et al. (2002). *J. Biol. Chem.*, 277(42): 39320–39326.

Maubon, D., Park, S., Tanguy, M., Huerre, M., et al. (2006). *Fungal. Genet. Biol.*, 43(5):366–375.

May, G. S., Xue, T., Kontoyiannis, D.P., & Gustin, M. C. (2005). *Med. Mycol.*, 43(Suppl. 1): S83–86.

McClelland, E. E., Bernhardt, P., & Casadevall, A. (2005). *PLoS Pathog.*, 1(4):e40.

Meier, A., Kirschning, C. J., Nikolaus, T., Wagner, H., et al. (2003). *Cell. Microbiol.*, 5(8): 561–570.

Monod, M., Capoccia, S., Lechenne, B., Zaugg, C., et al. (2002). *Int. J. Med. Microbiol.*, 292(5–6): 405–419.

Moss, R. B. (2005). *Med. Mycol.*, 43(Suppl. 1):S203–S206.

Mouyna, I., Henry, C., Doering, T. L., & Latge, J. P. (2004). *FEMS Microbiol. Lett.*, 237(2): 317–324.

Mullins, J., Harvey, R., & Seaton, A. (1976). *Clin. Allergy*, 6:209–217.

Nascimento, A.M., Goldman, G.H., Park, S., Marras, S.A., et al. (2003). *Antimicrob Agents Chemother.*, 47(5):1719–1726.

Nemecek, J. C., Wuthrich, M., & Klein, B. S. (2006). *Science*, 312(5773):583–588.

Netea, M.G, Warris, A., Van der Meer, J. W., Fenton, M. J., et al. (2003). *J. Infect. Dis.*, 188(2):320–326.

Nierman, W. C., Pain, A., Anderson, M. J., Wortman, J. R., et al. (2005). *Nature*, 438(7071): 1151–1156.

Odds, F. C. (2005). *Rev. Iberoam. Micol.*, 22(4):229–237.

Panepinto, J. C., Oliver, B. G., Fortwendel, J. R., Smith, D. L., et al. (2003). *Infect. Immun.*, 71(5): 2819–2826.

Paoletti, M., Rydholm, C., Schwier, E.U., Anderson, M.J., et al. (2005). *Curr Biol.*, 15(13): 1242–1248.

Paris, S., Debeaupuis, J. P., Crameri, R., Carey, M., et al. (2003). *Appl. Environ. Microbiol.*, 69(3):1581–1588.

Paris, S., Wysong, D., Debeaupuis, J. P., Shibuya, K., et al. (2003). *Infect. Immun.*, 71(6): 3551–3562.

Philippe, B., Ibrahim-Granet, O., Prevost, M. C., Gougerot-Pocidalo, M. A., et al. (2003). *Infect Immun.*, 71(6):3034–3042.

Pott, G. B., Miller, T. K., Bartlett, J. A., Palas, J. S., et al. (2000). *Fungal Genet. Biol.*, 31(1): 55–67.

Pringle, A., Baker, D. M., Platt, J. L., Wares, J. P., et al. (2005). *Evolution Int. J. Org. Evolution*, 59(9):1886–1899.

Reeves, E. P., Reiber K., Neville, C., Scheibner, O., et al. (2006). *FEBS J.*, 273(13):3038–3053.

Reiber, K., Reeves, E. P., Neville, C. M., Winkler, R., et al. (2005). *FEMS Microbiol. Lett.*, 248(1):83–91.

Rementeria, A., Lopez-Molina, N., Ludwig, A., Vivanco, A. B., et al. (2005). *Rev. Iberoam. Micol.*, 22(1):1–23.

Renwick, J., Daly, P., Reeves, E. P., & Kavanagh, K. (2006). *Mycopathologia*, 161(6):377–384.

Reyes, G., Romans, A., Nguyen, C. K., & G. S., May. (2006). *Eukaryot Cell*, 5(11) in press.

Rivera, A., Hohl, T., & Pamer, E. G. (2006). *Biol. Blood Marrow Transplant*, 12(1) (Suppl. 1):47–49.

Robson, G. D., Huang, J., Wortman, J., & Archer, D. B. (2005). *Med. Mycol.*, 43(Suppl. 1): S41–S47.

Roeder, A., Kirschning, C. J., Rupec, R. A., Schaller, M., et al. (2004). *Med. Mycol.*, 42(6):485–498.

Romano, J., Nimrod, G., Ben-Tal, N., Shadkchan, Y., et al. (2006). *Microbiology*,152 (Pt 7):1919–1928.

Ryckeboer, J., Mergaert, J., Coosemans, J., Deprins, K., et al. (2003). *J. Appl. Microbiol.*, 94(1): 127–137.

Rydholm, C., Szakacs, G., & Lutzoni, F. (2006). *Eukaryot Cell*, 5(4):650–657.

Sarfati, J., Monod, M., Recco, P., Sulahian, A., et al. (2006). *Diagn. Microbiol. Infect. Dis.*, 55(4):279–291.

Serrano-Gomez, D., Leal, J. A., & Corbi, A. L. (2005). *Immunobiology*, 210(2–4):175–183.

Shelton, B. G., Kirkland, K. H., Flanders, W. D., & Morris, G. K. (2002). *Appl. Environ. Microbiol.*, 68(4):1743–1753.

Shen, D. K., Noodeh, A. D., Kazemi, A., Grillot, R., et al. (2004). *FEMS Microbiol. Lett.*, 239(1): 87–93.

Sheppard, D. C., Doedt, T., Chiang, L. Y., Kim, H.S., et al. (2005). *Mol. Biol. Cell.*, 16(12): 5866–8679.

Schrettl, M., Bignell, E., Kragl, C., Joechl, C., et al. (2004). *J. Exp. Med.*, 200(9):1213–1219.

Slaven, J. W., Anderson, M. J., Sanglard, D., Dixon, G. K., et al. (2002). *Fungal Genet. Biol.*, 36: 199–206.

Smith, J. M., Davies, J. E., & Holden, D. W. (1993). *Mol. Microbiol.*, 9(5):1071–1077.

Steele, C., Rapaka, R. R., Metz, A., Pop, S. M., et al. (2005). *PLoS Pathog.*, 1(4):e42.

Steinbach, W. J., & Stevens, D. A. (2003). *Clin. Infect. Dis.* 37(Suppl. 3):S157–S187.

Sugui, J. A., Chang, Y. C., & Kwon-Chung, K. J. (2005). *Appl. Environ. Microbiol.*, 71(4): 1798–1802.

Svensson, A. A., Larsson, H., Emtenas, M., Hedenstrom, T., et al. (2001). *Chembiochem*, 12: 915–918.

Tanay, A., Steinfeld, I., Kupiec, M., & Shamir, R. (2005). *Mol. Syst. Biol.*, 1:2005.0002.

Tekaia, F. & Latge, J. P. (2005). *Curr. Opin. Microbiol.*, 8(4):385–392.

Thau, N., Monod, M., Crestani, B., Rolland, C., et al. (1994). *Infect. Immun.*, 62:4380–4388.

Thom, C. & K. B. Raper. (1945). *A Manual of the Aspergilli*, Waverly Press, Baltimore, MD.

Tillie-Leblond, I. & Tonnel, A. B. (2005). *Allergy*, 60(8):1004–1013.

Tobin, M. B., Peery, R. B., & Skatrud, P. L. (1997). *Gene*, 200:11–23.

Toews, M. W., Warmbold, J., Konzack, S., Rischitor, P., et al. (2004) *Curr. Genet.* 45(6): 383–389.

Tomee, J. F., Wierenga, A. T., Hiemstra, P. S., & Kauffman, H. K. (1997). *J. Infect. Dis.*, 176(1): 300–303.

Tomee, J. F. & Kauffman, H. F. (2000). *Clin. Exp. Allergy*, 30(4):476–484.

Tsai, H.F., Chang, Y.C., Washburn, R.G., Wheeler, M.H., & Kwon-Chung, K.J. (1998). *J. Bacteriol.*, 180(12):3031–3038.

Tsai, H.F., Wheeler, M.H., Chang, Y.C., & Kwon-Chung, K.J. (1999). *J. Bacteriol.*, 181(20): 6469–6477.

Tsai, H. F., Fujii, I., Watanabe, A., Wheeler, M. H., et al. (2001). *J. Biol. Chem.*, 276(31): 29292–29298.

Uematsu, S. & Akira, S. (2006). *J. Mol. Med.* [Epub ahead of print]

Verstrepen, K. J. & Klis, F. M. (2006). *Mol. Microbiol.*, 60(1):5–15.

Vicentefranqueira, R., Moreno, M. A., Leal, F., & Calera, J. A. (2005). *Eukaryot Cell*, 4(5): 837–848.

Walsh, T. J., Roilides, E., Cortez, K., Kottilil, S., et al. (2005). *Med. Mycol.*, 43(Suppl. 1): S165–S172.

Wang, J. E., Warris, A., Ellingsen, E. A., Jorgensen, P. F., et al. (2001). *Infect. Immun.*, 69(4): 2402–2406.

Weld, R. J., Plummer, K. M., Carpenter, M. A., & Ridgway, H. J. (2006). *Cell Res.*, 16(1): 31–44.

Wendland, J. (2001). *Fungal Genet. Biol.*, 34(2):63–82.

Wheeler, R. T., Kupiec, M., Magnelli, P., Abeijon, C., et al. (2003). *Proc. Natl. Acad. Sci. USA*, 100(5):2766–2770.

Wortman, J.R., Fedorova, N., Crabtree, J., Joardar, V., et al. (2006). *Med. Mycol.*, 44(Suppl. 1):3–7.

Xue, T., Nguyen, C. K., Romans, A., & May, G. S. (2004). *Eukaryot Cell*, 3(2):557–560.

Yu, J. H., Hamari, Z., Han, K. H., Seo, J. A., et al. (2004). *Fungal Genet. Biol.*, 41(11): 973–981.

Yu, J. H. & Keller, N. (2005). *Annu. Rev. Phytopathol.*, 43:437–458.

Youngchim, S., Morris-Jones, R., Hay, R. J., & Hamilton, A. J. (2004). *J. Med. Microbiol.*, 53(3): 175–181.

Zhao, W., Panepinto, J. C., Fortwendel, J. R., Fox. L., et al. (2006). *Infect. Immun.*, 74(8): 4865–4874.

Chapter 9
The Biology of the Thermally Dimorphic
Fungal Pathogen *Penicillium marneffei*

David Cánovas and Alex Andrianopoulos

Introduction

Fungal Pathogens and Dimorphism

Fungal pathogens are a significant public health problem, being responsible for a large and growing percentage of deaths from hospital-acquired infections (Asmundsdottir et al., 2002; Chakrabarti & Shivaprakash, 2005; McNeil, et al., 2001). A group of these pathogens are dimorphic, alternating between two different growth forms (Gow, 1995). The yeast form is unicellular, generally spherical, ellipsoid, or cylindrical in shape, and uninucleate due to coupled mitosis and cytokinesis (Figure 9.1). The filamentous form consists of highly polarized fibrillar cells, which elongate by apical growth, placing crosswalls called septa at regular intervals, exhibit incomplete cell separation, have the capacity to generate new grow foci by branching and are generally multinucleate.

The ability to alternate between the yeast and filamentous growth forms is a tightly regulated process known as dimorphic switching. The different growth forms represent cell types with different physiological properties, which are adapted to their respective environments. Interestingly, for most pathogenic dimorphic fungi only one growth form predominates during infection. Therefore, dimorphic switching is an intrinsic property of pathogenicity (Berman & Sudbery, 2002; Andrianopoulos, 2002; Lengeler et al., 2000).

A Distinct Penicillium Species

The *Penicillium* genus is comprised of more than 200 fungal species playing important roles in biotechnology, health, and the environment. Within the whole *Penicillium* genus and closely related genera, *P. marneffei* is the only species in which dimorphism has been described and the most prominent *Penicillium* species found in human infections.

K. Kavanagh (ed.), *New Insights in Medical Mycology.*
© Springer 2007

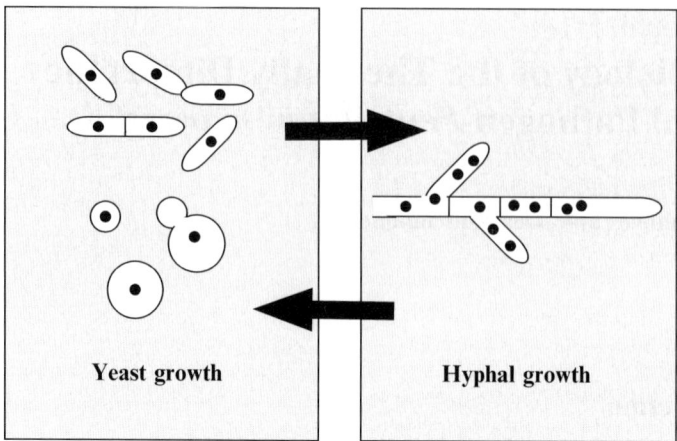

Figure 9.1 Diagrammatic representation of the two major fungal growth forms; unicellular yeast which are spherical, ellipsoidal, or cylindrical in shape and divide by budding or fission and multicellular hyphae consisting of cylindrical cells joined end-to-end which grow by apical extension and septation. Dimorphic fungi have the capacity to alternate between these two growth forms in response to extrinsic or intrinsic cues. Black filled circles in cells represent nuclei

Life Cycle

P. marneffei provides a novel and highly amenable system with which to address important aspects of fungal dimorphism and pathogenicity. *P. marneffei* is a thermally dimorphic human pathogenic fungus. The pathogenic form, expressed at 37°C, is an oval-shaped unicellular yeast, which divides by fission (Figure 9.2). As a saprophyte at 25°C, *P. marneffei* is filamentous in form and is morphologically similar to other *Penicillium* species (Andrianopoulos, 2002). Unlike some of the other pathogenic fungi, the growth form of *P. marneffei* is simple and easily controlled (Garrison & Boyd, 1973; McGinnis, 1994). In addition, the fact that morphologically extensive programmes for asexual development (conidiation) and dimorphic switching coexist in *P. marneffei* makes it an excellent system in which to study the relationship between these two developmental programmes.

The life cycle of *P. marneffei* can be divided into three stages: vegetative hyphal growth at 25°C, asexual development at 25°C, and yeast growth at 37°C (Figure 9.3). A conidium under the appropriate nutritional conditions germinates, initially expanding by isotropic growth until it reaches a certain size and then switching to polarized growth to produce a germ tube. This germ tube continues elongating by polarized growth at the apical tip. Cellular compartments are established behind the tip by septation. When grown *in vitro*, these first steps are common to both 25°C and 37°C. At 25°C, actively growing cellular compartments contain multiple nuclei, showing that nuclear and cellular division are uncoupled, while older subapical cells are usually uninucleate. Unlike the apical cells, subapical cells are capable of repolarizing to produce branches that continue growing by polarized elongation.

Asexual   Hyphae   Yeast
Development

Figure 9.2 Dimorphic switching in *P. marneffei*. Differential interference contrast microscopic images of the vegetative hyphal growth form evident at 25°C (centre panel), the vegetative pathogenic yeast growth form present at 37°C and in the host (right panel) and the asexual development structure (conidiophore, left panel) made up of multiple differentiated cell types which culminate in the production of the infectious conidia (arrowhead and inset)

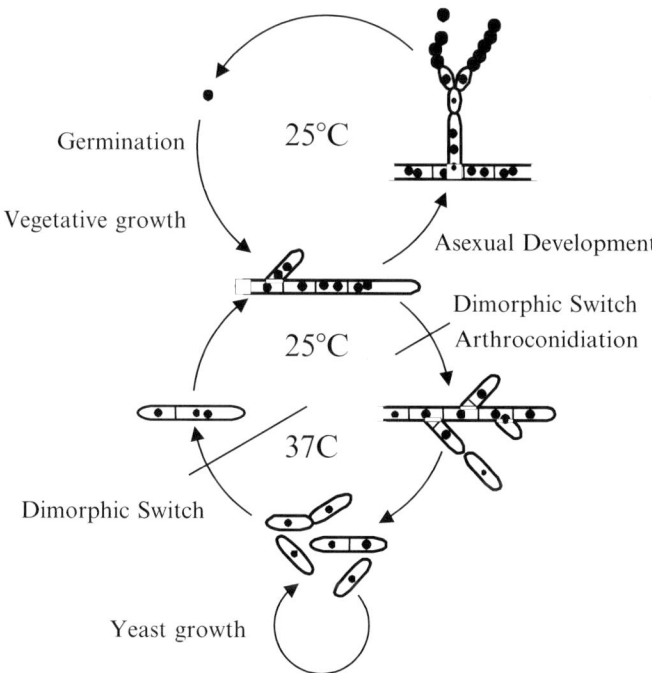

Figure 9.3 Diagrammatic representation of the *P. marneffei* life cycle. At 25°C, in the saprophytic growth phase, *P. marneffei* grows as multinucleate, septate, branched hyphae. These hyphae produce conidia, the infectious agent, from specialized multicellular structures termed conidiophores. Conidia can germinate under appropriate conditions to produce the vegetative hyphal growth form (top cycle). When switched to 37°C, *P. marneffei* undergoes a developmental process termed arthroconidiation. Cellular and nuclear division become coupled, double septa are laid down and hyphae fragment at these septation sites to liberate arthroconidia (centre cycle). Arthroconidia grow to become uninucleate yeast cells which consequently divide by fission (bottom cycle). The yeast cells are the pathogenic form. When shifted to 25°C, yeast cells will become highly polarized and produce the vegetative hyphal growth form

When *P. marneffei* is grown at 37°C *in vitro*, morphogenesis starts 48 h after germination with the onset of extensive hyphal branching and the coupling of nuclear and cellular division and gives rise to highly branched uninucleate hyphae. The cellular compartments of these branched hyphae are separated by double septa. Eventually the material between the double septa is degraded allowing for the complete separation of cells. These cells are called arthroconidia and the process is known as arthroconidiation. Arthroconidia grow apically and divide by fission to produce yeast cells, with a similar morphology to those of *Schizosaccharomyces pombe*. These yeast cells also divide by fission as long as the temperature remains at 37°C. Some strains of *P. marneffei* produce mixtures of yeast cells and arthroconidiating hyphae depending on the growth conditions. When hyphal cells growing at 25°C are switched to 37°C, they also undergo arthroconidiation. The hyphal to yeast dimorphic transition process is reversible such that decreasing the temperature from 37°C to 25°C leads to yeast cells becoming highly polarized to form hyphal cells and the uncoupling of nuclear and cellular division.

The morphology of *P. marneffei* yeast cells grown *in vitro* differs from that found in clinical specimens or when *P. marneffei* is growing inside macrophages in culture. In this case the yeast cells are substantially shorter in length and oval in shape. The search for *in vitro* culture conditions which can mimic the growth form evident in the host found that medium composed of 1% peptone produced the most morphologically similar yeast cells and these were produced directly from conidia without a hyphal intermediate stage (Tongchusak et al., 2004). What was also noted was that the growth rate on 1% peptone was significantly slower than on most other laboratory media.

In addition to dimorphic switching, *P. marneffei* is also capable of undergoing asexual development. Hyphal cells growing at 25°C require particular environmental cues to initiate this developmental programme. These conditions include light and an air interphase such as the growth on solid media. The programme starts after hyphal growth is fully established by the production of a multinucleate aerial stalk cell that grow away from the mycelium. In *P. marneffei*, the stalk cells are multinucleate, septate, and often branch to produce secondary stalk cells called rama. At the tip of the stalk cells, two sequential rounds of budding in which nuclear and cellular division are coupled, produces the two layers of sterigmata cells (metulae and phialides). Each stalk cell gives rise to 3–5 metulae and each metula produces 3–5 phialides by an acropetal mode of division. Phialides produce chains of asexual spores (conidia) by budding in a basipetal mode of division, such that the younger conidia displace the older ones, thereby forming a column of these cells. Mature conidia are pigmented with a pale grey–green colour (Andrianopoulos, 2002).

Pathology, Diagnosis, and Treatment

P. marneffei is an emerging fungal pathogen whose clinical prevalence has paralleled that of the increasing incidence of patients with human immunodeficiency virus (HIV).

Penicilliosis due to *P. marneffei* in Southeast Asia is the third most common opportunistic infection, after tuberculosis and cryptococcosis, in HIV patients in northern Thailand, and it is considered an AIDS-defining illness in the region. *P. marneffei* infection is endemic in Southeast Asia, with a particular focus in Thailand, southern China, Hong Kong, Vietnam, northeastern India, and Taiwan (Vanittanakom et al., 2006). Penicilliosis due to *P. marneffei* has been also reported in HIV patients residing outside of the endemic areas. The majority of these cases correspond to patients with a history of travel to the Southeastern Asia. Infection by *P. marneffei* has also been noted in other groups of immunocompromised patients, such as those having Hodgkin's disease, tuberculosis, or autoimmune disease (Lupi et al., 2005).

Infection by *P. marneffei* is presumed to occur by inhalation of conidia produced by the asexual development programme at 25°C, as has been noted for other dimorphic fungal pathogens such as *Histoplasma capsulatum* where infection occurs by inhalation of airborne conidia (Maresca & Kobayashi, 1989) or *Coccidioides immitis* due to inhalation of arthroconidial cells (Stevens, 1995). Once in the alveolar cavities of the lungs, alveolar macrophages, the primary defence response, engulf the fungal conidia. In immunocompromised individuals, *P. marneffei* is capable of colonizing and killing the macrophages, and disseminating throughout the host. Systemic penicilliosis presents symptoms which include fever, anaemia, weight loss, and skin lesions, lymphadenopathy, splenomegaly, and hepatomegaly, and it is indistinguishable from other mycoses such as histoplasmosis (Mootsikapun & Srikulbutr, 2006). Consequently, laboratory identification is required for the correct diagnosis of penicilliosis due to *P. marneffei*. A number of diagnostic tests have been developed that include culture and staining methods (Lim et al., 2006; Supparatpinyo et al., 1992b; Supparatpinyo & Sirisanthana, 1994; Chaiwun et al., 2002), serological tests (Cao et al., 1998, 1999; Pornprasert et al., 2005) and molecular techniques using polymerase chain reaction (Lindsley et al., 2001; Tsunemi et al., 2003).

Penicilliosis due to *P. marneffei* is fatal if untreated. Infections usually respond to treatment with 0.6 mg/kg/day intravenous amphotericin B for 2 weeks and 400 mg/day oral itraconazole for 10 weeks (Sirisanthana et al., 1998). Relapses have been reported 6 months after cessation of therapy (Supparatpinyo et al., 1992a), so secondary prophylaxis with oral itraconazole is highly recommended (Supparatpinyo et al., 1998).

Epidemiology and Clonality of *P. marneffei*

Epidemiology

One of the most significant issues surrounding *P. marneffei* infections is the source from which infections derive. Given that infections are thought to occur by inhalation of conidia, it is expected that source will be the hyphal form, which can undergo conidiation, growing in the environment as a saprophyte. Despite a number of extensive studies, the ecology of *P. marneffei* remains obscure. Several studies have suggested that bamboo rats may be an enzootic reservoir for *P. marneffei* (Chariyalertsak et al., 1996b; Gugnani et al., 2004) and a study using microsatellite typing revealed a genotypic correlation between *P. marneffei* is isolates from bamboo rats and humans (Fisher et al., 2005). However, the evidence for a zoonotic transmission is still lacking and is unlikely to explain all the cases of infection. Interestingly, infections by *Cryptococcus neoformans*, the causative agent of cryptococcosis, can be acquired by inhalation of either desiccated spores or yeast cells but aerosolized spores are 100 times more infectious than yeast cells. *C. neoformans* is readily found in the environment and in a mating-type dependent manner, can undergo monokaryotic fruiting to produce spores (Heitman, 2006 and references therein). Despite extensive sampling, isolation of *P. marneffei* from soil has seen limited success (Vanittanakom et al., 2006). As with *C. neoformans*, aerosolized yeast cells originated from patients or rats could be a means of infection by *P. marneffei*. Yeast cells grown in the laboratory were capable of being phago-cytosed by macrophage in culture, although it was media-dependent (Tongchusak et al., 2004). However, there is no evidence for human-to-human transmission to date. The main risk factors for infections remain associated with an HIV-positive status and an agricultural occupation. *P. marneffei* infection rates also show a seasonal variation, being more frequent during the rainy season (Chariyalertsak et al., 1996a).

Clonality and Adaptation to the Niche

An extensive study using multilocus microsatellite typing and 169 isolates representing a number of geographically distinct regions lead to the conclusion that *P. marneffei* is one of the most clonal species of fungus characterized (Fisher et al., 2005). Ecological studies performed in a number of opportunistic pathogens have revealed an apparent link between asexuality and pathogenicity. Fungi such as *C. neoformans, Candida albicans*, and *Aspergillus fumigatus*, together with *P. marneffei* show extensive levels of clonality in the environment with very low levels of recombination (Fisher et al., 2005; Heitman, 2006). The hypothesis is that clonality is a hallmark of organisms that are highly adapted to their environments. Despite this, all of these pathogens possess the genes which are required for mating and mating has been clearly demonstrated for *C. neoformans* and recently for *C. albicans* (Hull & Johnson, 1999; Lengeler et al., 2002; Paoletti et al., 2005). A recent report documented the presence of the *MAT1–1* α box and *MAT1–2* high-mobility group mating-type genes in *P. marneffei* (Woo et al., 2006). These genes are closely related to their homologues in *A. nidulans* and *A. fumigatus*. *P. marneffei* has a heterothallic arrangement of mating-type loci as is the case for *A. fumigatus*, which means that the strains require partners of the opposite mating-type to engage in sexual reproduction. In addition, previous studies have demonstrated that the *P. marneffei* STE12 homologue *stlA* can complement the sexual defects of a *steA* mutant in *A. nidulans* (Borneman et al., 2001), suggesting that some sexual genes in *P. marneffei* are indeed functional. Thus, the question is why has *P. marneffei*, an overwhelmingly clonal fungus, retained the sexual machinery. The simplest explanation is that *P. marneffei* has only recently lost the ability to mate. Alternatively, it has been suggested that limiting sexual reproduction appears to be a common virulence strategy, enabling the generation of clonal populations well adapted to host and environmental niches. As virulence is a polygenic trait, sexual reproduction might separate advantageous combinations of alleles (Heitman, 2006). Therefore, it is possible that by limiting sexual reproduction, *P. marneffei* has become restricted to the areas where it is well adapted and displays a high prevalence of infection in both human and bamboo rat populations for which it is adapted. Sexual reproduction between strains of the α mating-type is a contributor to the emergence of the hypervirulent clone of *C. gattii* on Vancouver Island in 1999 (Fraser et al., 2005; Kidd et al., 2005). A survey performed using a small number of clinical isolates has shown that *P. marneffei* isolates containing the *MAT1–1* α box mating-type gene are twice as frequent as their opposite mating-type counterparts (Woo et al., 2006). Whether this is significant remains to be tested.

Morphogenetic Programmes: Cell Types Suited for Pathogenesis

Fungal Dimorphism and the Environmental Cues

The environmental signals triggering dimorphic switching are species-specific, and the different cell types serve different purposes. For example, the ability of *C. albicans* to switch from yeast cells to the hyphal state is thought to be important for penetrating tissues and for escaping from phagocytic cells of the host innate immunity system. In addition, concomitant with the morphological changes are differences in the physiological state of the *C. albicans* cells including such things as cell wall composition. These changes can contribute significantly to the variety of host immune responses observed and the ability of *C. albicans* to escape the immune system (Whiteway & Oberholzer, 2004 and references therein).

Of the dimorphic fungi, a group which are known to respond to the simple cue of temperature, switching from the vegetative hyphal phase at ambient temperature (25°C) to the pathogenic yeast phase at body temperature (37°C). These thermally dimorphic fungi include *P. marneffei, H. capsulatum, Blastomyces dermatitidis, C. immitis, C. posadasii, Paracoccidioides brasiliensis*, and *Sporothrix schenkii*. For *P. marneffei*, the transition from hyphal to yeast grow occurs over an extended temperature range beginning at 32°C where few yeast cells are produced to 37°C where yeast cells are the predominant or exclusive growth form. Other dimorphic pathogens respond to more complex sets of signals such as *C. albicans* which switches from yeast to hyphal growth in response to a number of environmental signals such as temperature, pH, *N*-acetylglucosamine, and serum (Lengeler et al., 2000). *Ustilago maydis*, a dimorphic plant pathogen, switches from the non-pathogenic yeast form to the infective dikaryotic hyphal form after mating on the plant surface (Bolker, 2001). *Saccharomyces cerevisiae* is not capable of undergoing dimorphic switching, however diploid cells can undergo a morphogenetic programme leading to pseudohyphal growth in response to nitrogen limitation (for review see Lengeler et al., 2000).

Asexual Development

Under the appropriate environmental conditions, *P. marneffei* undergoes asexual development at 25°C resulting in the production of conidia. The molecular mechanisms which control conidiation are best understood in the monomorphic non-pathogenic fungus *A. nidulans* where a cascade of transcriptional regulators functions downstream of the inductive signals to effect the developmental programme (Clutterbuck, 1969): *brlA*, which encodes a C_2H_2 zinc finger protein (Adams et al., 1988), activates the expression of *abaA*, which encodes an ATTS/TEA protein (Andrianopoulos & Timberlake, 1994; Burglin, 1991) and in turn, AbaA activates the expression of *wetA*, which encodes a spore-specific protein (Marshall & Timberlake, 1991) and feedback regulates its own expression and that of *brlA* (Mirabito et al., 1989). In *P. marneffei*, asexual development is similarly regulated by BrlA and AbaA (Figure 9.4) (Borneman et al., 2000, 2002b). In addition, AbaA also plays an important role in yeast morphogenesis (Borneman et al., 2000).

Multiple controls impinge on the central cascade of transcriptional activators to modulate its activity (Adams et al., 1998). The *stuA* gene, encoding a member of the APSES family of transcriptional regulators, is required for the formation of sterigmata cells (metulae and phialides), but not for the production of conidia and is a regulator of *brlA* and *abaA* (Borneman et al., 2002a). The *tupA* gene, encoding a WD40 repeat protein homologous to the *S. cerevisiae TUP1* co-repressor gene, promotes filamentation and represses asexual development in a BrlA-dependent manner (Figure 9.4) (Todd et al., 2003). TupA activity is also required to repress yeast morphogenesis by an unknown mechanism (Todd et al., 2003). In addition to these transcriptional controls, morphogenesis of the cell types in the *P. marneffei* conidiophore is dependent on a Rac-like small GTPase encoded by *cflB* which controls cell polarization and separation. CflB also plays a role during hyphal, but not yeast, morphogenesis (Boyce et al., 2003).

Preceding the regulatory cascade are two Gα subunits of the heterotrimeric G protein complexes and a Ras small GTPase, all of which have multiple roles. The GasA Gasubunit plays a role in asexual development at 25°C by repressing *brlA* expression, possibly via a cAMP–PKA cascade, and thereby promoting vegetative growth (Zuber et al., 2002). It also plays a minor role in the regulation of secondary metabolism. In contrast, the GasC Gαsubunit has a significant effect on secondary metabolism and a minor effect during asexual development (Zuber et al., 2003). GasC also has a profound role in controlling conidial germination

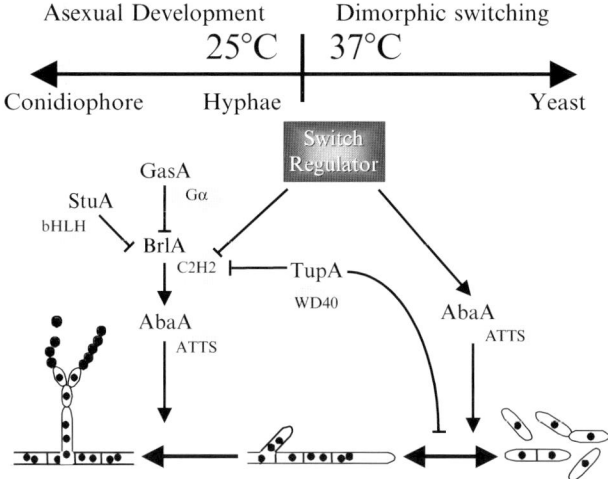

Figure 9.4 Proposed transcriptional regulatory circuit controlling asexual development and dimorphic switching in *P. marneffei*. When vegetative hyphae are exposed to the appropriate cues, asexual development is initiated. This requires removal of the negative influences of the Gα subunit GasA, the WD40-containing TupA proteins, and the predicted switch regulator which act on the *brlA* gene. BrlA (C$_2$H$_2$ transcriptional activator) activates AbaA (ATTS transcriptional activator) and, in concert, are required for differentiation of the conidiophore. When cells are grown at 37°C, the hyphal to yeast dimorphic switch is initiated by the predicted switch regulator and requires removal of the negative effects of TupA. Correct yeast cell morphogenesis requires the positive action of the AbaA

(Zuber et al., 2003). The RasA small GTPase functions at the onset of asexual development, controlling the timing and the extent of conidiation (Boyce et al., 2005). In addition, mutations in *rasA* affect conidial germination and hyphal and yeast cell morphogenesis (Boyce et al., 2005).

Yeast Morphogenesis

In *C. albicans*, dimorphic switching is controlled by a number of transcriptional regulators that integrate environmental signals. The Efg1 (APSES protein), Cph1 (Ste12 homologue), and Tec1 (ATTS protein) regulators promote filamentation while Tup1 represses filamentation (Lengeler et al., 2000; Schweizer et al., 2000). In *P. marneffei*, StuA (APSES protein) and StlA (Ste12 homologue) do not play a role during dimorphic switching (Borneman et al., 2001, 2002a). In contrast, the AbaA (ATTS protein) is required for the coupling of nuclear and cellular division during hyphal–yeast dimorphic switching, in addition to its role in asexual development, such that *abaA* deletion mutants produces multinucleate arthroconidiating hyphae and yeast cells (Figure 9.4) (Borneman et al., 2000). In *P. marneffei*, TupA is required to repress yeast morphogenesis at 25°C (Todd et al., 2003).

Morphogenetic changes during hyphal to yeast dimorphic switch require the CDC42-like small GTPase CflA and the Ras GTPases RasA in *P. marneffei*. Mutants affected in *cflA* function show significant defects in yeast cell polarization, shape, and division (Boyce et al., 2001), Mutations in the *rasA* gene of *P. marneffei* also affect yeast cell polarization, shape, and division and it is clear that this is predominantly due to RasA acting upstream of CflA (Boyce et al., 2005). However, there are subtle differences in some of the phenotypes seen in *rasA* and *cflA* mutants which suggest that RasA may also affect yeast morphogenesis via a second, CflA-independent, mechanism (Boyce et al., 2005).

Nutritional Supply and Pathogenesis: The Glyoxylate Cycle Case

Genes encoding enzymes of the glyoxylate cycle have been implicated in both fungal and bacterial pathogenesis (Munoz-Elias & McKinney, 2005; McKinney et al., 2000; Lorenz & Fink, 2001). The glyoxylate cycle involves two critical steps catalysed by the enzymes isocitrate lyase and malate synthase, which bypass the two decarboxylation steps of the TCA cycle. Therefore, the glyoxylate cycle is required for growth on gluconeogenic carbon sources, such as acetate and fatty acids, and it is activated under conditions of nutrient (glucose) deprivation. Such conditions are believed to occur inside macrophages and pose particular challenges to intracellular pathogens (Lorenz & Fink, 2002). In *C. albicans*, the isocitrate lyase gene *icl1* is upregulated during growth inside macrophages. Strains deficient for *icl1* are less virulent than wild-type strains in a mouse pathogenicity model (Lorenz & Fink, 2001). Therefore the ability to utilize fatty acids as carbon sources appears to be important for pathogenesis (Lorenz et al., 2004; Piekarska et al., 2006; Ramirez & Lorenz, 2006). Plant pathogens have also been shown to require the glyoxylate cycle for pathogenicity (Idnurm & Howlett, 2002; Solomon et al., 2004; Wang et al., 2003). However, this is not universal amongst all pathogens as it has been shown that loss of isocitrate lyase in *C. neoformans*, for example, has no effect in on pathogenicity (Rude et al., 2002). It is likely that host niche in which pathogens operates will dictate the requirement for the various nutrient assimilation pathways.

Despite the importance of the glyoxylate cycle in virulence, the molecular mechanisms which regulate the isocitrate lyase encoding genes in pathogens are poorly understood. In the non-pathogenic fungus *A. nidulans*, expression of the isocitrate lyase encoding *acuD* gene is regulated by a group of $Zn(II)_2Cys_6$ DNA-binding proteins: FacB, in response to acetate (Todd et al., 1997); FarA, in response to long-chain fatty acids; and ScfA and FarB, in response to short-chain fatty acids (Hynes et al., 2006). In *S. cerevisiae* the glyoxylate cycle genes are also activated by the $Zn(II)_2Cys_6$ DNA-binding transcriptional activators Cat8 and Sip4 in response to glucose limitation (Schuller, 2003). In addition to the acetate and fatty acid induction of the glyoxylate cycle genes, their expression is also subject to glucose-mediated repression by CreA in *A. nidulans* (Bowyer et al., 1994;

De Lucas et al., 1994) or Mig1 in *S. cerevisiae* (Schuller, 2003). In *P. marneffei, acuD* is independently regulated by acetate and temperature and a combination of both inducing conditions result in the highest level of expression, suggesting additive relationship (Figure 9.5) (Canovas & Andrianopoulos, 2006). *acuD* is induced at 37°C even in the presence of a repressing carbon source such as glucose. Regulation of *acuD* expression by temperature in *P. marneffei* is dependent on specific *cis* elements and *trans*-acting factors, as determined by reciprocal promoter exchange and heterologous expression studies using *P. marneffei* and *A. nidulans*, showing a unique evolutionary path for acetate and fatty acid regulation in this dimorphic pathogen. The $Zn(II)_2Cys_6$ DNA-binding motif transcriptional activator FacB is responsible for carbon source-dependent, but not temperature-dependent induction of *acuD* in *P. marneffei* (Canovas & Andrianopoulos, 2006). *acuD* is also independently regulated by the dimorphic switching developmental programme and part of this control is through the developmental transcriptional activator AbaA. However, deletion of *abaA* does not completely eliminate temperature-dependent induction, suggesting that *acuD* and the glyoxylate cycle are regulated by a complex network of factors in *P. marneffei* which may contribute to its pathogenicity (Canovas & Andrianopoulos, 2006). The *P. braziliensis acuD* orthologue was also shown to be differentially expressed during the dimorphic transition (Goldman et al., 2003) and is highly expressed in the pathogenic yeast growth form at 37°C (Felipe et al., 2005).

Recently it has been reported that the glyoxylate cycle genes in *C. albicans* are repressed by the physiological concentrations of glucose found in the bloodstream (Barelle et al., 2006). However, these genes are induced upon phagocytosis by macrophages or neutrophils, emphasizing the importance of carbon metabolism during pathogenesis. In *P. marneffei*, high concentrations of glucose are not enough

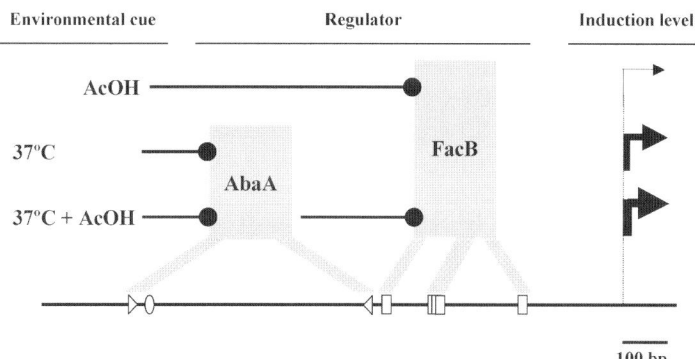

Figure 9.5 Transcriptional regulation of the glyoxylate bypass gene *acuD* encoding isocitrate lyase. The acuD promoter is shown (bottom) with the predicted DNA-binding sites depicted for the morphogenetic regulator AbaA (arrowhead), acetate regulator FacB (rectangle), and fatty acid regulators FarA/FarB (oval). Above the promoter are depictions of the AbaA and FacB regulators (centre) showing the environmental cues (left) which trigger their activity on the *acuD* promoter and the relative strength of these signals on transcriptional activation (right)

to repress the expression of *acuD* and temperature induction overrides glucose repression. After passing the primary line of defence (the macrophages), *P. marneffei* spreads throughout the body and yeast cells have been found in blood and fluids (Vanittanakom et al., 2006). Therefore, it is important to test the role of *P. marneffei acuD* during the early stages of infection and through progression to systemic infection in an appropriate pathogenic model.

Conclusions

P. marneffei is a member of a group of fungal pathogens which exhibit dimorphic switching, yet evolutionarily it is distinct from most of this group of organisms. It has a robust hyphal phase which can produce complex conidiophores, containing multiple cell types, that culminate in the production of infectious conidia at 25°C. At 37°C it produces the pathogenic yeast cell type which, unlike most other dimorphic pathogens, divides by fission. Consequently, it is a system which, when compared to the other dimorphic fungi, can provide important insights into the mechanism which underlie dimorphic switching, the ways in which dimorphic switching has evolved and the mechanisms which are central to pathogenicity. Understanding the biology of this emergent pathogen is in the early stages. Much work is required to understand the epidemiology and mode of transmission, the host–pathogen interactions and the molecular mechanisms controlling the morphogenetic programmes. The extensive set of genetic tools developed in this system make the molecular dissection of this intriguing pathogen highly amenable.

Acknowledgements The authors wish to acknowledge the current and past members of the laboratory for their contributions. The research was supported by grants from the Australian Research Council, the National Health and Medical Research Council, and the Howard Hughes Medical Institute to Alex Andrianopoulos. David Cánovas is a Marie Curie OIF fellow. Alex Andrianopoulos is a Howard Hughes Medical Institute International Scholar.

References

Adams, T. H., Boylan, M. T., & Timberlake, W. E. (1988). *Cell*, 54:353–362.
Adams, T. H., Wieser, J. K., & Yu, J. H. (1998). *Microbiol. Mol. Biol. Rev.*, 62:35–54.
Andrianopoulos, A. (2002). *Int. J. Med. Microbiol.*, 292:331–347.
Andrianopoulos, A. & Timberlake, W. E. (1994). *Mol. Cell. Biol.*, 14:2503–2515.
Asmundsdottir, L. R., Erlendsdottir, H., & Gottfredsson, M. (2002). *J. Clin. Microbiol.*, 40:3489–3492.
Barelle, C. J., Priest, C. L., Maccallum, D. M., Gow, N. A. R., Odds, F. C., & Brown, A. J. P. (2006). *Cell Microbiol.*, 8:961–971.
Berman, J. & Sudbery, P. E. (2002). *Nat. Rev. Genet.*, 3:918–930.
Bolker, M. (2001). *Microbiol.*, 147:1395–1401.
Borneman, A. R., Hynes, M. J., & Andrianopoulos, A. (2000). *Mol. Microbiol.*, 38:1034–1047.
Borneman, A. R., Hynes, M. J., & Andrianopoulos, A. (2001). *Genetocs*, 157:1003–1014.
Borneman, A. R., Hynes, M. J., & Andrianopoulos, A. (2002a). *Mol. Microbiol.*, 44:621–631.

Borneman, A. R., Hynes, M. J., & Andrianopoulos, A. (2002b). *Department of Genetics*, Melbourne, University of Melbourne.

Bowyer, P., De Lucas, J. R., & Turner, G. (1994). *Mol. Gen. Genet.*, 242:484–489.

Boyce, K. J., Hynes, M. J., & Andrianopoulos, A. (2001). *J. Bacteriol.*, 183:3447–3457.

Boyce, K. J., Hynes, M. J., & Andrianopoulos, A. (2003). *J. Cell Sci.*, 116:1249–1260.

Boyce, K. J., Hynes, M. J., & Andrianopoulos, A. (2005). *Mol. Microbiol.*, 55:1487–1501.

Burglin, T. R. (1991) *Cell*, 66:11–12.

Canovas, D. & Andrianopoulos, A. (2006). *Mol. Microbiol.*, 62:1725–1738.

Cao, L., Chan, C. M., Lee, C., Wong, S. S., & Yuen, K. Y. (1998). *Infect. Immun.*, 66:966–973.

Cao, L., Chan, K. M., Chen, D., Vanittanakom, N., Lee, C., Chan, C. M., Sirisanthana, T., Tsang, D. N., & Yuen, K. Y. (1999). *J. Clin. Microbiol.*, 37:981–986.

Chakrabarti, A. & Shivaprakash, M. R. (2005). *J. Postgrad. Med.*, 51:16–20.

Chaiwun, B., Khunamornpong, S., Sirivanichai, C., Rangdaeng, S., Supparatpinyo, K., Settakorn, J., Ya-In, C., & THORNER, P. (2002). *Mod. Pathol.*, 15:939–943.

Chariyalertsak, S., Sirisanthana, T., Supparatpinyo, K., & Nelson, K. E. (1996a). *J. Infect. Dis.*, 173:1490–1493.

Chariyalertsak, S., Vanittanakom, P., Nelson, K. E., Sirisanthana, T., & Vanittanakom, N. (1996b). *J. Med. Vet. Mycol.*, 34:105–110.

Clutterbuck, A. J. (1969). *Genetics*, 63:317–327.

De Lucas, J. R., Gregory, S., & Turner, G. (1994). *Mol. Gen. Genet.*, 243:654–659.

Felipe, M. S., Andrade, R. V., Arraes, F. B., Nicola, A. M., Maranhao, A. Q., et al., (2005). *J. Biol. Chem.*, 280:24706–24714.

Fisher, M. C., Hanage, W. P., De Hoog, S., Johnson, E., Smith, M. D., White, N. J., & Vanittanakom, N. (2005). *PLoS Pathog.*, 1:e20.

Fraser, J. A., Giles, S. S., Wenink, E. C., Geunes-Boyer, S. G., Wright, J. R., et al. (2005). *Nature*, 437:1360–1364.

Garrison, R. G. & Boyd, K. S. (1973). Dimorphism of *Penicillium marneffei* as observed by electron microscopy. *Can. J. Microbiol.*, 19:1305–1309.

Goldman, G. H., Dos Reis Marques, E., Duarte Ribeiro, D. C., De Souza Bernardes, et al. (2003). *Eukaryot. Cell*, 2:34–48.

Gow, N. R. (1995). *Yeast-Hyphal Dimorphism. The Growing Fungus*, London, Chapman & Hall.

Gugnani, H., Fisher, M. C., Paliwal-Johsi, A., Vanittanakom, N., Singh, I., & Yadav, P. S. (2004). *J. Clin. Microbiol.*, 42:5070–5075.

Heitman, J. (2006). *Curr. Biol.*, 16:R711–R725.

Hull, C. M. & Johnson, A. D. (1999). *Science*, 285:1271–1275.

Hynes, M. J., Murray, S. L., Duncan, A., Khew, G. S., & Davis, M. A. (2006). *Eukaryot. Cell*, 5:794–805.

Idnurm, A. & Howlett, B. J. (2002). *Eukaryot. Cell*, 1:719–724.

Kidd, S. E., Guo, H., Bartlett, K. H., Xu, J., & Kronstad, J. W. (2005). *Eukaryot. Cell*, 4:1629–1638.

Lengeler, K. B., Davidson, R. C., D'souza, C., Harashima, T., Shen, W. C., et al. (2000). *Microbiol. Mol. Biol. Rev.*, 64:746–785.

Lengeler, K. B., Fox, D. S., Fraser, J. A., Allen, A., Forrester, K., Dietrich, F. S., & Heitman, J. (2002). *Eukaryot. Cell*, 1:704–718.

Lim, D., Lee, Y. S., & Chang, A. R. (2006). *J. Clin. Pathol.*, 59:443–444.

Lindsley, M. D., Hurst, S. F., Iqbul, N. J., & Morrison, C. J. (2001) *J. Clin. Microbiol.*, 39:3505–3511.

Lorenz, M. C., Bender, J. A., & Fink, G. R. (2004). *Eukaryot. Cell*, 3:1076–1087.

Lorenz, M. C. & Fink, G. R. (2001). *Nature*, 412:83–86.

Lorenz, M. C. & Fink, G. R. (2002). *Eukaryot. Cell*, 1:657–662.

Lupi, O., Tyring, S. K., & Mcginnis, M. R. (2005). *J. Am. Acad. Dermatol.*, 53:931–951.

Maresca, B. & Kobayashi, G. S. (1989). *Microbiol. Rev.*, 53:186–209.

Marshall, M. A. & Timberlake, W. E. (1991). *Mol. Cell Biol.*, 11:55–62.

Mcginnis, M. R. (1994). *Clin. Microbiol. Newslett.*, 16:29–31.

Mckinney, J. D., Honer Zu Bentrup, K., Munoz-Elias, E. J., Miczak, A., et al. (2000). *Nature*, 406:735–738.

Mcneil, M. M., Nash, S. L. Hajjeh, R. A. Phelan, M. A., Conn, L. A., et al. (2001). *Clin. Infect. Dis.*, 33:641–647.

Mirabito, P. M., Adams, T. H., & Timberlake, W. E. (1989). *Cell*, 57:859–868.

Mootsikapun, P. & Srikulbutr, S. (2006). *Int. J. Infect. Dis.*, 10:66–71.

Munoz-Elias, E. J. & Mckinney, J. D. (2005). *Nat. Med.*, 11:638–644.

Paoletti, M., Rydholm, C., Schwier, E. U., Anderson, M. J., Szakacs, G., et al. (2005). *Curr. Biol.*, 15:1242–1248.

Piekarska, K., Mol, E., Van Den Berg, M., Hardy, G., Van Den Burg, J., Van Roermund, C., Maccallum, D., Odds, F., & Distel, B. (2006). *Eukaryot. Cell*, 5:1847–1856.

Pornprasert, S., Dettrairat, S., Vongchan, P., & Apichatpiyakul, C. (2005). *Southeast Asian J. Trop. Med. Public Health*, 36:966–969.

Ramirez, M. A. & Lorenz, M. C. (2006). *Eukaryot. Cell*, 8:8.

Rude, T. H., Toffaletti, D. L., Cox, G. M., & Perfect, J. R. (2002). *Infect. Immun.*, 70:5684–5694.

Schuller, H. J. (2003). *Curr. Genet.*, 43:139–160.

Schweizer, A., Rupp, S., Taylor, B. N., Rollinghoff, M., & Schroppel, K. (2000). *Mol. Microbiol.*, 38:435–445.

Sirisanthana, T., Supparatpinyo, K., Perriens, J., & Nelson, K. E. (1998). *Clin. Infect. Dis.*, 26:1107–1110.

Solomon, P. S., Lee, R. C., Wilson, T. J., & Oliver, R. P. (2004). *Mol. Microbiol.*, 53:1065–1073.

Stevens, D. A. (1995). *N. Engl. J. Med.*, 332:1077–1082.

Supparatpinyo, K., Chiewchanvit, S., Hirunsri, P., Baosoung, V., Uthammachai, C., Chaimongkol, B., & Sirisanthana, T. (1992a). *J. Med. Assoc. Thai.*, 75:688–691.

Supparatpinyo, K., Chiewchanvit, S., Hirunsri, P., Uthammachai, C., Nelson, K. E., & Sirisanthana, T. (1992b). *Clin. Infect. Dis.*, 14:871–874.

Supparatpinyo, K., Perriens, J., Nelson, K. E., & Sirisanthana, T. (1998). *N. Engl. J. Med.*, 339:1739–1743.

Supparatpinyo, K. & Sirisanthana, T. (1994). *Clin. Infect. Dis.*, 18:246–247.

Todd, R. B., Greenhalgh, J. R., Hynes, M. J., & Andrianopoulos, A. (2003). *Mol. Microbiol.*, 48:85–94.

Todd, R. B., Kelly, J. M., Davis, M. A., & Hynes, M. J. (1997). *Fungal Genet. Biol.*, 22:92–102.

Tongchusak, S., Pongtanalert, P., Pongsunk, S., Chawengkirttikul, R., & Chaiyaroj, S. C. (2004). *Asian Pac. J. Allergy Immunol.*, 22:229–235.

Tsunemi, Y., Takahashi, T., & Tamaki, T. (2003). *J. Am. Acad. Dermatol.*, 49:344–346.

Vanittanakom, N., Cooper, C. R., Jr., Fisher, M. C., & Sirisanthana, T. (2006). *Clin. Microbiol. Rev.*, 19:95–110.

Wang, Z. Y., Thornton, C. R., Kershaw, M. J., Debao, L., & Talbot, N. J. (2003). *Mol. Microbiol.*, 47:1601–1612.

Whiteway, M. & Oberholzer, U. (2004). *Curr. Opin. Microbiol.*, 7:350–357.

Woo, P. C., Chong, K. T., Tse, H., Cai, J. J., Lau, C. C., Zhou, A. C., Lau, S. K., & Yuen, K. Y. (2006). *FEBS Lett.*, 580:3409–3416.

Zuber, S., Hynes, M. J., & Andrianopoulos, A. (2002). *Eukar. Cell*, 1:440–447.

Zuber, S., Hynes, M. J., & Andrianopoulos, A. (2003). *Genetics*, 164:487–499.

Chapter 10
Immune Responses to Dermatophytoses

Roderick J. Hay

Introduction

Dermatophytes are exogenous pathogens that cause common superficial infections of the skin and keratinized structures arising from it, such as the hair and nails, known as tinea or dermatophytosis. While there are some fungi such as the lipophilic yeasts or *Malassezia* species that are part of the normal flora of the skin, dermatophytes are acquired from an external source. Dermatophytes are mould fungi which evolved from soil dwelling organisms, the keratinophilic fungi, that adapted to live in an environment where there was shed hair or skin, e.g. in the vicinity of animal homes or burrows. They adapted to invade keratin on living hosts and to cause infections on animals or humans. Dermatophytes are therefore known as geophilic, zoophilic, or anthropophilic depending on whether they originate from soil, animals, or humans, respectively.

Dermatophytes cause a variety of different clinical syndromes depending on the site of the infection. The commonest pattern is infection of the feet including the soles or web spaces, tinea pedis. Tinea capitis or dermatophyte infection of the scalp is endemic in many countries and, where the infecting fungi are anthropophilic, there may be high prevalence rates of infection reflecting the capacity of these organisms to be transferred in communities. Other forms of dermatophyte infection include tinea corporis (body) and tinea cruris (groin). A feature common to many dermatophyte infections is that some forms present with highly inflammatory lesions. This includes marked erythema and skin infiltration, but sometimes there is pustule formation. Others produce minimal clinical signs of abnormality such as mild scaling with barely discernible erythema. Prior to the availability of fungicidal treatment regimens these minimally inflamed infections were often persistent and refractory to treatment (Hay, 1982). Examples of such chronic infections include tinea pedis due to *Trichophyton rubrum*, tinea capitis due to *T. tonsurans* (Elewski, 1999), and tinea imbricate, a tropical fungal infection seen in well-defined areas of the tropics and caused by *T. concentricum* (Serjeantson & Lawerence, 1977). Nowadays, although they do not heal spontaneously, dermatophyte infections will clear completely with modern antifungals; recurrence is a common feature of many dermatophytoses (Elewski, 1999). Zoophilic organisms causing dermatophytosis,

K. Kavanagh (ed.), *New Insights in Medical Mycology.*
© Springer 2007

in man, are frequently highly inflammatory and infections are followed by resistance to further episodes of infection, e.g. *T. verrucosum* or cattle ringworm where recurrent infections are very uncommon (Hall, 1962); inflammatory tinea capitis, whatever the source of the organism, is also associated with recovery. As will be seen this behaviour, the development of an inflammatory response equating with recovery, is also reflected by the natural history of infection as seen in experimental infections in animals such as mice or guinea pigs. In humans there is a wider range of responses to infection.

Skin Invasion

The cells that are largely responsible for infection are arthrospores or thick-walled intercalary cells able to survive adverse conditions. Dermatophyte cells infect skin by a process of adherence to the cells of the epidermis followed by germination, growth, and penetration by fungal hyphae both within and between cells. The first phase of fungal attack on the stratum corneum, the outer layer of cornified cells, dead cells filled with the fibrous protein, keratin, depends on this process of inter-cellular adherence. Initial studies of this phenomenon utilized microconidia obtained from pure dermatophyte cultures; but these are not the natural propagules of infection (Zurita & Hay, 1987). Subsequently arthroconida which are the natural infecting particles have been generated *in vitro* by growth in a high concentration of carbon dioxide and used for these experiments. These studies have shown that there is no difference between different dermatophyte species or keratinocytes taken from different anatomical sites in adhesion rates between fungal cell and keratinocyte. This occurs after a time lag of approximately 2 h after initial contact with maximum adherence being achieved at 4 h. This process is usually accompanied by the appearance of fibrillar strands between the cells demonstrated by electron microscopy. The nature of the adhesin ligand bond is not known, although surface carbohydrates appear to play a key role (Esquenazi et al., 2003) and the process can be inhibited by sub-inhibitory concentrations of antifungals (Zurita & Hay, 1987). Adhering arthrospores swell during the interactive process.

Once adherence has occurred, the dermatophyte arthroconidia germinate and invasion of the outer layers of the stratum corneum proceeds (Kligman, 1952). Growing hyphae radiate from the initial cell contact but can be seen to grow towards and into holes in any underlying surface, mimicked *in vitro* by growth across and through a porous structure such as a Millipore membrane (Aljabre et al., 1992, 1993). This is likely to be mirrored *in vivo* by the ability of growing hyphae to take advantage of small breaks in the stratum corneum and penetrate between cells. Fungal growth however does not simply proceed by exploitation of epithelial defects as fungal hyphae can be seen within cells. In the case of hair shaft invasion they elaborate a specialized fungal cell, which apposes to the hair shaft and gives rise to a single modified hyphal strand known as the penetrating organ (Kligman, 1952). It is clear that cell penetration follows the production of enzymes and, in microscopy, a cleared area can be seen around the site of penetration in hair. The production of proteases, some of which are inducible in the presence of amino acid residues, is a key stage in the invasion of the skin by dermatophytes. In early studies at least three low-molecular weight proteases were isolated from *T. mentagrophytes* (Yu et al., 1971); a number of different proteases have also been extracted from *T. rubrum*. These range in size from 34, 77, and 105 kD. In addition, this dermatophyte produces a secreted metalloprotease with a molecular weight of approximately 200 kD which shows specificity for collagen and elastin. In *Microsporum canis* metalloproteases are produced by invading organisms at the site of hair shaft infection (Brouta et al., 2002). In addition it has been shown that the metalloproteases in *M canis* are members of the subtilisin family and to date three genes named sub1, 2, and 3 are linked to these.

In an experimental murine model of dermatophytosis, after successful invasion of the epidermis, a number of events can be seen. This starts with early recruitment of neutrophils which infiltrate the epidermis and upper dermis at the site of infection. In addition there is epidermal proliferation with an increased thickness of the epidermis and the presence of nucleated cells higher in the epithelium (Hay et al., 1988). This 'growth' spurt outlasts the infection and in mice the former infected area usually shows increased hair growth after resolution of the infection. In an experimental infection of BalbC mice with the mouse pathogen, *T. quinckeanum*, the inflammatory process maximizes between day 7 and 10 post infection with the formation of an epidermal crust containing fungal hyphae, epidermal debris, and neutrophils. This is similar in structure to the scutulum seen in the human scalp infection, favus, caused by *T. schoenleinii*. Thereafter the inflammation declines with fewer neutrophils and clustering of lymphocytes near the epidermo-dermal junction. Much later from day 25 onwards mast cells appear in the upper dermis. A similar limited histology, apart from the mast cell infiltrate, is seen in infections in humans (Graham & Barrosos, 1971). This suggests that there is an effective process, involving mobilisation of a cellular infiltrate, for limiting spread of infection which comes into operation during the course of a dermatophyte infection.

Defence Mechanisms

The principle means of defense against dermatophytes identified at present involve both non-immunological processes such as the interaction between fungi and unsaturated transferrin, activation of epidermal peptides, the inhibitory effect of fatty acids in sebum, and immunological processes including fungal killing by polymorphonuclear leucocytes attracted into the area of infection as well as the activation of T lymphocytes. The first of these are nonspecific mechanisms of defense against dermatophytosis.

The mode of action of fatty acids of medium chain length is unknown. However *in vitro* antifungal activity appears to depend on fatty acids of a chain length of 7–13 carbon residues (Abraham et al., 1975). Other members of the fatty acid families with similar structures, such as undecylenic acid, have antifungal inhibitory activity. These are found in a high concentration in post pubertal sebum, a fact that may account for the rarity of tinea capitis after puberty. Unsaturated transferrin inhibits the growth of dermatophytes by a direct mechanism involving its binding to the fungal cell membrane (King et al., 1975). As might be expected there is a similar effect of lactoferrin in experimental infections, although there is another possible pathway through an effect on mononuclear cell function (Wakabayashi et al., 2002). Epidermal peptides are also known to express antifungal activity. This is seen even in primitive vertebrates where frogs produce temporins and a species-specific peptide called ranatuerin, both of which have antifungal activity (Rollins-Smith et al., 2006). *In vitro* studies using the peptide produced by human epidermal cells, cathelicidin (Lopez-Garcia et al., 2006) showed that this is fungicidal to both *T. rubrum* and *T. mentagrophytes* in a concentration of 12.5 µM. Expression of the fungal inhibitor, detected by immunostaining for this peptide, was increased in infected versus uninfected human skin.

Accelerated epidermal growth may also aid clearance of fungi in dermatophytosis. This has been found in experimental dermatophytosis using guinea pig skin grafted onto athymic mice (Rollins-Smith et al., 2006). Epidermal proliferation occurs early, within the first 48 h after infection and in the absence of T lymphocytes. This suggests that there is a direct and T-cell-independent mechanism for increasing epidermal growth in response to fungal invasion. It is still possible, though, that immunological activation provides a means of amplifying this epidermal proliferative response. Similar increased epidermal cell growth has been detected in human infections

Neutrophils and macrophages of both murine and human origin are capable of killing dermatophytes. Of these the most important are the neutrophils, polymorphonuclear cells or PMN, and infiltration of the outer stratum corneum or around hair follicles can be demonstrated in the course of a dermatophyte infection. Human PMN have been shown to destroy up to 60% of dermatophyte germlings within 2 h; macrophages kill up to 20% in a similar time. Inhibitors of free radicals such as superoxide, including histidine or catalase inhibit killing which can proceed in the absence of ingestion (Calderon & Hay, 1987; Calderon & Shennan, 1987). The existence of a second non-oxidative method of phagocyte-mediated defense via a peptide mechanism, as occurs with *Candida*, for dermatophytes has not been investigated.

In common with other pathogenic fungi *T. rubrum* produces a catalase and a secreted superoxide dismutase, both of which may interfere with the outcome of phagocyte-mediated defense (Calderon & Shennan, 1987). Their role in blocking the effect of neutrophils *in vivo* remains unknown. Neutrophils are attracted to the site of infection both by production of epidermally derived chemotactic factors (from basal cells) such as leukotriene derivatives, adhesion molecules, and also by certain dermatophyte cell wall antigens which activate the alternate pathway of complement (Davies & Zaini, 1988).

The success of neutrophils in controlling dermatophyte invasion in all situations remains contestable as there are differences which emerge in the study of histology of experimental murine versus human infections. In experimentally infected mice, dermatophyte invasion appears to be limited by the migration of neutrophils to the site of infection (Hay et al., 1988). However, in man there is little evidence of early mobilization of neutrophils in some forms of dermatophytosis. Neutrophil accumulation is a feature of infections in which there is penetration of the hair follicle (Graham & Barrosos, 1971). However in infections of glabrous skin the histology often shows a striking lack of a cellular infiltrate. Human dermatophyte infections, as stated previously, are often viewed as either inflammatory or non-inflammatory; the former infections are often caused by zoophilic fungi, the latter by anthropophilic organisms.

However the development of an immune response and memory of previous infections is key to host defense in dermatophytosis which became apparent from the earliest studies of experimental infections. Keratinophilic fungi, such as dermatophytes, elicit appropriate responses via T-cell-mediated pathways and it is know that they interact with dendritic cells, the principle antigen detecting cells of the epidermis, through the DC–SIGN lectin (Serrano-Gomez et al., 2005) These antigen presenting cells have the capacity to process the antigen and present to lymphocytes in peripheral lymph nodes, possibly locally, thereby recruiting sensitized lymphocytes back to the site of infection. In an experimental mouse model of dermatophytosis using *T. quinckeanum*, the responses of lymphocytes from draining lymph nodes to lectins such as Concanavlin A (ConA) and phytohae-maglutinin (PHA) as well as an antigen from the dermatophyte were analysed during the course of infection (Calderon & Hay, 1984). These responses showed that at the peak of infection, at day 7, there was significant reduction in lymphocyte responsiveness (blastogenesis) to the polyclonal mitogens, PHA and ConA, although this reversed by the time of resolution of the infection by day 30. In contrast there were increased mitogenic reponses to the specific dermatophyte antigen. In secondary infections fungi were shed very quickly and at the outset mice had higher mitogenic responses to dermatophyte antigen than uninfected mice. Once again polyclonal mitogen responses were depressed at the peak of infection whereas responses to specific fungal antigen were augmented. Mixing of lymphocytes from infected and noninfected animals showed that a population of cells from infected animals with the Thy-1[+], Ly-2.2[+] phenotype were responsible for suppressing blastogenesis. Subsequently it was shown that transfer of T lymphocytes bearing the Thy-1[+], Ly-2.2[-] helper phenotype from immune animals to naive recipients is the key event in determining immunity (Calderon & Hay, 1984). Whereas immunity to infection can be transferred to irradiated naive animals with lymphocytes bearing this phenotype, it was not with those cells bearing the suppressor phenotype Thy-1[+], Ly-2.2 . Lymphocytes from uninfected donors and passive transfer of antibody would also not convey resistance to recipient animals.

A cohort of infected animals had prolonged infections of over 8 weeks without the normal resolution at 25–30 days. When spleen or lymph node-derived lymphocytes were transferred to naïve animals the recipients became immune to infection, illus-trating that cells from chronically infected mice transferred adoptive immunity (Calderon & Hay, 1984). However if these cells were transferred in the presence of autologous serum, i.e. serum from chronically infected animals, immunity was not transferred. It was subsequently shown that this effect could be mimicked by the additional of antigen to normal serum in similar transfer experiments (Calderon & Hay, 1984; Calderon, 1989). This suggested a critical role for antigen in blocking transfer of immunity.

Human Dermatophytosis

The majority of patients with dermatophytosis are not immunosuppressed and have no underlying disease. Previous studies have occasionally revealed a number of underlying diseases in patients with dermatophytosis, notably hereditary palmoplantar keratoderma (Elmros & Liden, 1983) and Raynaud's phenomenon (Hay, 1982). But the former is likely to reflect a major change in the local environment in which dermatophytes grow and the latter vascular disease. In fact it is possible to see patients with dermatophytosis on one foot where there is poor vascular perfusion detectable clinically but not on the other normally perfused foot (Hay, 1982).

Atopy is an underlying disease syndrome which has been consistently connected with dermatophytosis. A high proportion of chronically infected individuals, over 40% in some surveys, have hay fever, asthma, or atopic eczema either on personal or family history (Hay, 1982; Jones et al., 1973). In addition a high proportion of individuals seen in dermatological clinics with peripheral dermatophyte infection either have negative or immediate-type hypersensitivity to dermatophyte antigens on intradermal testing. Increased prevalence of immediate-type responses to intra-dermally injected antigens is also a feature of atopic subjects. There is also evidence that some individuals with persistent dermatophytosis may also have sensitivity to environmental moulds suggesting a modified (atopic) immunological response to a family of antigens.

The other group of patients showing altered responses to dermatophyte infections are those who have certain defects of immunity. Patients with chronic mucocutaneous candidosis (CMC) are particularly susceptible to widespread intractable dermato-phytosis as well as candidosis; human immunodeficiency virus (HIV)-infected individuals may also have chronic ringworm. These observations suggest that patients with impaired immunological defense mechanisms, in particular those affecting T-lymphocyte function, are prone to chronic or altered infection. However, in all these examples, whereas the clinical expression of infection may be modified by the patients' underlying condition, the prevalence of dermatophytosis in compromised patients is not significantly different to that seen in healthy subjects. In AIDS patients, the prevalence of infection has been no higher in some studies than in members of 'at risk' groups without HIV infection (Torssander et al., 1988) but there is an association with low CD4 counts (Munoz-Perez et al., 1998).

It is important to separate changes in prevalence, which also depends on the facility for transmission, from the clinical behaviour of the dermatophyte infection in immunosuppressed patients. Patients on chronic immunosuppressive therapeutic regimens, e.g. from solid organ transplant recipients, and patients with HIV/AIDS often have more widespread and difficult to treat dermatophyte infections. HIV positive women receiving antiretrovirals are less likely to have tinea pedis than those not receiving these drugs, irrespective of CD4 count (Maurer et al., 2004). There has been a report suggesting that patients receiving topical tacrolimus may develop widespread dermatophytosis (Siddaiah et al., 2004). These reflect the fact any prediposing condition that impairs T-cell function in the skin will affect the clinical expression of dermatophytosis in humans. But as stated previously, with the exception of atopy, underlying disease is not seen in the majority of patients with dermatophytosis (Svejgaard, 1985). This is in sharp distinction to patients with infections due to *Candida albicans* where there is usually some predisposing abnormality, ranging from occlusion of the skin surface to defects in neutrophil or T-lymphocyte function.

As stated previously, there is a correlation between inflammatory responses, T-lymphocyte activation, and recovery. In experimental infections, recovery occurs at the same time as lesions become inflamed and delayed-type hypersensitivity to the fungal antigen develops. Patients with persistent foot infections (*T. rubrum*) or tinea corporis (*T. concentricum*) show reduced levels of lymphocyte blastogenesis to dermatophyte antigen (Jones et al., 1974; Hay et al., 1983). In chronic infections caused by *T. rubrum* compared to those more inflamed infections due to *T. mentagrophytes*, for instance, the lymphocyte transformation responses of peripheral blood lymphocytes from infected patients are low. There is little clinical inflammatory reaction and relapse or persistence of infection is common. In less than 15% of cases is there evidence of an underlying disease affecting the immune system that explains these findings. There is also evidence that some immune responses, such as absent delayed-type hypersensitivity to trichophytin, can be reversed by successful therapy (Elewski et al., 2002). In the rare disseminated forms of dermatophytosis where there is involvement of internal organs there may also be evidence of disturbance of immune function (Allen et al., 1977; Liautaud & Marill, 1984).

Modulation of the Immune Response in Dermatophytosis

While immunological mechanisms provide potential methods of defense, the persistence of infection in many apparently healthy individuals suggests that these are either ineffective or inoperative in some patients. It has been shown that some patients with persistent dermatophytosis have defective lymphocyte blastogenesis to T-cell mitogens and dermatophyte antigen and that this can be reversed either by substituting heterologous (foetal calf) for autologous serum or after successful antifungal treatment (Mayou et al., 1987). This suggests that an inhibitory factor(s) is present in serum. If this parallels the situation described in mice referred to previously then dermatophyte antigen is a potential candidate as a blocking agent (Calderon, 1989). Dermatophyte antigen has been identified in such infected serum employing an immunoradiometric assay using the mouse antidermatophyte IgM monoclonal TQ-1 (Calderon et al., 1987). Antigen derived from *Trichophyton* species, containing TQ-1 reactive epitopes, increases susceptibility of Balb/C mice to dermatophyte infection and interferes with T-cell-mediated immunity (Calderon, 1989). TQ-1 antibody reacts with phosphoryl-choline (PC) and immunoreactivity can be abrogated by pretreatment with PC. Phosphoryl-choline hapten is found in other parasites including filaria and this also affects expression of immune responses in these infections. Another, probably different, factor has also been identified. Oligosaccharides derived from glycoproteins present in the dermatophyte cell wall may interfere with both T- and B-lymphocyte activation. This is a reversible process *in vitro*, but pre-exposure of lymphocytes to dermatophyte inhibitory factor (DIF) will prevent proliferation of T cells in response to mitogens such as phytohaemagglutinin (PHA), as well as dermatophyte antigens (Dahl, 1994; McGregor et al., 1992).

It is clear that within the complex of antigen epitopes that make up the phenotype of a dermatophyte there are antigenic components that potentially stimulate as well as some that block effective immunity. There are also antigenic components that are more likely to trigger an 'allergic' or immediate-type hypersensitivity response, although these are usually the same as those that, under different circumstances, elicit effective immunity and delayed-type hypersensitivity (Woodfolk, 2005). The fungal components which are most closely identified with host resistance via T-cell activation are glycopeptides. However other emerging candidates are heat-shock proteins of the HSP60 family (Raska et al., 2004) and metalloproteinases or subtilases (Mignon, 2005). There is also evidence that these can induce protection against experimental infection in animals.

The possibility that there is interference with the process of immunological activation in the skin is supported by immunohistochemical studies of biopsies from chronically infected skin. In acute dermatophyte infections, immunophenotypic techniques can be used to demonstrate the presence of numbers of effector lymphocytes in the vicinity of the infection (Brasch & Sterry, 1992). Work has now shown that the dermal infiltrate mainly contains cells which are Leu2a positive (viz. T-helper cells) (Brasch & Sterry, 1992; Leibovici et al., 1995). Conversely few express CD-8 (T-suppressor markers). HLA-DR is strongly expressed and most biopsies show Langerhans cells using a variety of different cell markers. In other inflammatory processes cells of the epidermis produce certain cytokines. If fungi including the dermatophyte *T. mentagrophytes* are co-cultured with human keratinocytes produce IL8 and TNF but other cytokines were undetectable (Kano, 2004). It has been shown by co-incubation of dermatophytes such as *T. tonsurans* that cultured keratinocytes of the PHK16–0b cell line produce cytokines. With *Arthoderma benhamiae* (the perfect state of the *T. mentagrophytes* complex a wide range of cytokines, chemokines and immunostimulatory cytokines are secreted; this involved upregulation of genes encoding arrange of cytokines including IL-1, Il-2, IL-4, Il-6, IL-13, IL15, and IL-16 as well as interferon γ. However *T. tonsurans* only triggered release of IL-8 and Il-16 with upregulation of IL-1 and IL-16. This suggests that there are very different responses to different dermatophytes (Shiraki et al., 2006). It is of further interest that an organism that normally causes a highly inflammatory reaction in man, *A. benhamiae (T. mentagrophytes)* triggers a wider range of inflammatory cytokines whereas *T. tonsurans* associated with chronic and less inflammatory disease stimulated a limited range. At present there is no evidence that this occurs *in vivo*.

Despite the presence of an infiltrate in chronically infected patients, adhesion molecules, such as ICAM-1, are poorly expressed in epidermis (Schectman et al., 1993). This may reflect suppression of the expression of these integrins in the epidermis despite intact cell-mediated immunity. Suppression of this aspect of immune activation by DIF produced by dermatophytes *in situ* is a possible explanation, which may account for the success of dermatophytes, such as *T. rubrum*, in causing persistent infections.

A recent study of the interaction between macrophages and *T. rubrum* conidia and hyphae showed that, in the presence of fungus, macrophages produce TNF and IL10 but not IL12 or nitric oxide. There was also evidence that the dermatophytes downregulated the production of co-stimulatory molecules CD80 and CD50. Phagocytosis is inhibited by addition of crude fungal exoantigen or mannan to the medium (Campos et al., 2006) and the viability of macrophages is reduced ingestion of dermatophyte spores

Unsolved Issues

There are a number of unexplained findings resulting from published studies, in particular the relationship between atopy and infection. Atopic subjects have been reported to be more susceptible to persistent infections, particularly if they are not involved in occupations such as coal mining or heavy industry where exposure to infection occurs frequently (Hay et al., 1983). This may simply reflect the fact that a proportion of *T. rubrum* infections become persistent and that the increased opportunity for spread in an industrial setting overrides normal transmission. These studies generally have used a clinical definition of atopy based on family history and, in some instances, prick testing to other antigens. It should also be noted the relationship between persistence of infection and atopy has not been seen in all studies.

Patients with persistent foot and groin infections due to *T. rubrum* are more likely to have positive immediate-type immune responses to intradermal tests with trichophytin (Jones et al., 1973). A similar finding has been recorded for chronic tinea imbricata infections, i.e. a different species affecting large areas of the trunk and limbs (Hay et al., 1983). This is supported by elevated levels of IGE specific to different dermatophytes in patients with dry type *T. rubrum* infections and tinea imbricata. Currently it is thought that this is most likely to reflect a Th2 switch occurring in the development of dermatophytosis. This interpretation is supported by a study which has shown that high IgE levels are accompanied by raised specific IgG4. In the same study, a higher level of delayed-type hypersensitivity to intradermal

trichophytin, but similar lymphocyte transformation responses, were seen in
T. rubrum patients compared to controls (Koga et al., 2001a). It is clear that IFN γ
producing cells (Koga et al., 2001b) are found in the infiltrate of dermatophyte
lesions. However at present a correlation between dry or chronic type infection and
defective IFN γ production is not established. Variations in IgE production in
response to successful therapy would, once again, indicate a fluctuating relationship
between Th1 and Th2 pathways (Escalante et al., 2000).

Once again there are some parallels here with the chronic infection, CMC. For
although this is beyond the scope of a review of defense against dermatophytosis
there is a relevance as a proportion of patients with CMC have persistent and hyper-
keratotic dermatophytosis of the feet or scalp.

There have been recent advances in our understanding of the relationship
between this disorder and immunological function (Atkinson et al., 2001). It
appears that in some patients with CMC associated with the autoimmune polyen-
docrinopathy syndrome there are mutations in the autoimmune regulator (AIRE)
gene. Two siblings who had CMC and hypoparathyroidism were found to have
specific mutations in the AIRE gene accompanied by altered expression of the T
cell V beta receptor (Kogawa et al., 2002). Here there is a direct link between T-cell
functional regulation and the risk of chronic dermatophytosis. Previous studies of
CMC have demonstrated a wide variety of immunological defects (Kirpatrick et al.,
1971; Lilic & Gravenor, 2001). These include negative delayed-type hypersensitivity
both to *Candida* as well as other antigens, defective lymphocyte transformation and
defects in neutrophil killing of yeasts, leucocyte migration, and cytokine production as well as selective antibody deficiency (Kalfa et al., 2003). In one study (Lilic
et al., 1996), production of IFN γ was variable, with some patients producing little
or none at all. As with dermatophytosis some authors have shown reversal of the
immune defect with successful antifungal chemotherapy (Drouhet & Dupont, 1983;
Kennedy et al., 1981). However there is a change in T-cell immunoregulation which
appears to be associated with defective memory responses to antigens. In some
patients there are increased levels of IL-4 and a decrease in Th1 mediated
cell-mediated immunity (Kobrynski et al., 1996). As reversal of immune defects,
such as absent delayed-type hypersensitivity to *Candida* antigens, has been seen
with successful clearance of candidosis in CMC patients, it is possible that some
immunological changes may be secondary to the infection itself. There are intriguing
parallels here with the situation described with dermatophytosis and the potential
for immunomodulation by fungal antigen.

Other constituents of superficial fungi are believed to affect expression of an
effective immune response. Other immunomodulators described in superficial
fungi include a lipid and melanin. *Malassezia* species elaborate a lipid rich layer
external to the cell wall, removal of which is associated with significant improvement
in vitro of T-cell-mediated responses (Kesavan et al., 2000). Melanin is an essential
constituent of the fungal cell wall that has been described in many pathogenic
species such as *Cryptococcus neoformans* and *Histoplasma capsulatum*. The type
of melanin varies although it is commonly Dopa or pentaketide melanin. Melanized
fungal cells show enhanced capacity to resist T-cell mechanisms and neutrophil

attack largely through neutralizing the effect of oxidative products such as superoxide or reactive oxygen. In *C. neoformans* defective melanin biosynthesis correlates with reduced virulence. At this stage it would be premature to abstract these data to interpret the failure of immune responses to dislodge other organisms. However it is worth noting that a number of superficial pathogens, such as *Scytalidium* species and some dermatophytes, have now been shown to contain melanin. These include anthrophilic fungal species such as *T. rubrum* (Morris-Jones et al., 2004).

Summary

In summary it is possible to connect these findings by proposing the following hypothesis for the development and persistence of dermatophyte infections.

• Dermatophyte infections are normally eliminated through a largely Th1 path involving effector mechanisms that range from accelerated epidermal turnover to production of adhesion molecule-directed neutrophil trafficking in the epidermis at the site of infection and subsequent phagocyte-mediated fungal cell destruction.
• A number of different factors can delay elimination of the organisms. These include gross thickening of the epidermis as seen in patients with forms of keratoderma, immunosuppression that affects the function of T lymphocytes (therapeutic or disease-related), but fungal antigen-mediated modulation of T-cell responses may also play a role in blunting host defense.
• Persistence of these organisms in the face of immune-recognition is a key factor involved in triggering a Th2 switch, which in turn leads to a defective defense mechanism. This switch is more likely to occur in individuals with atopic background but is not exclusive to them – this explains the higher prevalence of dermatophyte infections seen in some, but not all, studies in atopic subjects.

Whatever the truth of this hypothesis it is clear that dermatophytosis provides an intriguing model for the adaptation of effective immune mechanisms in both human and experimental infections.

References

Abraham, A., Mohapatra, L. N., & Kandhari, K. C. (1975). *Dermatologica*, 151:144–149.
Aljabre, S. H., Richardson, M. D., Scott, E. M., & Shankland, G. S. (1992). *J. Med. Vet. Mycol.*, 30:145–152.
Aljabre, S. H., Richardson, M. D., Scott, E. M., Rashid, A., & Shankland, G. S (1993). *Clin. Exp. Dermatol.*, 18:231–235.
Allen, D. E., Snyderman, R., Meadows, L., et al. (1977). *Am. J. Med.*, 63:991–1000.
Atkinson, T. P., Schaffer, A. A., Grimbacher, B., et al. (2001). *Am. J. Hum. Genet.*, 69:791–803.
Brasch, J. & Sterry, W. (1992). *Acta. Dermato. Venereol.*, 72:345–347.

Brouta, F., Descamps, F., Monod, M., et al. (2002) *Infect. Imm.*, 70:5676–5683.

Calderon, R. A. (1989). *Crit. Revs. Microbiol.*, 16:339–368.

Calderon, R. A. & Hay, R. J. (1984). *Immunology*, 53:457–464.

Calderon, R. A. & Hay, R. J. (1984). *Immunology*, 53:465–472.

Calderon, R. A. & Hay, R. J. (1987). *Immunology*, 61:289–296.

Calderon, R. A., Hay, R. J., & Shennan, G. I. (1987). *J. Gen. Microbiol.*, 133:2699–2705.

Calderon, R. A. & Shennan, G. I. (1987). *Immunology*, 61:283–288.

Campos. M. R., Russo, M., Gomes, E., & Almeida, S. R. (2006). *Microbes and Infect.*, 8:372–379.

Dahl, M. V. (1994). *J. Am. Acad. Dermatol.*, 31:S34–S41.

Davies, R. R. & Zaini, F. (1988). *Clin. Exp. Dermatol.*, 13:228–231.

Drouhet, E. & Dupont B. (1983). *Am. J. Med.* 74 (Suppl. 1B):30–45.

Elewski, B. (1999). *J. Am. Acad. Dermatol.*, 40:S27–S30.

Elewski, B. E., El Charif, M., Cooper, K. D., et al. (2002). *J. Am. Acad. Dermatol.*, 46:371–375.

Elmros, T. & Liden, S. (1983). *Acta. Derm. Venereol.*, 63:254–257.

Escalante, M. T., Sanchez-Borges, M., Capriles-Hulett, A., et al. (2000). *J. Allerg. Clin. Immunol.*, 105:547–551.

Esquenazi, D., de Souza, W., Alviano, C. S., & Rozental, S. (2003). *FEMS Immunol. Med. Microbiol.*, 35:113–123.

Graham, J. H. & Barrosos-Tobila, C. (1971). Dermatophytosis. In: *The Pathologic Anatomy of the Mycoses*, Baker, R. D., Ed., Springer, Berlin, pp. 211–235.

Hall, F. R. (1962). *Arch. Dermatol.*, 94:35–37.

Hay, R. J. (1982). *Br. J. Dermatol.*, 106:1–9.

Hay, R. J., Reid, S., Talwat, E., & MacNamara, K. (1983). *Br. J. Dermatol.*, 108:581–586.

Hay, R. J., Calderon, R. A., & Mackenzie, C. D. (1988). *Br. J. Exp. Pathol.*, 45:56–63.

Hay, R. J., Campbell, C. K., Wingfield, R., & Clayton, Y. M. (1983). *Br. J. Indust. Med.*, 40:353–355.

King, R. D., Khan, H. A., Foye, J. C., et al. (1975). *J. Lab. Clin. Med.*, 86:204–212.

Koga, T., Duan, H., Urabe, K., et al. (2001a). *Eur. J. Dermatol.*, 11:105–107.

Koga, T., Shimizu, A., & Nakayama, J. (2001b). *Med. Mycol.*, 39:87–90.

Jones, H. E., Reinhardt, J. H., & Rinaldi, M. G. (1973). *Arch. Dermatol.*, 108:61–68.

Jones, H. E., Reinhardt, J. H., & Rinaldi, M. G. (1974). *Arch. Dermatol.*, 110:369–374.

Kalfa, V. C., Roberts, R. L., & Stiehm, E. R. (2003). *Ann. Allergy Asthma Immunol.*, 90:259–264.

Kano, R. (2004). *Jap. J. Med. Mycol.*, 45:131–136.

Kennedy, C. T., Valdimarsson, H., & Hay, R. J. (1981). *J. R. Soc. Med.*, 74:158–162.

Kesavan, S., Holland, K. T., & Ingham, E. (2000). *Med Mycol.*, 38:239–247.

Kirkpatrick, C. H., Rich, R. B., & Bennett, J. E. (1971). *Ann. Intern. Med.*, 74:955–978.

Kligman, A. M. (1952). *J. Invest. Dermatol.*, 18:231–246.

Kobrynski, L. J., Tanimune, L., Kilpatrick, L., et al. (1996). *Clin. Diagnost. Lab. Immunol.*, 3:740–745.

Kogawa, K., Kudoh, J., Nagafuchi, S., et al. (2002). *Clin. Immunol.*, 103:277–283.

Leibovici, V., Evron, R., Axelrod, O., et al. (1995). *Clin. Exp. Dermatol.*, 20:390–394.

Liautaud, B. & Marill, F. G. (1984). *Bull. Soc. Fr. Pathol. Exotique.*, 77:637–648.

Lilic, D. & Gravenor, I. (2001). *J. Clin. Pathol.*, 54:81–83.

Lilic, D., Cant, A. J., Abinun, M., et al. (1996). *Clin. Exp. Immunol.*, 105:205–212.

Lopez-Garcia, B., Lee. P. H., & Gallo, R. L. (2006). *J. Antimicrob. Chemother.*, 57:877–882.

McGregor, J. M., Hamilton, A. J., & Hay, R. J. (1992). *Br. J. Dermatol.*, 127:233–238.

Maurer, T., Rodrigues, L. K., Ameli, N., Phanuphak, N., & Gange, S. J., DeHovitz, J., French, A. L., Glesby, M., Jordan, C., Khalsa, A., & Hessol, N. A. (2004). *Clin. Infect. Dis.*, 38:579–584.

Mayou, S. C., Calderon, R. A., Goodfellow, A., & Hay, R. J. (1987). *Clin. Exp. Dermatol.*, 12:358–388.

Mignon, B. (2005). *Bull. Mem. Acad. Royale Med. Belgique.*, 160:270–275.

Morris-Jones, R., Youngchim, S., Hextall, J. M., Gomez, B. L., Morris-Jones, S. D., Hay, R. J., Casadevall, A., Nosanchuk, J. D., & Hamilton, A. J. (2004). *J. Clin. Microbiol.*, 42:3789–3794.

Munoz-Perez, M. A., Rodriguez-Pichardo, A., Camacho, F., et al. (1998). *Br. J. Dermatol.*, 139:33–39.

Raska, M., Zemanova, E., Kafkova, L., Belakova, J., Vudattu, N. K., Kopecek, P., & Weigl, E. (2004). *Mycoses*, 47:482–490.

Rollins-Smith, L. A., Woodhams, D. C., Reinert, L. K., Vredenburg, V. T., Briggs, C. J., Nielsen, P. F., & Conlon, J. M. (2006). *Develop. Comp. Immunol.*, 30:831–842.

Schectman, R. C., Allen, M. H., McGregor, J. M., & Hay, R. J. (1993). *J. Med. Vet. Mycol.*, 31:459–462.

Serjeantson, S. & Lawrence, G. (1977). *Lancet*, 1:13–15.

Serrano-Gomez, D., Leal, J. A. C., & Angel, L. (2005). *Immunobiology*, 210:175–183.

Shiraki, Y., Ishibashi, Y., Hiruma, M., Nishikawa, A., & Ikeda, S. (2006). *J. Med. Microbiol.*, 55:1175–1185.

Siddaiah, N., Erickson, Q., Miller, G., & Elston, D. M. (2004). *Cutis*, 73:237–238.

Svejgaard, E. (1985). *Sem. Dermatol.*, 4:201–221.

Torssander, J., Karlsson, A., Morfeldt-Mason, L., et al. (1988). *Acta. Derm. Venereol.*, 68:53–59.

Wakabayashi, H., Takakura, N., Yamauchi, K., et al. (2002). *J. Med. Microbiol.*, 51:844–850.

Woodfolk, J. A. (2005). *Clin. Microbiol. Revs.*, 18:30–43.

Yu, R. J., Harmon, S. R., et al. (1971). *J. Invest. Dermatol.*, 56:27–32.

Zurita, J. & Hay, R. J. (1987). *J. Invest. Dermatol.*, 89:529–534.

Chapter 11
Insights in *Paracoccidioides brasiliensis* Pathogenicity

Luiz R. Travassos, Gustavo Goldman, Carlos P. Taborda, and Rosana Puccia

Introduction

Paracoccidioides brasiliensis is the agent of the human systemic disease paracoccidioidomycosis (PCM) which affects individuals in endemic areas extending from Argentina to Central America. It is the most prevalent deep mycosis in this vast region. Infected people include rural workers but also dwellers of urban centers located on the route of migration movements. Ten million people may be infected and up to 2% of them may develop the disease (Restrepo, 1985; McEwen et al., 1995; Camargo & Franco, 2000). PCM is the eighth most common cause of death among chronic/recurrent infections and parasitic diseases in Brazil (Coutinho et al., 2002). It is also recognized as one of the imported fungal infections in Japan. More than 15 cases of PCM have been reported in that region with the diagnosis methods including the amplification of *P. brasiliensis* genes (Sano et al., 2001).

The infection is most probably acquired by inhalation of conidia which transform into yeast forms that give rise to an asymptomatic lung infection, or progress to cause acute, subacute (juvenile type), and chronic (adult type) clinical forms (McEwen et al., 1987). Experimental intratracheal (i.t.) infections with cultured yeast cells of *P. brasiliensis* have been obtained (Castañeda et al., 1987) but it is unclear whether yeast forms arising at environmental temperatures of 33–38°C, common in tropical habitats, particularly in the Amazon, remain viable for a while sufficient to infect human beings. Generally, *P. brasiliensis* is a restricted saprobe growing in nature with low production of infective propagulae (Bagagli et al., 2006). Natural infection of wild and domestic animals has been reported based on intradermal and serological tests. The fungus has frequently been isolated from the nine-banded armadillo *Dasypus novemcinctus* in Brazil (Bagagli et al., 2003) and from the naked-tailed armadillo *Cabassous centralis* in Colombia (Corredor et al., 2005). It was found in 75–100% of armadillos from hyperendemic PCM areas (Bagagli et al., 2003; Restrepo et al., 2000), and was recovered from young and older adult animals with no sign of disease. Granulomas with fungal cells were eventually observed in lymph nodes, liver, and lungs of a few animals (Bagagli et al., 2003). IgM and IgG antibodies to two *P. brasiliensis* major antigens (gp43 and gp70) were detected in a survey of 47 armadillos (Fernandes et al., 2004).

K. Kavanagh (ed.), *New Insights in Medical Mycology*.
© Springer 2007

These animals rove through areas of disturbed and humid vegetation near water sources. Isolation of *P. brasiliensis* from habitats of armadillo roving has been unsuccessful but specific fungal DNA has been amplified from soil samples collected inside the animal's burrow (Theodoro et al., 2005). The increase in the number of PCM cases in the Amazon has been attributed to soil churning and other practices by developmental projects in the region.

Malnutrition is a recognized factor favoring PCM in rural workers. Men are more frequently affected than women who seem to be more resistant owing to estrogenic hormone production (Restrepo et al., 1984). In fact, the transition from mycelium to yeast in *P. brasiliensis* is inhibited by 17-β-estradiol. It seems that the host defense mechanisms are primarily effective at the portal of entry of the fungus where conidia transform into infective yeast forms. Local phagocytes, leukocyte migration, natural killer (NK) cells, and complement are elements of the host natural defense that attempt to destroy the yeast cells (Camargo & Franco, 2000) which have a unique capacity to form multibudding structures (Figure 11.1). Alveolar macrophages when in contact with *P. brasiliensis* yeast forms expressed threefold more MHC class II, intercellular adhesion molecule-1, and B7-2 molecules. Cultured macrophages produced IL-6, TNF-α, and MIP-1α (Fornazim et al., 2003). As determined by Schall et al. (1993), HuMIP-1α at 100 pg/mL attracts B cells and cytotoxic T cells, whereas at higher concentrations (10 ng/mL), the migration of these cells appears diminished, and the migration of CD4$^+$ T cells is enhanced. Conidia ingested by resident macrophages transform into yeast cells that grow intracellularly by budding (Cano et al., 1992). When monocyte-derived macrophages were treated for 3 days with recombinant human IFN-γ (300 U/mL) or cytokines from concanavalin-A-stimulated mononuclear cells, the multiplication of *P. brasiliensis* was inhibited by 65% and 95%, respectively, as compared with control macrophages (Moscardi-Bacchi et al., 1994).

It seems then that natural defense mechanisms are apt to recruit and activate cells that will destroy the fungus at the start of infection. Effective granuloma formation could contain the fungal cells in the lung thwarting the infection progress. Well-resolved lesions might still contain detectable fungal remains such as specific DNA. Detection of *P. brasiliensis* DNA in tissue samples from infected mice is now possible based on a highly sensitive and specific nested PCR assay (Bialek et al., 2000). The gp43 gene was selected for the test aiming at a highly specific response.

Expression of *P. brasiliensis* virulence is directly related to the host resistance but there are fungal isolates that show increased pathogenicity based on a number of attributes. Generally, host resistance is translated by granuloma formation along with a robust cell-mediated immune response. If the original infection evolves into a clinical form, it is benign, localized, with a compact epithelioid granuloma enclosing few fungal elements. In the absence of an effective cell-immune response, which corresponds to the anergic pole of the disease, disseminated infection is observed with mixed suppurative and loose granulomatous inflammation, extensive areas of necrosis and large numbers of multiplying fungi (reviewed in Camargo & Franco, 2000). Experimentally, a similar picture could be reproduced in athymic

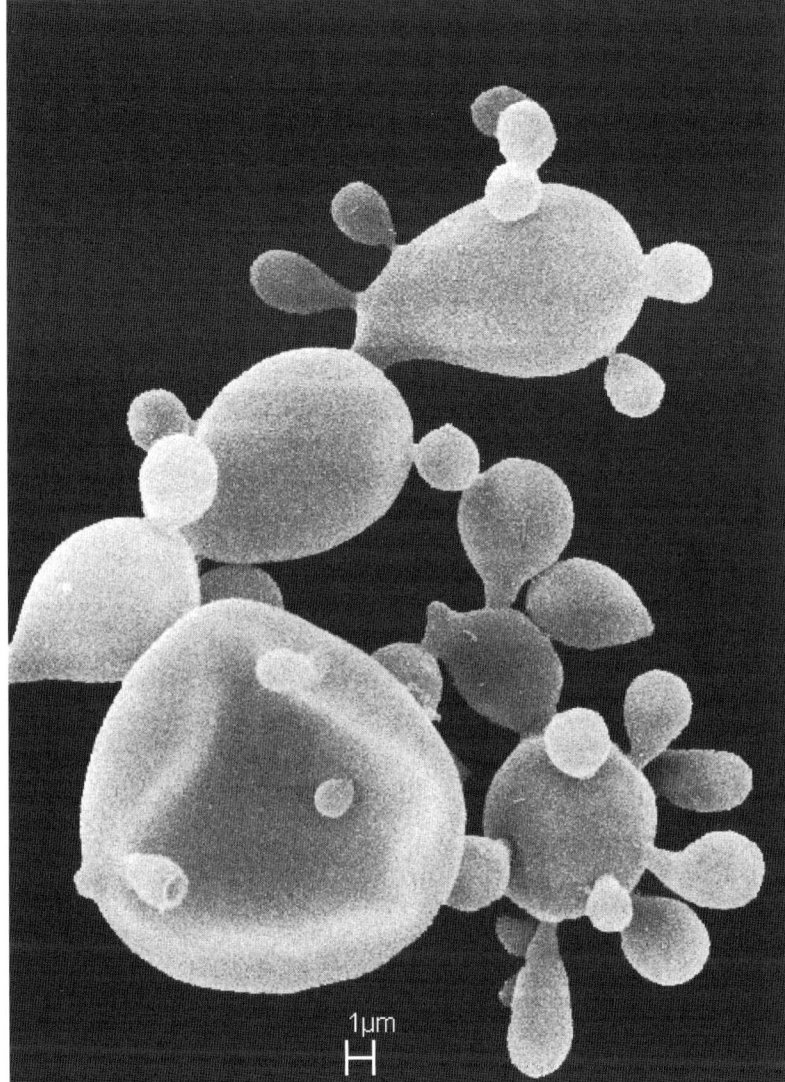

Figure 11.1 Scanning electron micrograph of *P. brasiliensis* (Pb18) yeast cells

(nu/nu) mice that showed high mortality, with few granulomas and many viable fungi (Miyagi & Nishimura, 1983). Apparently, the absence of T lymphocytes led to unstimulated macrophages that lost their fungicidal activity. IFN-γ is the key cytokine mediating macrophage activation and the organization of granulomas. Depletion of IFN-γ by specific monoclonal antibodies exacerbated lung infections and resulted in increased dissemination of yeast forms to other organs (Cano et al., 1998). Mice resistant and susceptible to PCM have long been explored by Calich research group using Pb18 isolate (Calich et al., 1994). Production of IFN-γ and IL-2 and a predominant secretion of IgG2a antibodies are associated with resistance. Low levels of IFN-γ and early secretion of IL-5 and IL-10, eosinophilia, and production of IgG2b and IgA isotypes are found in susceptible animals with progressive disease (Kashino et al., 2000). In mice homozygous for the null mutation of the gene encoding the IFN-γ receptor, and which were highly susceptible to *P. brasiliensis* infection, typical epithelioid granulomas were not formed. Infiltrates of neutrophils with numerous isolated yeast cells were seen. Upon i.t. challenge with virulent yeasts 100% mortality occurred after 3–4 weeks in groups of IFN-γ, IFN-γ-R, and IRF-1 knockout (–/–) mice but not in IFN-α-R and IFN-β-R knockouts (Travassos et al., 2004a). Moreover, IFN-γ, TNF-α, and IL-12-dependent expression of ICAM-1 was upregulated in T lymphocytes after infection with *P. brasiliensis*. ICAM-1 is effectively involved in cellular migration and organization of the granulomatous lesion (Moreira et al., 2006). In Balb/c mice infected with conidia, ICAM-1 was present in bronchiolar epithelium, type II pneumocytes, and macrophages as well as on vascular endothelium. The decrease in CFUs on the early stage postchallenge coincided with upregulation of ICAM-1, VCAM-1, CD18, and Mac-1 expression by cells of the inflammatory process (Gonzalez et al., 2005b).

An association of KC and MIP-1alpha (CCL3) and neutrophil infiltration in the lungs of infected mice was observed during the early acute phase of infection. In IFN-γ KO mice, both KC and MIP-1alpha had increased expression simultaneously with chronic neutrophilia. High levels of RANTES/CCL5, MCP-1/CCL2, EP-10/CXCL10, and Mig/CXCL9 were observed along with mononuclear cell infiltration in the lungs of infected mice. Infected cells from wild-type mice preferentially migrated in response to IP-10 (CXCR3 ligand), whereas those from IFN-γ KO mice migrated in response to eotaxin/CCL11 (CCR3 ligand) (Souto et al., 2003). These results show that IFN-γ by modulating the expression of chemokines and their receptors strongly influences the kind of cells that infiltrate the lungs in mice infected with *P. brasiliensis*.

As a summary for the overall series of events that characterize the inflammatory response in PCM, lymphokines produced by activated lymphocytes of helper and cytotoxic subsets attract and activate macrophages that kill *P. brasiliensis*, produce cytokines and further differentiate into epithelioid and giant cells. Neutrophils, NK cells, and eosinophils are effector cells that concentrate around the fungal cells. Around the tubercles there are B lymphocytes and plasma cells which produce antibodies that diffuse into the granuloma to interact with fungal antigens. The pathology arises as a consequence of the insufficient macrophage and neutrophil activity which depends on optimal activation of T lymphocytes and production of cytokines. In the circumstance of a Th-2/B cell predominant response with high levels of specific antibodies, increased production of IL-4, IL-5 and IL-10 is associated with host susceptibility to infection. With a predominant Th-1 response, IFN-γ, IL-2, and IL-12 are associated with host resistance. Recently, the persistence of *P. brasiliensis* infection in the host has been attributed to natural regulatory T (Treg) cells that could control systemic and local immune response in patients with PCM. A high frequency of CD4⁺ CD25⁺ T cells expressing CTLA4, glucocorticoid-inducible

TNFR, membrane-bound TGF-β, and forkhead-box 3 was found in PBMC from patients exhibiting a strong suppressive activity (Cavassani et al., 2006).

Dimorphic fungi which cause systemic mycoses, *P brasiliensis, Blastomyces dermatitidis, Histoplasma capsulatum*, and *Coccidioides immitis*, have been classified in the Onygenaceae family *sensu lato*, order Onygenales and phylum Ascomycota, although not all the telomorphic forms have been discovered. Analyses of rRNA divided Onygenaceae into several clades. The new Ajellomycetaceae family Untereiner, Scott and Sigler fam. nov. was introduced for a distinct monophyletic clade that includes the genus *Ajellomyces* and species *dermatitidis* (anamorph *B. dermatitidis* Gilchrist and Stokes), *capsulatus* (anamorph *H. capsulatum* Darling), and *crescens* (anamorph *Emmonsia crescens*), as well as other genera and species, *Emmonsia* spp., *Paracoccidioides* (*P. brasiliensis*) and *Lacazia* (*L. loboi*) (Untereiner et al., 2004). *P. brasiliensis, A. dermatitidis*, and *A. capsulatus* grow in the yeast phase *in vivo* and *in vitro* at 35–37°C, therefore are called thermodimorphic, with the mycelial phase developing at lower temperatures. None of the members of this clade display keratinolytic activity which is a mycological phenotype of many fungal species within Onygenales.

A recent study of ten clinical and environmental isolates demonstrated that *P. brasiliensis* has either haploid or aneuploid genome, whose size is in the 26.3–35.5 Mb range per uninucleated yeast cell (Almeida et al., 2006). The authors used flow cytometry and Syber Green I-stained cells to determine the DNA content values in R_1 uninucleated cell subpopulations (Almeida et al., 2005). Ploidy was estimated by comparing these values with those previously reported for the same isolates using weighted sum of pulsed field gel electrophoresis chromosomal bands, which totalize 4–5 with molecular masses between 2 and 10 Mb (Cano et al., 1998; Feitosa et al., 2003; Montoya et al., 1997, 1999).

Recently, three distinct phylogenetic groups of *P. brasiliensis* have been identified following analysis of multilocus genealogies of eight regions from five different loci (Matute et al., 2006a). A total of 65 clinical and environmental isolates from endemic regions of Latin America were studied. Most of the samples were classified as species S1, distributed in Brazil, Argentina, Paraguay, Peru, and Venezuela, while phylogenetic species PS3, which was strongly supported by the α-tubulin genealogy, included 21 Colombian isolates. Phylogenetic species PS2 comprised five Brazilian and one Venezuelan isolate. The genealogies of four regions strongly supported PS2 as an evolutionary independent lineage, which is, however, in sympatry with S1, suggesting barriers to gene flow other than geographic isolation. Additional sequencing analysis by Matute et al. (2006a) also indicates for the first time that *P. brasiliensis* might undergo recombination in Nature. A complementary survey showed that the fungal phylogenetic species can easily be identified with a microsatellite marker system that unequivocally discriminates between S1 and PS2 (Matute et al., 2006b).

Attempts at transformation of *P. brasiliensis in vitro* have only recently been successful. Yeast cells of strain Pb01 were transformed to hygromycin B resistance using plasmid pAN7.1. The integrated *HPH* sequence in the transformed genomes had a degree of instability probably due to the multinuclearity of *P. brasiliensis* cells (Soares et al., 2005). A similar transformation of *P. brasiliensis* was achieved by coculturing with *Agrobacterium tumefasciens* (Leal et al., 2004).

Major Antigenic Molecules of *P. brasiliensis*

The 43 kDa glycoprotein (gp43) is the major diagnostic antigen of *P. brasiliensis*. Discovered in 1986 (Puccia et al., 1986), this glycoprotein represents approximately 80% of the exoantigen protein as examined by SDS-PAGE (reviewed in Camargo & Franco, 2000). By using Western blotting, anti-gp43 antibodies from patients in recovery after chemotherapy had lower reactivity in contrast to patients with relapsing disease who showed higher reactivity relative to the patients with installed disease. A decrease of anti-gp43 IgG, IgA, and IgM correlated well with clinical improvement (Giannini et al., 1990). Since anti-gp70 in parallel with anti-gp43 antibodies also decreased in patients undergoing antimycotic therapy, both molecules are markers of human PCM (Camargo et al., 1989). Different isoforms of the gp43 with pIs ranging from 5.8 to 8.5 were not all recognized by patients' sera in capture immunoassays with bound anti-gp43 monoclonal antibodies. The pI 8.5 gp43 was the least recognizable isoform by both sera from acute and chronic patients (Souza et al., 1997). We determined early in 1991 that epitopes in gp43 that elicited a strong antibody response were peptidic in nature and therefore the patients' sera reacted intensely with the deglycosylated molecule (Puccia & Travassos, 1991).

Heat-shock proteins have also been shown to elicit antibody responses in patients with PCM. Recombinant hsp60 was recognized by sera from 72/75 patients. Overall the sensitivity and specificity of assays with hsp60 were high (97.3% and 92.5%, respectively) (Cunha et al., 2002). Also, sera from PCM patients recognized recombinant HSP70 from *P. brasiliensis* (Bisio et al., 2005). The 87 kDa antigen can be detected in the sera of infected patients. It is homologous with heat-shock proteins and a mAb raised against *H. capsulatum* 80 kDa hsp cross-reacted with purified 87 kDa, predominantly yeast-form antigen (Diez et al., 2002). A 45 kDa formamidase detected in both *P. brasiliensis* yeast and mycelium reacted with sera from PCM patients and not with sera from healthy individuals (Borges et al., 2005). Another recombinant antigen which also reacted with sera from patients with 87.5% specificity, although at 58.7% compared to patients with other mycoses, is the 27 kDa protein used at 1 μg/well. Cross-reactivity with sera from patients with aspergillosis and histoplasmosis was observed (Ortiz et al., 1998).

Gp43 is regularly isolated from yeast culture supernatants by affinity chromatography using monoclonal antibody (mAb) 17c. Intracellularly this antigen is found in cytoplasmic vacuoles and within lomasomes connected to the plasma membrane (Straus et al., 1996). It then diffuses in the cell wall but is secreted at restricted sites as drop-like aggregates. Therefore gp43 is responsible in part for the surface reactivity of yeast cells *in vitro* and *in vivo* and, as a secreted product, for the interaction with macrophages and B cells, and with specific antibodies. Yeast cells adhered to laminin-1 from mouse sarcoma and the interaction seemingly involved gp43. *In vitro*, laminin-1 bound to gp43 with a Kd of 3.7 nM. *In vivo*, laminin-coated yeast cells had increased virulence in hamsters challenged by testicular injection (Vicentini et al., 1994). Different anti-gp43 mAbs, recognizing different epitopes in the gp43, also showed different effects in the hamster infection model. Some mAbs were protective, inhibited granuloma formation and tissue destruction whereas another mAb did not control infection and increased histopathogenicity (Gesztesi et al., 1996). Laminin-1-coated yeast cells were also administered intratracheally in mice. No differences in CFUs were found in the lungs, livers, and spleens of these animals (Andre et al., 2004) in comparison with untreated yeasts. Apparently, the internal basement lamina containing laminin is less accessible for homotypic binding to externally present laminin-coated yeasts. The latter interact with the alveolar epithelium cells and macrophages and are equally infective compared to uncoated cells. Moreover, *P. brasiliensis* produces a serine-thiol proteinase that degrades laminin, fibronectin, collagen IV, and proteoglycan (Carmona et al., 1995; Puccia et al., 1998) which can be construed as a virulence factor allowing fungal cells to degrade the basal membrane, thus favoring tissue invasion and hematogenous dissemination. The gp43 mediated phagocytosis of yeast cells and anti-gp43 F(ab) fragments inhibited this activity (Almeida et al., 1998b). The interaction of gp43 with macrophages involved in part the mannose/fucose receptor which agrees with the single high-mannose N-linked $Hex_{13}GlcNAc_2$ oligosaccharide chain identified in the glycoprotein (Almeida et al., 1996). Gp43 can be presented by dendritic cells or by B cells resulting in Th1 or Th2-type immune responses, respectively, with the corresponding major cytokines and implications in the outcome of *P. brasiliensis* infection (Almeida et al., 1998a; Ferreira et al., 2003).

Gp43 gene was cloned, sequenced, and expressed in *Escherichia coli* as a recombinant fusion protein (Cisalpino et al., 1996). It encodes a polypeptide of 416 amino acids with a leader sequence of 35 residues and is present in one copy per genome as shown by Southern blotting and chromosomal mega-restriction analysis. The open-reading frame has two exons and one intron. As mentioned, the mature protein has a single N-glycosylation site. Sequences containing B cell epitopes are still under study chiefly by examining the reactivity of anti-gp43 mAbs. The T cell epitope responsible for delayed-type hypersensitivity reactions, and CD4[+] T cell proliferation, has been mapped to a 15-mer peptide called P10 with the sequence: QTLIAIHTLAIRYAN (Taborda et al., 1998). The hexapeptide HTLAIR with varying flanking regions is essential for the immune cellular response. Both gp43 and P10 protect mice against the i.t. challenge with virulent *P. brasiliensis*. Gp43 has 54–60% homology and 50% identity with the exo-β-1, 3-D-glucanase sequences from different fungi: *Aspergillus oryzae, Blumeria graminis, Schizosaccharomyces pombe, Pichia augusta, Saccharomyces cerevisae*, and *Candida albicans*. The sequences in these proteins that could contain homologous peptides to P10 or gp43 (181–195) showed 60% identity of P10 with peptide NTLRAIQALAERYAP from β-1, 3-glucanase of *B. graminis* (Genbank accession number Q96V64). For comparison, the identity of the corresponding sequences from β-1,3-glucanases of dimorphic fungi related to *P. brasiliensis, H. capsulatum* and *B. dermatitidis*, and of *A. nidulans* (hypothetical

protein with endoglucanase region) and *C. albicans* to gp43 (179–199) that contains
P10, was of 57.1%, 57.1%, 42.8%, and 28.5%, respectively (Blast, NCBI assessments
XP661656, P29717 and website http://genome.wustl.edu/). Gp43 has a mutated
catalytic peptide (NKP instead of NEP) which is essential for glucanase activity
(Figure 11.2).

Polymorphism in gp43 was suggested owing to the identification of isoforms
with distinct isoelectric points (Souza et al., 1997). The gp43 precursor genes of 17
P. brasiliensis isolates were aligned and compared with a consensus sequence. The
genotypic types showed 1–4 or 14–15 informative substitution sites, localized
between 578 and 1166 bp (Morais et al., 2000). Nucleotide differences were
consistent with a second isoelectric point for the deduced protein. The most polymorphic sequences encoded basic gp43 isoforms. The nucleotides encoding the
P10 sequence were conserved, reinforcing the possibility of a peptide based vaccine
(Travassos et al., 2004a, b).

The promoter region and exon 2 of the *PbGP43* locus have been included in the
multilocus studies of Matute et al. (2006). Nucleotide polymorphism in the Pb*GP43*
gene characterized by Morais et al. (2000) has later been analyzed for the promoter
region as well (Carvalho et al., 2005). The authors observed the existence of six
genotypes by comparing sequences of two cloned PCR fragments of the whole
gene, where substitutions were concentrated on exon 2 and tended to lead to amino
acid changes. The core of P10 containing the T cell epitope and glycosylation site
in gp43 are conserved. The sequences from isolates Pb2, Pb3, and Pb4, recovered
from adult patients with PCM, were highly polymorphic, hence phylogenetically

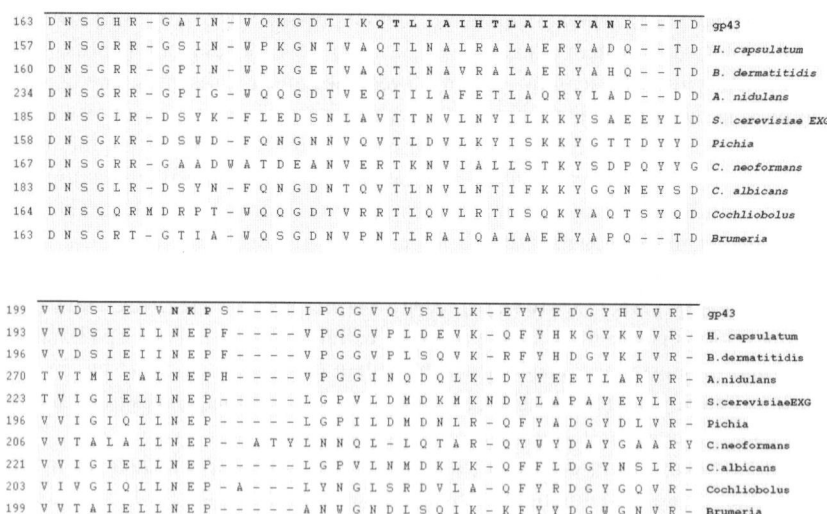

Figure 11.2 Sequence alignment of fragments of *P. brasiliensis* gp43 containing P10 and the
tripeptide *NKP*, with homologous sequences of fungal β-1, 3-glucanases or hypothetical proteins
with endoglucanase domains, showing the catalytic **NEP** motiv

distant from the others. Similar sequences were later found in an armadillo (T10F1) and another clinical (Bt84) sample (Hebeler-Barbosa et al., 2003), where the translated gp43 has basic isoelectric point in contrast with the neutral to acidic protein resulting from the other genotypes. Not coincidentally, the isolates bearing highly polymorphic Pb*GP43* sequences are in the PS2 group defined by Matute et al. (2006a). Therefore, the Pb*GP43* genotypes are reflecting global differences in genetic background. Although there was no association between Pb*GP43* and PCM clinical forms (Morais et al., 2000), differences in the outcome of experimental PCM have been observed in animals inoculated with isolates from the S1 and PS2 species. Using the susceptible B10.A mouse model and organ-recovered yeasts from i.t. and intravenous (i.v.) infections, Carvalho et al. (2005, unpublished data) observed that Pb2, Pb3, and Pb4 (phylogenetic group PS2) evoked few deaths and were recovered from the lungs at significantly fewer colony forming units (CFU) than five isolates from the S1 species, including the highly virulent Pb18 strain. Infection was followed for 30, 60, and 120 days. Anti-gp43 antibody responses elicited by i.t.-inoculated Pb2, Pb3, and Pb4 were rich in IgG2a, IgG2b, and IgG3, while the other sera were richer in IgG1 and IgA. Further i.t. infections, with representative isolates that were adapted *in vivo* before inoculation, showed that mice infected with Pb3, whose CFU in the lungs declined after 120 days infection, secreted increasing amounts of IFN-γ, whereas IL-10 could only be detected after 30 days of infection. The animals infected with S1 isolates had progressive infection and higher CFU counts by day 120, when IFN-γ was undetectable. In these animals, IL-10 was detected at all time points. It is relevant to point out that a reliable correlation between genetic background and virulence should be made using comparable animal models, inoculation routes, inoculum size, and fungal growth status and conditions. In that respect, armadillo isolate T10F1 from species PS2 has been reported as very aggressive in the hamster testicular model (Hebeler-Barbosa et al., 2003), in contrast with the results mentioned above for other PS2 members in the mouse model.

A Peptide Vaccine Against Paracoccidioidomycosis

As mentioned above gp43, the major diagnostic antigen of *P. brasiliensis*, elicits a vigorous IFN-γ-mediated T-CD4$^+$ response that is protective against the i.t. challenge by virulent yeasts of this fungus. The P10 sequence contains the T cell epitope that is presented by MHC class II molecules from three different mouse haplotypes (Taborda et al., 1998). The promiscuous nature of P10 was also shown with HLA-DR molecules, since this peptide and the analogous gp43 (180–194) without C-terminal asparagine residue (N-glycosylation site in the original gp43) and with N-terminal lysine, bound to the 9 most prevalent HLA-DR molecules (Iwai et al., 2003). Furthermore, four additional peptides from gp43 all identified by the TEPITOPE algorithm were also promiscuous with respect to HLA-DR binding. Gp43 (180–194) was recognized by 53% of patients with treated PCM and the other promiscuous peptides were recognized by 32–47% of patients; 74% of patients recognized the combination of 5 promiscuous gp43 peptides (Iwai et al., 2003). An expansion of this study with 10 more patients (total of 29) showed that 79% of PCM patients recognized at least 1 peptide, and by pooling these peptides, the recognition frequency increased to 86%. The TEPITOPE algorithm scanned 25 Caucasian HLA-DRs and P10 and neighboring peptides were predicted to bind to 90% or more of these molecules. These results are the basis for devising a peptide vaccine against PCM that could be used at first as an immunological adjuvant to chemotherapy. The latter may involve a prolonged time and relapses are frequent.

To test this hypothesis, immunization with peptide 10 and chemotherapy were used together in an attempt to improve treatment of experimental PCM and prevent relapses. Two protocols were used. Mice were infected intratracheally with yeast cells of *P. brasiliensis* and drug treatment was started after 48 h or 30 days of the infection. The treatment continued for 30 days, during which groups of mice received intraperitoneal doses of itraconazole, fluconazole, ketoconazole, sulfame-thoxazole, or trimethoprim-sulfamethoxazole every 24 h. Amphotericin B was administered every 48 h. Immunization with P10 was carried out weekly for 4 weeks, once in complete Freund's adjuvant and three times in incomplete Freund's adjuvant (Marques et al., 2006). There was a significant reduction in the fungal load (measured as CFUs) in both groups immunized with P10 only, or treated with antifungal drugs, but an additive protective effect was observed with the combination of both. In animals treated with sulfamethoxazole, an early protection was followed by relapse, but the association of sulfamethoxazole and P10 vaccina-tion succeeded in controlling the disease. In the second protocol that aimed at reproducing the condition of established infection, the fungal burden was examined after 60 and 120 days of infection. An additive protective effect of P10 immuniza-tion and drug treatment was also observed, with 60–80% reduction of lung CFUs compared with untreated infected animals. Lung homogenates were examined for cytokines in all groups. Generally, chemotherapy led to a predominant Th2 response with increased production of IL-4 and IL-10. P10 vaccination stimulated a Th1 response, rich in IFN-γ and IL-12 without suppressing the Th2 component of the immune response (Marques et al., 2006).

In most cases of acute or subacute forms of paracoccidioidomicosis (most aggressive forms of PCM) the specific humoral immune response is preserved with the patients presenting high antibody titers, accompanied by severe depression of cell-immunity (Del Negro et al., 1994). In an attempt to reproduce experimental acute or subacute disease, Balb/c mice were treated with dexamethasone-21 phosphate added in the drinking water. After 30 days of treatment animals were anergic, as shown by negative DTH reactivity. This was previously determined in infected animals, treated with 0.15 mg/kg of dexamethazone and injected with *P. brasiliensis* crude antigen. Mice were then infected with a virulent *P. brasiliensis* isolate. Fifteen days after infection, animals were treated with trimethoprim-sulfamethoxazole or itraconazole, were immunized with P10 or received the association of both treatments (drug and peptide). Mice were killed 45 or 90 days after infection and CFU determination showed that treatment with drugs or P10 immunization resulted in significant reduction of the fungal burden in the lung, spleen and liver. The association of drugs and P10 immunization conferred additional protection. There was a significant increase of IL-12 and IFN-γ and decrease of IL-4 and IL-10 in groups of mice immunized with P10 alone or in association with antifungal drugs. These results suggest that P10 immunization can be effective in the case of anergic patients.

The question of antigen delivery using formulations that do not require CFA has been addressed in two ways. A plasmid DNA construct including the gp43 gene was used to immunize mice followed by i.t. challenge with *P. brasiliensis* yeast forms. A mixed Th-1/Th-2 immune response was elicited which protected Balb/c mice against fungal infection (Pinto et al., 2000). When lymphocytes from mice, which had been immunized with plasmid over a 60-day period were tested *in vitro* with gp43 or ConA, only IL-2 and IFN-γ but not IL-4 and IL-10 were produced. Moreover, the lung clearance observed with plasmid immunization was very similar to that obtained with P10 immunization. Immunization with plasmid DNA with the P10 minigene, associated or not with a plasmid carrying a murine IL-12 DNA insert, was tested in Balb/c mice intratracheally infected with a virulent isolate (Pb18) of *P. brasiliensis*. Similarly, as described above, such immunization elicited significant reduction of fungal burden in lung, spleen and liver. Co-immunization with P10/IL-12 plasmids rendered increased protection with virtually no detection of dissemination to spleen and liver. A mixed Th1/Th2 immune response was also detected. A significant production of IL-12 and IFN-γ and reduction of IL-4 levels in lung homogenates were observed with DNA vaccination with plasmid-P10 or -IL-12 individually or combined.

Another approach to deliver the T cell epitope involved a multiple antigen peptide (MAP) construction. Four equal LIAIHTLAIRYAN chains were synthesized on a branched lysine core containing glycine at the C-terminal position and the product was called M10 because of its analogy with P10. Control MAPs had the same lysine core attached to truncated peptides derived from P10, with 5–10 amino acids long. Proliferation of lymph node cells from P10 or M10-sensitized mice was identical when stimulated *in vitro* with either P10 or M10. Mice immunized once with M10 with no adjuvant and challenged intratracheally with *P. brasiliensis* showed significantly fewer lung, spleen and liver CFUs and few or no yeasts in lung histopathological sections (Taborda et al., 2004).

Further association of P10 with a gp43 peptide eliciting protective antibody response is still being considered because no defined linear sequences have been found so far reacting with antibodies. A report on the characterization of anti-gp70 mAbs protective against experimental PCM (Mattos Grosso et al., 2003) stimulated the parallel study with gp43. Both gp43 and gp70 are markers monitoring the regression of PCM through their decreased antigenemia and specific antibody response in patients submitted to chemotherapy (Marques da Silva et al., 2004; Silva et al., 2004).

The effect of anti-gp43 mAbs on the i.t. infection by *P. brasiliensis* was examined in Balb/c mice injected intraperitoneally with 6 IgG2a and IgG2b different antibodies 24h before the challenge. Analysis of the CFU in the lungs of animals 15 days after infection showed that all tested antibodies, except mAb 32H, were partially protective. Thirty days after infection, however, mAbs 19G, 10D, and 3E still showed a protective effect, with reduced lung CFUs, while mAb 32H, 17D, and 21F allowed full infection, with lung CFUs similar to control animals treated only with irrelevant mAbs. In a phagocytosis assay with peritoneal and alveolar macrophages, all monoclonal antibodies, except 32H, increased significantly the phagocytosis index (Buissa-Filho et al., 2005). Studying the reactivity of mAb 3E, the most efficient in the reduction of fungal burden and phagocytosis, against a panel of gp43-derived peptides it is suggested that the epitope recognized by mAb 3E lies within the sequence NHVRIPIGYWAV shared with *A. fumigatus, A. oryzae* and *B. graminis* internal sequences of β-1,3-glucanases.This peptide might play a role, together with P10, as another candidate for a peptide vaccine against PCM. A scheme showing the network association of data on gp43 and P10 aiming at a human vaccine is shown in Figure 11.3.

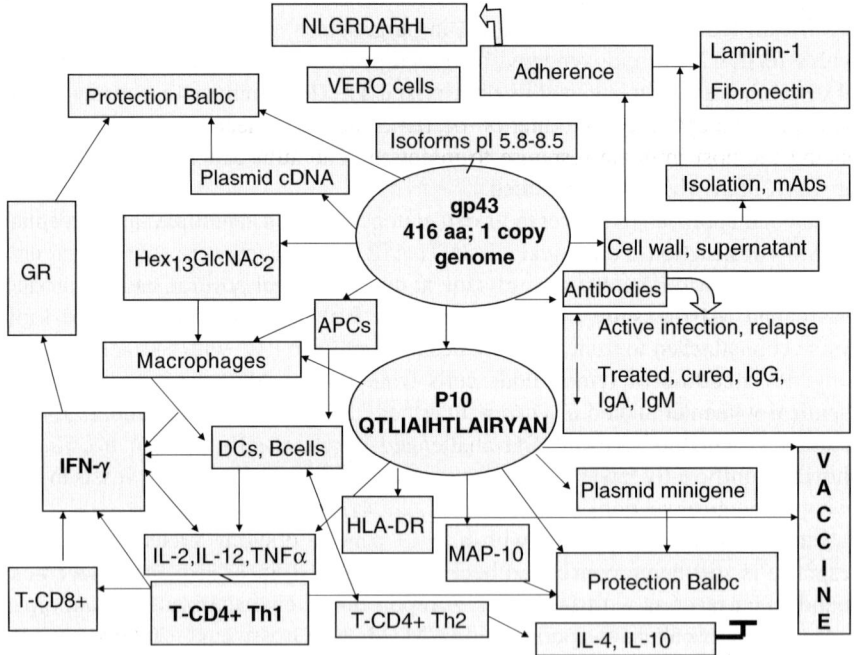

Figure 11.3 Network associating the diverse studies with gp43 and derived peptide P10, aiming at a human vaccine. All interactions and materials used are described in the text. APCs, antigen presenting cells; DCs, dendritic cells; MAP, multiple antigen peptide; GR, granuloma; HLA-DR, human lymphocyte antigen, DR alleles; Hex$_{13}$GlcNAc$_2$, N-linked oligosaccharide of gp43, Hex$_{13}$ = 12 α-mannosyl and 1 β-galactofuranosyl, residues, NAc$_2$Glc = 2 N-acetylglucosaminyl residues

Virulence Factors: A Phenotypic Study

Melanin and Laccase in P. brasiliensis

Melanin pigments are multifunctional polymers ubiquitous in Nature produced by all biological kingdoms, including bacteria, fungi, plants, helminthes, and animals (reviewed in Nosanchuk & Casadevall, 2006). They are amorphous, negatively charged, hydrophobic pigments of high molecular weight, insoluble in aqueous and organic solvents, resistant to concentrated acid, susceptible to bleaching by oxidizing agents and not amenable to either solution or crystallographic structural studies and, consequently, a definitive chemical structure by conventional biochemical and biophysical techniques is a difficult project to pursue (Hamilton & Gomez, 2002; Nosanchuk & Casadevall, 2006). Melanins are composed of polymerized phenolic and/or indolic compounds, usually dark brown or black, and consist of stable organic free radicals (Enochs et al., 1993). In mammals, melanin synthesis is catalyzed by tyrosinase (Sanchez-Ferrer et al., 1995) while microbes generally synthesize melanin via various phenoloxidases such as tyrosinases, laccases, or catacholases and/or by the polyketide synthase pathway (reviewed in Wheeler and Bell, 1988). Many diverse functions have been attributed to melanins and have been associated with virulence in a variety of pathogenic microbes.

Melanin has been associated with virulence also in *P. brasiliensis* (Gomez et al., 2001). It is believed to contribute to microbial virulence by reducing the fungal pathogen susceptibility to killing by host antimicrobial mechanisms and by influencing the host immune response to infection. The authors recovered pigmented particles after chemical and enzymatic treatment of conidia and yeasts grown *in vitro* (on water agar and in the presence of L-DOPA, respectively). Melanin-like particles were also obtained from infected mouse tissue after chemical and enzymatic treatments. They were reactive with antibody to melanin and had yeast-like morphology. A laccase-like activity was detected in protein extracts providing strong evidence for the enzymatic synthesis of melanin in yeast cells. On the other hand, evidence for *in vitro* conidial melanization is even stronger since conidia become pigmented when suspended in water, indicating a capacity to synthesize melanin in the absence of L-DOPA (Gomez et al., 2001).

The ability of melanin to protect microbes from host defenses is relevant to antimicrobial therapy because the clinical efficacy of some antimicrobial drugs has to be complemented by host immune defenses (Nosanchuk & Casadevall, 2006). Melanized *P. brasiliensis* yeast cells showed increased resistance to phagocytosis. Since melanins are charged polymers their presence in the cell wall can alter the fungal cell surface charge as in *C. neoformans* (Nosanchuk & Casadevall, 1997) and thus may contribute to inhibition of phagocytosis. In addition to reducing ingestion, melanization protects *P. brasiliensis* against killing by macrophages (Silva et al., 2006). In fact, melanization protects *P. brasiliensis* against injury mediated by nitrogen- and oxygen-derived free radicals and enhances protection against H_2O_2 and hypochlorite similarly to *C. neoformans, Aspergillus* spp., *S. schenckii* (reviewed in Nosanchuk & Casadevall, 2003). Melanins are highly effective scavengers of free radicals (Sichel et al., 1991). Electron transfer from free radical species generated in solution to *C. neoformans* melanin has been demonstrated by electron spin resonance spectroscopy (Wang et al., 1996). Similar spectra were generated with melanins from *P. brasiliensis, H. capsulatum, S. schenckii*, and *Pneumocystis* spp. (Nosanchuk & Casadevall, 2003).

Melanin is an antifungal resistance factor, given its ability to reduce the susceptibility of melanized cells to antifungal drugs (Ikeda et al., 2003). Melanization of *P. brasiliensis* yeast cells (Figure 11.4) did not affect the cytotoxicity of amphotericin B, ketoconazole, fluconazole, itraconazole, sulfamethoxazole as measured by the standard broth macrodilution procedures for assessing the susceptibility of yeast cells to antifungal drugs (Silva et al, 2006). Similar results have previously been observed with melanized in comparison with nonmelanized *C. neoformans* and *H. capsulatum* (Van Duin et al., 2002). By using a killing assay, however, the increased resistance of *P. brasiliensis* melanized cells to amphotericin B was observed. This effect was less pronounced with the azoles (Silva et al, 2006). The killing assay with melanized *C. neoformans* and *H. capsulatum* also showed less susceptibility to amphotericin B and caspofungin, but melanization did not affect cell resistance to fluconazole and itraconazole (Van Duin et al., 2002).

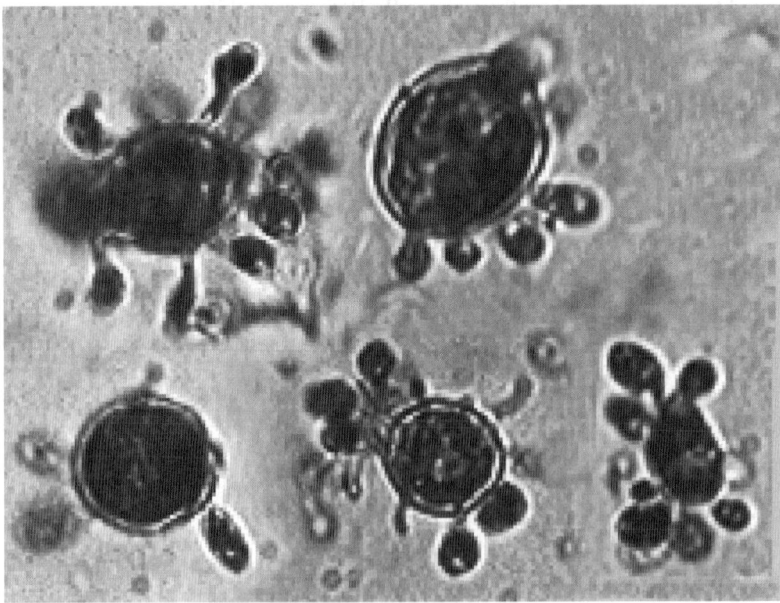

Figure 11.4 Light microscopy of melanized *P. brasiliensis* (Pb18) yeast cells grown in liquid medium with L-DOPA. 400X

Laccase is an important virulence factor in several fungal infections. Laccase production in *P. brasiliensis* has been described by Gomez et al. (2001) and Silva et al. (2006) by two different methods. Laccase in *C. neoformans* has been associated with virulence through melanin synthesis, iron uptake, interference with oxidative burst, and a role in extrapulmonary dissemination. The enzyme was found in the cytoplasm, cell wall and capsule of fungal cells grown *in vivo* (Garcia-Rivera et al., 2005). Further studies are necessary to define the role of laccase as a virulent factor in *P. brasiliensis*.

Adherence to Host Cells and Extracellular Matrix

Cell–cell adhesion and adhesion to extracellular matrix (ECM) proteins play an important role in invasive fungal disease. The first glycoprotein described to be involved in *P. brasiliensis* adherence to ECM proteins was gp43, as described previously. These data would place gp43 as a virulence factor but it is clear that binding to laminin, which can be demonstrated *in vitro*, highly depends on the model used for fungal infection *in vivo*. Another report claimed that gp43 displayed an inhibitory effect on peritoneal macrophages but no clear definition of the cell types present was done in this study (Popi et al., 2002). B1-b cells that are abundant in the peritoneal and pleural cavities were shown by the same group to be antigen presenting cells expressing high levels of MHC class II and inducing an efficient proliferation of gp43-sensitized T lymphocytes (Vigna et al., 2002). Furthermore, activated B1-b cells significantly produce IL-10 that can inhibit macrophage functions. In an *in vitro* model of granuloma assembly it was shown that granulomas were formed when isolated B-1 cells were added to macrophage cultures and that gp43 strongly stimulated this response (Vigna et al., 2006). These results and the intense humoral and cellular immune responses to gp43, which combined, are protective in infected mice, do not support a role for gp43 as a virulence factor in the human or experimental lung infection. The same happened with gp43-plasmid vaccination (Pinto et al., 2000). Different responses, however, probably related to the genetic background of mice have been observed. Protection experiments by gp43 and P10 were mainly obtained in Balb/c mice. By using B10.A susceptible mice, it was shown that administration of dendritic cells incubated with gp43 (10 μg/mL) and LPS caused an increase in lung CFUs of infected animals (Ferreira & Almeida, 2006). Macrophages from susceptible mice, however, produced high and persistent levels of NO throughout the fungal infection and incubation of resident macrophages with *P. brasiliensis* also resulted in NO production. Over expression of NO can induce T cell immunosuppression during infection, inhibition of TNFα production, and less killing of the fungus (Nascimento et al., 2002). It is still unclear whether gp43 acts in B10.A mice as a stimulant of NO production beyond the level that is needed for anti-fungal activity as observed in A/Sn resistant mice. Gp43 and most certainly the high molecular weight glycoconjugate of *P. brasiliensis* (Matsuo et al., 2006) may, however, stabilize the serine-thiol proteinase activity of yeast cells which degrade components of the ECM, conceivably related to fungal dissemination.

An *in vitro* assay for the adhesion of *P. brasiliensis* to cultured mammalian epithelial cells has established a correlation between adherence and virulence, with virulent strain (Pb18) having a greater capacity to adhere. Yeast cells were shown to attach to Vero cells via a narrow tube similar to a germ tube, which subsequently led to the yeast cells making close and extensive contact with the membrane of the host cell, inducing cytoskeletal changes before being taken up or invading the cell (Hanna et al., 2000). Both adhesion that depended on gp43 and cell invasion induced apoptosis demonstrated by TUNEL, DNA fragmentation and Bak and Bcl-2 expression (Mendes-Giannini et al., 2004).

Dectin-1 is a major β-glucan receptor highly expressed in alveolar monocyte/ macrophages and particularly in inflammatory cells (Taylor et al., 2002). *A. fumigatus* stimulated alveolar macrophages to produce IFN-γ and mediate a Th-1 response that was dectin-1 dependent. Curiously, IL-10 synthesis was also dectin-1 dependent (Steele et al., 2005). Macrophages from mice transgenic for pentraxin-3 expression had improved phagocytosis of zymosan particles and yeast forms of *P. brasiliensis*. Increased microbicidal activity and nitric oxide production was observed. Blockade of dectin-1 receptor inhibited phagocytosis in this system (Diniz et al., 2004), suggesting that *P. brasiliensis* yeast forms expose β-glucan despite the fact that α-1,3-glucan predominates in this morphological phase. As provocative hypothesis conidia reaching the alveoli could bind to alveolar macrophages via dectin-1, be internalized and differentiate into infective yeast forms similarly to the model of early pathogenesis of *H. capsulatum* (Procknow et al., 1960; Newman et al., 1990). Adhesion and invasion of nonprofessional phagocytes (epithelial cells) by *P. brasiliensis* depends on signal-transduction involving protein tyrosine kinase (PTK). Treatment with genistein significantly inhibited fungal invasion (Monteiro da Silva et al., 2006).

The adhesion of yeasts to Vero cells (Hanna et al., 2000) was inhibited by anti-gp43 antiserum and a pool of sera from patients with PCM. Mendes-Giannini et al. (2006) showed that ECM proteins bind to the surface of *P. brasiliensis* yeast cells in distinct qualitative patterns. Extracts from Pb18 isolate (virulent), exhibited differential adhesion to ECM components, before and after animal inoculation. After animal inoculation, Pb18 had a higher capacity to adhere to ECM proteins. Laminin was the most adherent component followed by type I collagen, fibronectin (fragment of 120 kDa), and type IV collagen. In addition to gp43 that reacted with both laminin and fibronectin, type I collagen was recognized by 47 and 80 kDa proteins. A peptide derived from gp43 (NLGRDAKRHL) contributed most to the adhesion of *P. brasiliensis* to Vero cells and synthetic peptides derived from peptide YIGRS of laminin or from RGD of both laminin and fibronectin inhibited the adhesion of gp43 to Vero cells.

Other molecules in *P. brasiliensis* were found to interact with laminin, fibrinogen, and fibronectin. Two peptides of 19 and 32 kDa bound to all three proteins and the purified 32 kDa component as well as a mAb against it inhibited adherence of conidia to the ECM proteins in a dose dependent manner (Gonzalez et al., 2005a).

Another 30 kDa protein, with pI 4.9, was described with properties of an adhesin. Laminin but none of the other ECM components bound specifically to this fungal protein (Andreotti et al., 2005).

The glyceraldehyde-3-phosphate dehydrogenase (GAPDH) from *P. brasiliensis*, was identified by immunoproteomic sequencing and microsequencing of peptides in two isoforms of 36 kDa (pIs 6.8 and 7.0). The protein is expressed in the cytoplasm and on the cell wall of the yeast phase, as an adhesin. Recombinant GAPDH bound to fibronectin, laminin, and type I collagen in ligand far-Western blotting assays. Treatment of yeast cells with anti-GAPDH polyclonal antibody and incubation of pneumocytes with the recombinant protein inhibited adherence and internalization of *P. brasiliensis* to those *in vitro*-cultured cells (Barbosa et al., 2006).

The ability of microorganisms to adhere to host cells and to ECM components, resist the immune response, grow and invade tissues and the blood stream for a disseminated infection, are hallmarks of microbial pathogenicity. The research on *P. brasiliensis* virulence has made a significant advance towards the understanding of all these interactions and events at the molecular level.

Virulence Factors: Genes and Transcription

At present, there are two databases of *P. brasiliensis* expression sequences that differ in the isolate used, fungal phase, status, and growth conditions. The transcriptome reported by Goldman et al. (2003) analyzed mRNA from Pb18 (S1 species) yeasts, which had been recovered from the spleen of Balb/c mice infected intraperitoneally and then grown under aerobic conditions in rich medium. The authors obtained 13,490 EST sequences from both 5' and 3'ends (http://143.107.203.68/pbver2/default.html), among which 5,121 singlets and 1,397 contigs have been identified from a total of 4,692 expressed sequences. Felipe et al. (2003, 2005) obtained a total of 19,718 EST sequences from both mycelium and yeast phases of *in vitro*-adapted Pb01 isolate grown in semisolid rich medium. These sequences yielded 6,022 assembled sequences from 2,655 contigs and 3,367 singlets that are deposited in the http://www.biomol.unb.br/Pb database. The details concerning bioinformatics of this project can be found in Brígido et al. (2005), including the subtraction of yeast and mycelium ESTs that was carried out to determine differentially expressed genes. The information obtained with the Pb01 EST sequencing project was discussed in a series of reviews on the annotated transcripts following gene categories related to RNA biogenesis (Albuquerque et al., 2005), therapeutic targets (Amaral et al., 2005), metabolism (Arraes et al., 2005), hydrolytic enzymes (Benoliel et al., 2005; Parente et al., 2005), oxidative stress (Campos et al., 2005), GPI-anchored proteins (Castro et al., 2005), transporters and drug resistance (Costa et al., 2005), cell signaling (Fernandes et al., 2005), molecular chaperones (Nicola et al., 2005), DNA-related events (Reis et al., 2005), translation and protein fate (Souza et al., 2005), virulence (Tavares et al., 2005), and cell wall (Tomazett et al., 2005).

By random sequencing seven loci ranging between 4,439 and 11,093 kb from *P. brasiliensis* Venezuelan isolate IVICPb73, Reinoso et al. (2005) found 20 ORFs at an estimated density of one gene per 3.0–3.7 kb, which would average 9,166 expressed genes for an estimated genome size between 25 and 30 Mb. Therefore, more than 50% of the expressed genes by *P. brasiliensis* may be represented in each EST database described above. The complete genome of *P. brasiliensis* will probably be available in a couple of years, following the efforts of the Broad Institute, with support of the Dimorphic Fungal Genomes Consortium, that is currently developing a comparative genomics project on dimorphic fungal pathogens. The project is centered on *C. immitis*, whose genome will be compared with that of closely related *P. brasiliensis, H. capsulatum, B. dermatitidis*, and *L. loboi*. This study will allow the identification, among many other features, of shared and individual genetic determinants of pathogenicity and virulence in dimorphic fungi. The *P. brasiliensis* isolates included in this study are Pb18, representing major S1 group and virulence, Pb3 from phylogenetic cryptic species PS2, and Pb01 as a molecular model. It has recently been shown that Pb01 alone belongs to a phylogenetic group distinct from those previously identified.

The *P. brasiliensis* EST databases were screened for clusters similar to virulence genes from *C. albicans* and other pathogenic fungi (Goldman et al., 2003; Felipe et al., 2005; Tavares et al., 2005). Although *C. albicans* might not be the best choice for comparison with *P. brasiliensis* in terms of fungal biology and disease outcome, it is still the best-studied fungal pathogen, with the relevance of many genes being assessed using gene-directed mutagenesis. Virulence gene homologs have been recognized in groups of metabolic, cell wall, signal transduction, detoxi-fication, and other genes encoding secreted proteins, suggesting that they could be important for *P. brasiliensis* pathogenesis as well. Another approach to finding virulence-related genes involved detection of differentially expressed transcripts in the yeast phase and during mycelium-to-yeast transition. Large-scale strategies to achieve that purpose included Pb01 database subtraction, macro and microarray hybridization, notably using *P. brasiliensis* biochip carrying 4,962 expressed genes from the Pb18 database (Marques et al., 2004; Felipe et al., 2005; Nunes et al., 2005; Ferreira et al., 2006). Genes with increased mRNA expression in the yeast phase have originally been identified using differential display (Venancio et al., 2002) and later, by high-throughput suppression subtraction hybridization (SSH) analysis, where yeast mRNA was the tester and mycelium mRNA was the driver population (Marques et al., 2004). The latter study added 163 new Pb18 clusters to the EST database, while 20 gene homologs were more expressed in the yeast phase of *P. brasiliensis* grown in three different culture media. Since temperature shift is used to promote the dimorphic transition in all these studies, genes that are exclusively committed to phase transition are hardly distinguishable. Fungal isolate, growth phase, aeration, and culture media are other features that could change the metabolic pattern independently from phase specificity, hence results that do not test these points should be cautiously interpreted. Bailão et al. (2006) aimed at detecting *P. brasiliensis* genes differentially expressed by Pb01 in host conditions using SSH (or RDA for representational difference analysis). The authors used as tester mRNA from yeasts either briefly cultured *in vitro* after isolation from mouse liver or incubated for 10 and 60 min in human blood. The cDNA population obtained from *P. brasiliensis* yeasts exposed to host conditions was subtracted from that isolated from *in vitro*-adapted yeasts grown in solid rich medium. A total of 490 ESTs were identified in the library representing survival in liver and 417 ESTs represented early fungal response upon contact with human blood, but new genes were scarce. In all the studies mentioned above, selected genes were validated for differential expression using Northern blotting, semiquantitative RT-PCR and/or real time RT-PCR.

Increased Transcriptions and Pathogenicity

Since extensive description of *P. brasiliensis* yeast-expressed and phase-transition related genes can be found in the various reports listed above, this section discusses representative genes that seem to be experimentally related to the fungal interaction with the host.

So far the best approach to studying *P. brasiliensis* gene transcriptional response to host cell contact is the one by Tavares et al. (2005) who studied regulation of 1,152 selected fungal genes after 6 h of coculture with peritoneal murine macrophages, with 90% of yeast cells internalized or adherent to phagocytes. Microarray hybridization revealed differential expression of 152 Pb01 transcripts considering a twofold variation in comparison with transcription during *in vitro* growth. Apparently, *P. brasiliensis* reacted to macrophage nitrogen and oxygen radicals by increasing transcription of genes related to oxidative stress. In this sense, *SOD3* (Cu–Zn superoxide dismutase), which encodes a predicted GPI-anchored plasma membrane protein (Castro et al., 2005), was highly expressed, and *HSP60* and *QCR8* (subunit VIII of citocrome oxidase c) mRNAs were also above the twofold cutoff. Histone-related and ribosomal subunit genes were also upregulated, while the β-1,3-glucan synthase gene was downregulated. It should be pointed out that the α-1,3-glucan synthase gene is highly expressed in yeast cells (Marques et al., 2004).

The most abundant transcripts detected by Bailão et al. (2006) in yeasts recovered from liver were *HSP30* and *ZTR1* (high-affinity zinc/iron permease), followed by *CRT3* (high-affinity copper transporter) and *GADPH* (glyceraldehyde 3-phosphate dehydrogenase). While *ZTR1* and *CRT3* transcripts might have been upregulated in response to iron limitation in the organ, GADPH as discussed above adheres to ECM-associated proteins (Barbosa et al., 2004), and may contribute to fungal invasion. Two serine proteinase gene homologs more expressed in host conditions (*PR1H* and *SP1*) may contribute to the process by cleaving local proteins.

It should be pointed out that the glyoxylate cycle-specific genes *ICL1* and *MLS1*, encoding isocitrate lyase and malate synthase, have not been differentially detected in *P. brasiliensis* cultivated with macrophages or human blood, in contrast to *C. albicans* (Fradin et al., 2003). These enzymes are *C. albicans* virulence factors and potential drug targets due to their absence in humans (Lorenz et al., 2001).

During temperature-driven mycelium-to-yeast transition of Pb18 in rich medium, a total of 2,583 genes from a 4,692-element microarray were modulated, showing the complexity of a phenomenon that involves amino acid catabolism, signal transduction, protein synthesis, cell wall metabolism, genome structure, heat and oxidative stress response, growth control, and development (Nunes et al., 2005). The most interesting gene clusters involved transcripts whose levels increased sharply after either 5 or 48 h of temperature shift and remained high until 120 h, when over 80% of the cells were transformed into yeasts. Together, these two clusters involved 30 genes, among which there were two related to amino acid catabolism, encoding a homolog of 4-hydroxyphenyl-pyruvate dehydrogenase (4-HPPD), the *C. immitis* protective T-cell stimulatory antigen, and a branched chain α-keto acid dehydrogenase. The relevance of amino acid catabolism to mycelium-to-yeast conversion and yeast-phase maintenance was determined by the specific inhibitory effect of nitisinone in the dimorphic transition to yeast and yeast growth. Nitisinone, or NTBC (2-(2-nitro-4-trifluoromethylbenzoyl)-cyclohexane-1,3-dione), is a commercially available inhibitor of 4-HPPD used in treatment of hereditary type I tyrosinemia, and could be a completely new alternative in the treatment of PCM. More importantly, Nunes et al. (2005) tested a series of new NTBC derivatives and found that compound 8 is 40 times more potent than nitisinone in the inhibition of yeast cells development.

Included in the two clusters mentioned above (Nunes et al., 2005) are also Pb*GP43, METR*, encoding a zinc-finger transcription factor related to induction of genes of the sulfate assimilation pathway (Ferreira et al., 2006), Pb*Y20*, and gene homologs encoding an alcohol dehydrogenase III and a formamidase. The formamidase gene (Pb*FMD*) has recently been characterized (Borges et al., 2005) and the antigenic properties of the gene product were mentioned in a previous section. Upon contact with macrophages, however, Pb*FMD* was downregulated. The Pb*Y20* gene has recently been characterized (Daher et al., 2005). It encodes a member of the flavodoxin-like WrbA family of quinone reductases, which might be involved in protection against oxygen intermediates. PbY20 protein, together with enolase, has originally been recognized as yeast-preferential by comparing 2-D profiles of total proteins extracted from fully formed mycelium and yeast cells (Cunha et al., 1999). The transcript and the protein are poorly expressed in mycelium cells (Daher et al., 2005), which start to overexpress Pb*Y20* mRNA only after 48 h of temperature shift for transition to yeast cells that express the transcript in large amounts (Felipe et al., 2005; Nunes et al., 2005). The protein was mainly localized to large cytoplasmic vacuoles and in the cell wall. PbY20 might act as an antioxidant *in vivo*, but this function remains to be proved. A 61 kDa catalase, possibly peroxisomal, has been detected in yeast fungal extracts with PCM patients' sera (Fonseca et al., 2001; Moreira et al., 2004). Both transcript and protein were upregulated during transition to yeast and an increase in protein levels was already apparent 20 min after H_2O_2 addition, suggesting that it has an antioxidant role. We have recently characterized two adjacent genes, Pb*LON* and Pb*MDJ1*, encoding mitochondrially sorted ATP-dependent proteinase and an Hsp40, respectively (Barros et al., 2001; Batista et al.,

2006a). The genes share a promoter region that bears 3 mapped heat-shock elements and one AP-1-binding motif directed to Pb*LON*, which upregulates fivefold upon addition of H_2O_2 (Batista et al., 2006b).

References

Albuquerque, P., Baptista, A. J., Derengowsky, L. S., Procopio, L., et al. (2005). *Genet. Mol. Res.*, 4(2):251–272.

Almeida, A. J., Martins, M., Carmona, J. A., Cano, L. E., et al. (2006). *Fungal. Genet. Biol.* 43(6): 401–409.

Almeida, A. J., Matute, D. R., Carmona, J. A., Martins, M., et al. (2006). *Fungal. Genet. Biol.*, Jul 29 (Epub ahead of print).

Almeida, I. C., Neville, D. C., Mehlert, A., Treumann, A., et al. (1996). *Glycobiology*, 6(5): 507–515.

Almeida, S. R., Moraes, J. Z., Camargo, Z. P., Gesztesi, J. L., et al. (1998a). *Cell. Immunol.*, 190(1):68–76.

Almeida, S. R., Unterkircher, C. S., & Camargo, Z. P. (1998b). *Med. Mycol.*, 36(6):405–411.

Amaral, A. C., Fernandes, L., Galdino, A. S., Felipe, M. S., et al. (2005). *Genet. Mol.*, 4(2): 430–449.

Andre, D. C., Lopes, J. D., Franco, M. F., Vaz, C. A. C., & Calich, V. L. G. (2004). *Microbes. Infect.*, 6(6):549–558.

Andreotti, P. F., Monteiro da Silva, J. L., Bailao, A. M., Soares, C. M., et al. (2005). *Microbes. Infect.*, 7(5–6):875–881.

Arraes, F. B., Benoliel, B., Burtet, R. T., Costa, P. L., et al. (2005). *Genet. Mol. Res.*, 4(2): 290–308.

Bagagli, E., Bosco, S. M. G., Theodoro, R. C., & Franco, M. (2006). *Infect. Genet. Evol.*, 6(5): 344–351.

Bagagli, E., Franco, M., Bosco, SMG, Hebeler-Barbosa, F., et al. (2003). *Med. Mycol.*, 41(3): 217–223.

Bailão, A. M., Schrank, A., Borges, C. L., Dutra, V., et al. (2006). *Microbes. Infect.*, 8(12–13): 2686–2697.

Barbosa, M. S., Cunha Passos, D. A., Felipe, M. S., Jesuino, R. S., et al. (2004). *Fungal. Genet. Biol.*, 41(7):667–675.

Barbosa, M. S., Bao, S. N., Andreotti, P. F., Faria, F. P., et al. (2006). *Infect. Immun.*, 74(1): 382–389.

Barros, T. F. & Puccia, R. (2001). *Yeast*, 18(10):981–988.

Batista, W. L., Matsuo, A. L., Ganiko, L., Barros, T. F., et al. (2006a). *Eukaryot. Cell*, 5(2): 379–390.

Batista, W. L., Barros, T. F., Goldman, G. H., Morais, F. V., & Puccia, R. (2006b). *Fungal. Genet Biol.*, Dec 11 (Epub ahead of print).

Benoliel, B., Arraes, F. B., Reis, V. C., Siqueira, S. J., et al. (2005). *Genet. Mol. Res.*, 4(2): 450–461.

Bialek, R., Ibricevic, A., Aepinus, C., Najvar, L. K., et al. (2000). *J. Clin. Microbiol.*, 38(8): 2940–2942.

Bisio, L. C., Silva, S. P., Pereira, I. S., Xavier, M. A., et al. (2005). *Med. Mycol.*, 43(6):495–503.

Borges, C. L., Pereira, M., Felipe, M. S., Faria, F. P., et al. (2005). *Microbes. Infect.*, 7(1):66–77.

Brigido, M. M., Walter, M. E., Oliveira, A. G., Inoue, M. K., et al. (2005). *Genet. Mol. Res.*, 4(2): 203–215.

Buissa-Filho, R., Marques, A. F., da Silva, M. B., Puccia, R., et al. (2005). *IX International Meeting on Paracoccidioidomycosis*, Aguas de Lindóia, São Paulo. Revista do Instituto de Medicina Tropical, São Paulo, Vol. 47, p. 41.

Calich, V. L., Singer-Vermes, L. M., Russo, M., Vaz, C. A. C., & Burger, E. (1994). In: *Paracoccidioidomycosis*, Franco, M., Lacaz, C. S., Restrepo-Moreno, A., & Del Negro, G., Eds., CRC Press, Boca Raton, FL, pp. 151–173.

Camargo, Z. P. & Franco, M. F. (2000). *Rev. Iberoam. Micol.*, 17(2):41–48.

Camargo, Z. P., Unterkircher, C. S., & Travassos, L. R. (1989). *J. Med. Vet. Mycol.*, 27(6): 407–412.

Campos, E. G., Jesuino, R. S., Dantas, A. S., et al. (2005). *Genet. Mol. Res.*, 4(2):409–429.

Cano, L. E., Brummer, E., Stevens, D. A., & Restrepo, A. (1992). *Infect. Immun.*, 60(5): 2096–2100.

Cano, L. E., Kashino, S. S., Arruda, C., Andre, D., et al. (1998). *Infect. Immun.*, 66(2):800–806.

Cano, M. I., Cisalpino, P. S., Galindo, I., Ramirez, J. L., et al. (1998). *J. Clin. Microbiol.*, 36(3): 742–747.

Carmona, A. K., Puccia, R., Oliveira, M. C., Rodrigues, E. G., et al. (1995). *Biochem. J.*, 309 (Pt 1):209–214.

Carvalho, K. C., Ganiko, L., Batista, W. L., Morais, F. V., et al. (2005). *Microbes Infect.*, 7(1): 55–65.

Castañeda, E., Brummer, E., Pappagianis, D., & Stevens, D. A. (1987). *J. Med. Vet. Mycol.*, 25(6): 377–387.

Castro, N. S., Maia, Z. A., Pereira, M., & Soares, C. M. (2005). *Genet. Mol. Res.*, 4(2):326–345.

Cavassani, K. A., Campanelli, A. P., Moreira, A. P., Vancim, J. O., et al. (2006). *J. Immunol.*, 177(9): 5811–5818.

Cisalpino, P. S., Puccia, R., Yamauchi, L. M., Cano, M. I. N., et al. (1996). *J. Biol. Chem.*, 271(8): 4553–4560.

Corredor, G. G., Peralta, L. A., Castano, J. H., Zuluaga, J. S., et al. (2005). *Med. Mycol.*, 43(3): 275–280.

Costa, C. S., Albuquerque, F. C., Andrade, R. V., Oliveira, G. C., et al. (2005). *Genet. Mol. Res.*, 4(2): 390–408.

Coutinho, Z. F., Silva, D., Lazera, M., Oliveira, R. M., et al. (2002). *Cad. Saude. Publica.*, 18(5): 1441–1454.

Cunha, A. F., Sousa, M. V., Silva, S. P., Jesuino, R. S., et al. (1999). *Med. Mycol.*, 37(2): 115–121.

Cunha, D. A., Zancope-Oliveira, R. M., Sueli, M., Felipe, S., et al. (2002). *Clin. Diagn. Lab. Immun.*, 9(2):374–377.

Daher, B. S., Venancio, E. J., de Freitas, S. M., Bao, S. N., et al. (2005). *Fungal Genet. Biol.*, 42(5): 434–443.

Del Negro, G., Lacaz, C. S., Zamith, V. A., & Siqueira, A. M. (1994). In: *Paracoccidioidomycosis*, Franco, M., Lacaz, C. S., Restrepo-Moreno, A., & Del Negro, G., Eds., CRC Press, Boca Raton, FL, pp. 225–232.

Diez, S., Gomez, B. L., Restrepo, A., Hay, R. J., & Hamilton, A. J. (2002). *J. Clin. Microbiol.*, 40(2): 359–365.

Diniz, S. N., Nomizo, R., Cisalpino, P. S., Teixeira, M. M., et al. (2004). *J. Leukoc. Biol.*, 75(4): 649–656.

Enochs, W. S., Nilges, M. J., & Swartz, H. M. (1993). *J. Neurochem.* 61(1):68–79.

Feitosa, L. S., Cisalpino, P. S., Santos, M. R., Mortara, R. A., et al. (2003). *Fungal Genet. Biol.*, 39(1): 60–69.

Felipe, M. S., Andrade, R. V., Arraes, F. B., Nicola, A. M., et al. (2005). *J. Biol. Chem.*, 280(26): 24706–24714.

Felipe, M. S., Andrade, R. V., Petrofeza, S. S., Maranhão, A. Q., et al. (2003). *Yeast*, 20(3): 263–271.

Felipe, M. S., Torres, F. A., Maranhao, A. Q., Silva-Pereira, I., et al. (2005). *FEMS Immunol. Med. Microbiol.*, 45(3):369–381.

Fernandes, G. F., Deps, P., Tomimori-Yamashita, J., & Camargo, Z. P. (2004). *Med. Mycol.*, 42(4): 363–368.

Fernandes, L., Araujo, M. A., Amaral, A., Reis, V. C., et al. (2005). *Genet. Mol. Res.*, 4(2): 216–231.

Ferreira, K. S. & Almeida, S. R. (2006). *Immunol. Lett.*, 103(2):121–126.

Ferreira, K. S., Lopes, J. D., & Almeida, S. R. (2003). *Scand. J. Immunol.*, 58(3):290–297.

Ferreira, M. E., Marques, E. R., Malavazi, I., et al. (2006). *Mol. Genet. Genomics*, 276(5): 450–463.

Fonseca, C. A., Jesuino, R. S., Felipe, M. S., Cunha, D. A., et al. (2001). *Microbes. Infect.* 3(7): 535–542.

Fornazim, M. C., Balthazar, A., Quagliato, R., Jr., Mamoni, R. L., et al. (2003). *Eur. Respir. J.*, 22(6): 895–899.

Fradin, C., Kretschmar, M., Nichterlein, T., Gaillardin, C., et al. (2003). *Mol. Microbiol.* 47(6): 1523–1543.

Garcia-Rivera, J., Tucker, S. C., Feldmesser, M., Williamson, P. R., & Casadevall, A. (2005). *Infect. Immun.*, 73(5):3124–3127.

Gesztesi, J. L., Puccia, R., Travassos, L. R., Vicentini, A. P., et al. (1996). *Hybridoma*, 15(6): 415–422.

Giannini, M. J., Bueno, J. P., Shikanai-Yasuda, M. A., Stolf, A. M., et al. (1990). *Am. J. Trop. Med. Hyg.*, 43(2):200–206.

Goldman, G. H., Marques E. R., Ribeiro, D. C. D., Bernardes, L. A. S., et al. (2003). *Eukaryot. Cell*, 2(1):34–48.

Gómez, B. L., Nosanchuk, J. D., Díez, S., Youngchim, S., et al. (2001). *Inf. Immun.*, 69(9): 5760–5767.

Gonzalez, A., Gomez, B. L., Restrepo, A., Hamilton, A. J., & Cano, L. E. (2005a). *Med. Mycol.*, 43(7): 637–645.

Gonzalez, A., Lenzi, H. L., Motta, E. M., Caputo, L., et al. (2005b). *Microbes. Infect.*, 7(4): 666–673.

Hamilton, A. J. & Gomez, B. L. (2002). *J. Med. Microbiol.*, 51(3):189–191.

Hanna, S. A., Monteiro da Silva, J. L., & Giannini, M. J. (2000). *Microbes. Infect.*, 2(8): 877–884.

Hebeler-Barbosa, F., Morais, F. V., Montenegro, M. R., Kuramae, E. E., et al. (2003). *J. Clin. Microbiol.*, 41(12):5735–5737.

Ikeda, R., Sugita, T., Jacobson, E. S., & Shinoda, T. (2003). *Microbiol. Immunol.*, 47(4): 271–277.

Iwai, L. K., Yoshida, M., Sidney, J., Shikanai-Yasuda, M. A., et al. (2003). *Mol. Med.*, 9(9–12): 209–219.

Kashino, S. S., Fazioli, R. A., Cafalli-Favati, C., Meloni-Bruneri, L. H., et al. (2000). *J. Interf. Cytok. Res.*, 20(1):89–97.

Leal, C. V., Montes, B. A., Mesa, A. C., Rua, A. L., et al. (2004). *Med. Mycol.*, 42(4):391–395.

Lorenz, M. C. & Fink, G. R. (2001). *Nature*, 412(6842):83–86.

Marques da Silva, S. H., Queiroz-Telles, F., Colombo, A. L., Blotta, M. H., et al. (2004). *J. Clin. Microbiol.*, 42(6):2419–2424.

Marques, A. F., da Silva, M. B., Juliano, M. A., Travassos, L. R., & Taborda, C. P. (2006). *Antimicrob. Agents Chemother.*, 50(8):2814–2819.

Marques, E. R., Ferreira, M. E., Drummond, R. D., Felix, J. M., et al. (2004). *Mol. Genet. Genomics*, 271(6):667–677.

Matsuo, A. L., Tersariol, I. L. S., Kobata, S. I., Travassos, L. R., et al. (2006). *Microbes. Infect.*, 8(1): 84–91.

Matute, D. R., McEwen, J. G., Puccia, R., Montes, B. A., et al. (2006a). *Mol. Biol. Evol.*, 23(1): 65–73.

Matute, D. R., Sepulveda, V. E., Quesada, L. M., Goldman, G. H., et al. (2006b). *J. Clin. Microbiol.*, 44(6):2153–2157.

McEwen, J. G., Bedoya, V., Patiño, M. M., Salazar, M. E., & Restrepo, A. (1987). *J. Med. Vet. Mycol.*, 25(3):165–175.

McEwen, J. G., Garcia, A. M., Ortiz, B. L., Botero, S., & Restrepo, A. (1995). *Arch. Med. Res.*, 26(3): 305–306.

Mendes-Giannini, M. J., Andreotti, P. F., Vincenzi, L. R., Silva, J. L., et al. (2006). *Microbes. Infect.*, 8(6):1550–1559.

Mendes-Giannini, M. J., Hanna, S. A., Silva, J. L., Andreotti, P. F., et al. (2004). *Microbes. Infect.*, 6(10): 882–891.

Miyagi, M. & Nishimura, K. (1983). *Mycopathologia*, 82(3):129–141.

Monteiro da Silva, J. L., Andreotti, P. F., Benard, G., Soares, C. P., et al. (2006). *Anton. van Leeuw.*, Nov 21 (Epub ahead of print).

Montoya, A. E., Alvarez, A. L., Moreno, M. N., Restrepo, A., & McEwen, J. G. (1999). *Med. Mycol.*, 37(3):219–222.

Montoya, A. E., Moreno, M. N., Restrepo, A., & McEwen, J. G. (1997). *Fungal. Genet. Biol.*, 21(2): 223–227.

Morais, F. V., Barros, T. F., Fukada, M. K., Cisalpino, P. S., & Puccia, R. (2000). *J. Clin. Microbiol.*, 38(11):3960–3966.

Moreira, A. P., Campanelli, A. P., Cavassani, K. A., Souto, J. T., et al. (2006). *Am. J. Pathol.*, 169(4): 1270–1281.

Moreira, S. F., Bailao, A. M., Barbosa, M. S., Jesuino, R. S., et al. (2004). *Yeast*, **21**(2):173–182.

Moscardi-Bacchi, M., Brummer, E., & Stevens, D. A. (1994). *J. Med. Microbiol.*, 40(3): 159–164.

Nascimento, F. R. F., Calich, V. L. G., Rodriguez, D., & Russo, M. (2002). *J. Immunol.*, 168(9): 4593–4600.

Newman, S. L., Bucher, C., Rhodes, J., & Bullock, W. E. (1990). *J. Clin. Invest.*, 85(1): 223–230.

Nicola, A. M., Andrade, R. V., & Silva-Pereira, I. (2005). *Genet. Mol. Res.*, 4(2):346–357.

Nosanchuk, J. D. & Casadevall, A. (1997). *Infect. Immun.*, 65(5):1836–1841.

Nosanchuk, J. D. & Casadevall, A. (2003). *Cell. Microbiol.*, 5(4):203–223.

Nosanchuk, J. D. & Casadevall, A. (2006). *Antimicrob. Agents Chemother.* 50(11):3519–3528.

Nunes, L. R., Oliveira, R. C., Leite, D. B., Silva, V. S., et al. (2005). *Eukaryot. Cell*, 4(12): 2115–2128.

Ortiz, B. L., Diez, S., Uran, M. E., Rivas, J. M., et al. (1998). *Clin. Diagn. Lab. Immun.*, 5(6): 826–830.

Parente, J. A., Costa, M., Pereira, M., & Soares, C. M. (2005). *Genet. Mol. Res.*, 4(2):358–371.

Pinto, A. R., Puccia, R., Diniz, S. N., Franco, M. F., & Travassos, L. R. (2000). *Vaccine*, 18(26): 3050–3058.

Popi, A. F. F., Lopes, J. D., & Mariano, M. (2002). *Cell Immunol.*, 218(1–2):87–94.

Procknow, J. J., Page, M. I., & Loosli, C. G. (1960). *Arch. Pathol.*, 69(4):413–426.

Puccia, R. & Travassos, L. R. (1991). *Arch. Biochem. Biophys.*, 289(2):298–302.

Puccia, R., Carmona, A. K., Gesztesi, J. L., Juliano, L., & Travassos, L. R. (1998). *Med. Mycol.*, 36(5): 345–348.

Puccia, R., Schenkman, S., Gorin, P. A., & Travassos, L. R. (1986). *Infect. Immun.*, 53(1): 199–206.

Reinoso, C., Nino-Vega, G., San Blast, G., & Dominguez, A. (2005). *Med. Mycol.*, 43(8): 681–689.

Reis, V. C., Torres, F. A., Pocas-Fonseca, M. J., Souza, M. T., et al. (2005). *Genet. Mol. Res.*, 4(2): 232–250.

Restrepo, A., Salazar, M. E., Cano, L. E., Stover, E. P., et al. (1984). *Infect. Immun.*, 46(2): 346–353.

Restrepo, A. (1985). *J. Med. Vet. Mycol.*, 23(5):323–334.

Restrepo, A., Baumgardner, D. J., Bagagli, E., Cooper, C. R., et al. (2000). *Med. Mycol.*, 38 (Suppl. 1): 67–77.

Sanchez-Ferrer, A., Rodriguez-Lopez, J. N., Garcia-Canovas, F., & Garcia-Carmona, F. (1995). *Biochimica. Biophysica. Acta*, 1247(1):1–11.

Sano, A., Yokoyama, K., Tamura, M., Mikami, Y., et al. (2001). *Nippon Ishinkin Gakkai Zasshi*, 42(1): 23–27.

Schall, T. J., Bacon, K., Camp, R. D. R., Kaspari, J. W., & Goeddel, D. V. (1993). *J. Exp. Med.*, 177(6): 1821–1826.

Sichel, G., Corsaro, C., Scalia, M., Di Bilio, A. J., & Bonomo, R. P. (1991). *Free Rad. Biol. Med.* **11**(1): 1–8.

Silva, M. B., Marques, A. F., Nosanchuk, J. D., Casadevall, A., et al. (2006). *Microbes. Infect.*, 8(1): 197–205.

Silva, S. H., Mattos Grosso, D., Lopes, J. D., Colombo, A. L., et al. (2004). *J. Clin. Microbiol.*, 42(10): 4480–4486.

Soares, R. B., Velho, T. A., Moraes, L. M., Azevedo, M. O., et al. (2005). *Med. Mycol.*, 43(8): 719–723.

Souto, J. T., Aliberti, J. C., Campanelli, A. P., Livonesi, M. C., et al. (2003). *American J. Pathol.*, 163(2): 583–590.

Souza, D. P., Silva, S. S., Baptista, A. J., Nicola, A. M., et al. (2005). *Genet. Mol. Res.*, 4(2): 273–289.

Souza, M. C., Gesztesi, J. L., Souza, A. R., Moraes, J. Z., et al. (1997). *J. Med. Vet. Mycol.*, 35(1): 13–18.

Steele, C., Rapaka, R. R., Metz, A., Pop, S. M., et al. (2005). *PLoS Pathog.* 1(4):323–334.

Straus, A. H., Freymuller, E., Travassos, L. R., & Takahashi, H. K. (1996). *J. Med. Vet. Mycol.*, 34(3): 181–186.

Taborda, C. P., Nakaie, C. R., Cilli, E. M., Rodrigues, E. G., et al. (2004). *Scand. J. Immunol.*, 59(1): 58–65.

Taborda, C. P., Juliano, M. A., Puccia, R., Franco, M., & Travassos, L. R. (1998). *Infect. Immun.*, 66(2): 786–793.

Tavares, A. H., Silva, S. S., Bernardes, V. V., Maranhao, A. Q., et al. (2005). *Genet. Mol. Res.* 4(2): 372–389.

Taylor, P. R., Brown, G. D., Reid, D. M., Willment, J. A., et al. (2002). *J. Immunol.*, 169(7): 3876–3882.

Theodoro, R. C., Candeias, J. M. G., Araujo, J. P., Jr., Bosco, S. M. G., et al. (2005). *Med. Mycol.*, 43(8): 725–729.

Tomazett, P. K., Cruz, A. H., Bonfim, S. M., Soares, C. M., & Pereira, M. (2005). *Genet. Mol. Res.*, 4(2):309–325.

Travassos, L. R., Taborda, C. P., Iwai, L. K., Cunha-Neto, E., & Puccia, R. (2004a). In: *The Mycota XII, Human Fungal Pathogens*, Domer, J. E. & Kobayashi, G. S., Eds., Springer, Berlin, Heildeberg, pp. 279–296.

Travassos. L. R., Casadevall, A., & Taborda, C. P. (2004b). In: *Pathogenic Fungi: Host Interactions and Emerging Strategies for Control.*, San-Blas, G. & Calderone, R.A., Eds., Caister Academic Press, Norfolk, England, pp. 241–283.

Untereiner, W. A., Scott, J. A., Naveau, F. A., Sigler, L., et al. (2004). *Mycologia*, 96(4): 812–861.

Van Duin, D., Casadevall, A., & Nosanchuk, J. D. (2002). *Antimicrob. Agents Chemother.*, 46(11): 3394–3400.

Venancio, E. J., Kyaw, C. M., Mello, C. V., Silva, S. P., et al. (2002). *Med. Mycol.*, 40(1):45–51.

Vicentini, A. P., Gesztesi, J. L., Franco, M. F., Souza, W., et al. (1994). *Infect. Immun.*, 62(4): 1465–1469.

Vigna, A. F., Almeida, S. R., Xander, P., Freymuller, E., et al. (2006). *Microbes. Infect.*, 8(3): 589–597.

Vigna, A. F., Godoy, L. C., Almeida, S. R., Mariano, M., & Lopes, J. D. (2002). *Immunol. Lett.*, 83(1): 61–66.

Wang, Y., Aisen, P., & Casadevall, A. (1995). *Infect. Immun. Infect. Immun.*, 63(8):3131–3136.

Wheeler, M. H. & Bell, A. A. (1988). *Current Topics in Med. Mycol.* 2:338–387.

Chapter 12
Fusarium and *Scedosporium*: Emerging Fungal Pathogens

Emmanuel Roilides, John Dotis, and Aspasia Katragkou

The Organisms

Fusarium spp. and *Scedosporium* spp. have emerged as important fungal pathogens during the last decades causing significant morbidity and mortality especially in immunocompromised patients. The two fungal genera possess several biological and clinical characteristics in common, most notably the very high mortality of the diseases caused by them, and thus they are discussed together.

Fusarium spp. are non-dematiaceous hyaline molds (Ajello, 1986). The main characteristic of the genus *Fusarium* is the production of microconidia in the early phase of growth followed by the production of multiseptate sickle-shaped macroconidia, with a more or less pronounced foot cell in later phase (Boutati & Anaissie, 1997). The genus currently contains over 20 species, of which *Fusarium solani* is the most frequent human pathogen (almost half of the cases). Other medically relevant species are *F. oxysporum, F. moniliforme* and less commonly, *F. anthophilum, F. chlamydosporum, F. dimerum, F. equiseti, F. lichenicola, F. napiforme, F. proliferatum, F. Semitecum*, and *F. verticilloides* (Nelson et al., 1994; Dignani & Anaissie, 2004).

Fusarium spp. are ubiquitus fungi commonly found in soil, water, decomposing organic matter, and plants. *Fusarium* spp. can colonize pharyngeal specimens of healthy adults and also the conjunctival sac, especially in diseased eyes. However, in immunologically competent hosts, *Fusarium* spp. infections are very rare involving mainly skin around surgical wounds, burns, deep ulcers, nails, or cornea. Less commonly, these organisms have been documented as aetiological agents in localized tissue infections, including septic arthritis, endophthalmitis, cystitis, peritonitis, and brain abscesses (Guarro & Gene, 1995; Nucci & Anaissie, 2002; Jensen et al., 2004).

In contrast, disseminated infection due to *Fusarium* spp. often occurs in severely immunocompromised patients. *Fusarium* spp. have emerged as the second most frequent filamentous fungal pathogens, after *Aspergillus* spp., in high-risk patients with haematological malignancies, recipients of solid organ and allogeneic bone marrow, or stem cell transplants (Boutati & Anaissie, 1997; Walsh & Groll, 1999; Sampathkumar & Paya, 2001).

K. Kavanagh (ed.), *New Insights in Medical Mycology.*
© Springer 2007

In the latter patient population, the distribution of fusariosis is bimodal, with peaks observed before and several weeks after engraftment. Besides, the epidemiological distribution of fusariosis is not homogeneous, since most of the cases have been reported in the USA. While the incidence in Europe has remained stable over the last 20 years, it has significantly increased in some US institutions (Boutati & Anaissie, 1997). Although it is unclear whether this reflects a unique phenomenon, it has been found that the incidence of *Fusarium* spp. infections can be affected by the environmental fungal load or the contamination of hospital water distribution system (Raad et al., 2002; Anaissie et al., 2001).

Infections due to *Fusarium* spp. can be divided into localized or disseminated. In immunocompetent hosts *Fusarium* infections are more frequently localized, whereas in immunocompromised patients they are disseminated. They usually present as persistent fever refractory to antibacterial and antifungal agents combined with sinusitis and/or rhinocerebral infection, cellulitis of toe or finger, painful skin lesions, myositis, pneumonia, and infection of the central nervous system (Boutati & Anaissie, 1997). Skin lesions, the cornerstone of disseminated fusariosis, occur in 70–90%, followed by lungs and sinuses in about 70% (Bodey et al., 2002).

Fusarium spp. are a frequent cause of corneal damage and are one of the most frequent causes of fungal keratitis. The main predisposing factor is ocular trauma due to the implantation of a contaminated material (Gopinathan et al., 2002). An additional risk factor for *Fusarium* keratitis is the presence of a pathological corneal condition and concomitant therapy with topical steroids combined with antibiotics (Nelson et al., 1994). However, the most frequently noted predisposing factor for keratitis has been improper lens care, which has led to contamination of contact lens paraphernalia. Contamination of the lens can also occur in windy conditions. This fungus can penetrate the matrix of the soft contact lens with increasing microbial growth in lenses. *Fusarium* keratitis can also develop among users of daily disposable soft contact lenses. Fusarial endophthalmitis is a rare complication of keratitis or following surgical and non-surgical trauma (Dignani & Anaissie, 2004).

Fusarium spp. can cause localized infections of the nails and/or the skin after soil contamination. The most common clinical presentations include proximal subungual onychomycosis with or without paronychia and white superficial onychomycosis (Tosti et al., 2000). Skin infections due to *Fusarium* spp. can be either primary or metastatic. Among immunocompetent hosts, skin lesions typically are localized and develop after trauma at the site of infection (Bodey et al., 2002). Cutaneous infections in these immunocompetent individuals present most commonly as necrotic lesions that complicate extensive wounds, cellulitis after onychomycosis, or chronic ulcers and abscesses. By contrast, the majority of immunocompromised and particularly neutropenic patients have disseminated skin lesions. The most common lesions in these patients are multiple painful erythematous papules or nodules with necrosis, frequently associated with cultures of blood and skin positive for *Fusarium* spp. and resulting in death (Rippon et al., 1988). Probably, the preexisting skin damage allows the entry of these pathogenic fungi into the underlying soft tissues, leading to the development of cellulitis and subsequent dissemination in a manner reminiscent of bacterial infections (Nucci & Anaissie, 2002).

In addition, *Fusarium* spp. can less frequently cause peritonitis following continuous ambulatory peritoneal dialysis, catheter-associated fungemia, and other single organ infections. Specifically, rare cases of osteomyelitis, arthritis, otitis, sinusitis, and brain abscess due to *Fusarium* spp. have been reported (Walsh and Groll, 1999).

Disseminated fusariosis is the most important manifestation within the wide spectrum of *Fusarium* spp. infections (Martino et al., 1994; Nucci et al., 2004). The clinical features of disseminated fusariosis mimic those of disseminated aspergillosis. However, distinct differences can be seen, such as an increased incidence of skin lesions and positive blood cultures associated with disseminated fusariosis in contrast to invasive aspergillosis (Martino et al., 1994). Among patients undergoing solid organ transplantation, fusarial infections tend to be more localized, occur later after transplantation and have a better outcome than among neutropenic patients with haematological malignancies or bone marrow transplantation recipients. Skin lesions involve practically any skin site, with predominance on the lower extremities (Bodey et al., 2002; Nucci & Anaissie, 2002).

Early identification of fusariosis is an important factor for a successful outcome. Furthermore, diagnosis requires the demonstration of hyphae in pathological samples; however, hyphae of *Aspergillus, Scedosporium,* and *Fusarium* are difficult to discriminate. Although diagnosis of fusariosis is not always easy, in patients with severe immunosuppression, a high suspicion for disseminated fusariosis should be raised when mold fungemia is reported. In addition, when preceding or concomitant toe cellulitis or cutaneous or subcutaneous lesions are present, *Fusarium* spp. infection must be suspected.

Positive cultures are needed for a definitive diagnosis of a *Fusarium* spp. infection and in about 40% of patients fusariosis can be diagnosed only by blood cultures. The rate of positive blood cultures increases to over 60% in the presence of disseminated skin lesions, while fungemia is extremely rare in patients with localized skin infections (Dignani & Anaissie, 2004). *Fusarium* spp. cultures can be confused with other genera such as *Acremonium, Verticilium,* and *Cylindrocarpon* (Walsh & Groll, 1999).

To enhance diagnostic potential, tissue diagnosis of fusariosis can be made by immunohistological staining, using polyclonal fluorescent antibody reagents that distinguish *Fusarium* from *Aspergillus* spp. (Kaufman et al., 1997). In addittion, *in situ* hybridization may also help to distinguish *Fusarium* from *Aspergillus* and *Scedosporium* spp. in tissue sections with a very high positive predictive value (Hayden et al., 2003).

Nowadays, molecular methods including polymerase chain reaction (PCR) analysis can be used in addition to conventional techniques to provide an earlier and better diagnosis. Different molecular methods have been used to differentiate *Fusarium* spp., such as amplified fragment length polymorphism (AFLP), multilocus sequence typing (MLST) (O'Donnell et al., 2004), restriction fragment length polymorphism (RFLP) (Hopfer et al., 1993), random amplified polymorphic DNA (RAPD) (Ouellet & Seifert, 1993), automated repetitive sequence-based PCR (rep-PCR) (Healy et al., 2004), cleaved amplified polymorphic sequences (CAPS), derived CAPS (dCAPS), and simple sequence repeats (SSR) analysis (O'Donnell et al., 2004). These PCR methods can detect *Fusarium* spp. even to the species level and for that reason PCR-based assays have the potential for use as an adjunct to conventional methods of diagnosis. Developing a standardized PCR-based assay should be a priority, as the complexes that exist in *Fusarium* along with the numerous anamorphic species provide a very dynamic and at times confusing nomenclature.

The 'ideal' methodology for determining *in vitro* susceptibility of molds to antifungal agents is not currently standardized. However, with the present methods, *Fusarium* spp. are only moderately susceptible or resistant to most of the antifungal agents (Pujol et al., 1997). Comparative pharmacodynamic data on their susceptibility to various antifungal agents indicate low susceptibility to flucytosine, miconazole, ketoconazole, itraconazole, and fluconazole and variable susceptibility to amphotericin B (Anaissie et al., 1991; Espinel-Ingroff, 2001; Espinel-Ingroff et al., 2001). Notably, *in vitro* susceptibility or resistance to these antifungal agents may not predict the clinical outcome of *Fusarium* infection.

A discrepancy has been reported between microdilution MICs and E test endpoints for amphotericin B and more extensive investigations are required to establish which method predicts the clinical outcome after treatment best. However, amphotericin B appears to be the most active agent against *Fusarium* spp. infections *in vitro* and *in vivo*. About 50% of isolates are susceptible to this drug and MICs for 90% of the strains tested range from 1 to 4 µg/mL (Pfaller et al., 2002; Lewis et al., 2005). Unfortunately the correlation between *in vitro* values and clinical efficacy is low and many patients remain unresponsive to treatment despite *in vitro* susceptibility.

In addition to amphotericin B the newer triazoles including voriconazole, posaconazole and ravuconazole are promising. Although these triazoles show variable activity, they do not exert significant fungicidal effects on *F. solani* unlike amphotericin B (Espinel-Ingroff, 2001; Paphitou et al., 2002; Pfaller et al., 2002; Lewis et al., 2005). *F. solani* seems to be somewhat more susceptible to amphotericin B but less susceptible to voriconazole than *F. oxysporum* (Paphitou et al., 2002).

Fusarium spp. resistance to echinocandins is probably due to the particularity of the cell wall structure of this mold. Specifically, a possible mechanism is that *Fusarium* spp. possess less 1,3-β-glucans, the target of echinocandins, than other non-1,3-β glucans (Arikan et al., 2001). Thus, caspofungin, anidulafungin, and micafungin have no *in vitro* activity against *Fusarium* spp. (Espinel-Ingroff, 2003). However, combination of caspofungin plus amphotericin B *in vitro* can result in the reduction of caspofungin MIC (Arikan et al., 2002).

Due to the *in vitro* resistance of *Fusarium* spp., combinations of antifungal agents with other agent categories have been used. Although amphotericin B plus rifampicin have shown *in vitro* synergy, the results have not been encouraging in human fusariosis. In addition, when amphotericin B was combined with flucytosine and rifampicin in mice with disseminated *F. solani* infection, no synergy was demonstrated (Guarro et al., 1999).

Despite the equivocal *in vitro* susceptibility results and numerous treatment failures, amphotericin B with its lipid formulations (liposomal, lipid complex, and colloidal dispersion amphotericin B) has remained the drug of choice for the treatment of disseminated fusarial infections (Jensen et al., 2004). Lipid formulations of amphotericin B seem to be more effective *in vivo* than conventional amphotericin B [deoxycholate amphotericin B (DAMB)], with a response rate approaching 85% for amphotericin B lipid complex (ABLC) (Walsh et al., 1998). Several individual case reports of invasive fusariosis also favour the use of a lipid formulation of amphotericin B; however, in some cases higher doses were required (Walsh et al., 1999).

Newer triazoles appear to be an alternative therapeutic option for patients with fusariosis who are either intolerant of, or have infection refractory to standard antifungal therapy. Voriconazole, posaconazole, and ravuconazole are active against *Fusarium* spp., with variable and species-dependent fungistatic activity (Ghannoum & Kuhn, 2002; Torres et al., 2005). In addition, voriconazole is approved for the treatment of fusariosis in patients intolerant of, or with infection refractory to other drugs.

In immunosuppressed patients with disseminated fusariosis mortality rates exceed 75% (Boutati & Anaissie, 1997). In general, most failures occurred in patients who had undergone hemopoietic stem cell transplantation, had disseminated fusariosis or had persistent neutropenia. Due to increased mortality, different antifungal combinations have been used. A combined approach of liposomal amphotericin B, voriconazole, surgery and granulocyte transfusions in a patient suffering from disseminated *F. oxysporum* infection with a relapse of acute B cell leukemia during induction chemotherapy resulted in a favourable outcome (Durand-Joly et al., 2003).

Hemopoietic growth factors and cytokines have been used as adjuvant therapy *in vitro* and *in vivo*. Granulocyte colony-stimulating factor (G-CSF) as well as granulocyte-macrophage colony-stimulating factor (GM-CSF) enhance the antifungal activities of phagocytes against *Fusarium* spp. (Hubel et al., 2002; Gaviria et al., 1999a). Other potential adjuvant agents, such as interleukin-15 (IL-15), have been shown to enhance neutrophil-induced damage of hyphae of *Fusarium* spp. suggesting a positive effect on the immune response against *Fusarium* spp. infections (Winn et al., 2005).

The genus *Scedosporium* includes two hyphomycetes of emerging medical importance, *Scedosporium apiospermum* and *S. prolificans*. *Pseudallescheria boydii* is the teleomorph (sexual state) distinguished from its anamorph (asexual state) *S. apiospermum* by the presence of dark round cleistothecia (the organ of sexual reproduction) (Walsh et al., 2004). During the past decades, both states have undergone several sequential name changes having been referred to as *Petriellidium boydii*, *Allescheria boydii*, *P. Sheari*, and *Monosporium apiospermum* (Tadros et al., 1998). In culture, *P. boydii* and *S. apiospermum* appear as branching septate hyphae with parallel cell walls and terminal anneloconidia. However, it is not usually possible to distinguish *P. boydii* and *S. apiospermum* from other filamentous fungi such as *Aspergillus* or *Fusarium* because terminal conidia are usually absent in tissue sections.

S. prolificans, formerly known as *S. inflatum* due to the inflated appearance of their phialides, is morphologically and physiologically similar to *S. apiospermum*.

However, *S. prolificans* was established as a distinct species owing to its more rapid growth on standard culture media and the formation of annellides with swollen bases (Salkin et al., 1988). *S. prolificans* is not known to have a teleomorph (Rainer & De Hoog, 2006). However, molecular studies have demonstrated a close relationship of *S. prolificans* with the genus *Petriella* regarded as its sexual state (Walsh et al., 2004; Issakainen et al., 1997). *S. prolificans* is a relatively new fungus since infections due to this fungal species have been noticed after mid-1980s (Guarro et al., 1991; Marin et al., 1991). It is differentiated from *S. apiospermum* by the bottle-shaped annellides and the lack of growth on Sabouraud agar with cyclohex-imide. One typical feature useful in distinguishing between *S. apiospermum* and *S. prolificans* is their susceptibility profiles with the majority of *S. apiospermum* isolates being more susceptible to antifungals (Steinbach & Perfect, 2003).

Recently, studies have shown that high genetic variation exists within this species leading to different pathogenicity profiles. Initially, Bell has found differences in virulence between a strain from a patient with subcutaneous infections and one from the environment (Bell, 1978). De Hoog et al. recognized, on the basis of nuclear DNA (nDNA) homology, three infraspesific ecological and clinical groups. They also noted that one of the groups has a relatively high frequency of airway infections and systemic infections (de Hoog et al., 1994). Rainer et al., found five different small-subunit rRNA gene sequence lengths (Rainer et al., 2000). Further, RAPD studies reported the existence of numerous and varying genotypes, while no clustering according to geographic origin of the isolates was noted (Defontaine et al., 2002). While, in several cases, strains isolated from a single region proved to be genetically different, some lineages appeared to have a regional distribution (Rainer et al., 2000). All these data suggest that *P. boydii* is a 'species complex' that includes a high number of phylogenetic species though only a few of them can be recognized morphologically (Gilgado et al., 2005). Overall, an evolution process seems to take place with the persistence of groups and species with increased virulence to humans (Rainer & De Hoog, 2006).

S. apiospermum is widely distributed in soil, sewage, and contaminated water. Its presence in polluted water is responsible for the fungus dissemination into the CNS in near-drowning victims the majority of whom are otherwise healthy adults or children (Dworzack et al., 1989; Buzina et al., 2006; Katragkou et al., 2007). The clinical spectrum of *S. apiospermum* infection involves localized disease after trauma (mycetoma, arthritis, osteomyelitis, endophthalmitis, onychomycosis, lymphocutaneous infection) occurring mainly in immunocompetent individuals, asymptomatic or symptomatic colonization of cavities (sinusitis, otitis, pulmonary fungus ball) observed in patients with predisposing pulmonary disorders like cystic fibrosis and systemic invasive disease (pneumonia, endocarditis, CNS infection) found primarily in immunocompromised patients (Guarro et al., 2006).

Infections of *S. prolificans* are not homogeneously distributed around the world (Ortoneda et al., 2002b). While disseminated infections have been relatively frequently reported in Spain and Australia, in the USA most of the cases have been localized osteoarticular infections and a very small number of disseminated infections have been reported even in high-risk groups (cancer and leukaemia patients) (Wood et al., 1992). However, this irregular distribution is not found in infections caused by *S. apiospermum* and other species of this genus (Ortoneda et al., 2002b).

Numerous studies have shown that both clinically important *Scedosporium* spp. are generally resistant to amphotericin B while some of the newer triazoles like voriconazole and posaconazole exhibit promising results (Berenguer et al., 1997; Espinel-Ingroff, 1998; Carrillo & Guarro, 2001; Gilgado et al., 2006). Enhanced antifungal activity has been demonstrated *in vitro* for combinations of amphotericin B plus azoles (miconazole, itraconazole, and fluconazole) (Walsh et al., 1995). The optimal management of scedosporiosis is not apparent. Therapy usually includes extensive surgical debridement whenever possible and other salvage treatment options like high concentrations of antifungals administered locally (Mursch et al., 2005; Wilson et al., 1990). Among adjunctive therapeutic modalities aiming to the reconstitution of the patient's immune status, administration of certain cytokines, such as interferon gamma (IFN-γ) and GM-CSF, has been proposed (Gil-Lamaignere et al., 2005). However, there are not enough clinical data supporting these findings.

Pathogenesis

Two modes of acquisition of *Fusarium* spp. seem likely: (i) respiratory, following inhalation of conidia and (ii) cutaneous (via trauma, burns, catheters, or other foreign bodies), as in cases of cellulitis, catheter-related infections, and burns. The predominant mode of infection is inhalation of conidia into the respiratory system and establishment in the lungs. It is well known that host defence mechanisms [especially macrophages, neutrophils, and monocytes (MNCs)] influence the manifestation and severity of fungal infections; thus, clinical forms of the disease depend on a patient's immune response. If the host is unable to effectively clear the primary pulmonary infection, widespread hematogenous dissemination may occur. In addition, *Fusarium* spp. possess several cellular and molecular attributes that may confer different degrees of inherent virulence on these organisms. The combination of these virulence factors and the immunocompromised status of the host contribute to the development of invasive fusarial infections. Risk factors for disseminated fusariosis include severe immunosuppression (mainly patients with haematological malignancies) in addition to colonization and tissue damage. Specifically, conditions like neutropenia, lymphopenia, corticosteroid therapy or any other immunosuppressive treatment, graft versus host disease, receipt of a graft from an HLA mismatched or unrelated donor are considered risk factors for disseminated fusariosis (Guarro & Gene, 1995; Marr et al., 2002; Raad et al., 2002).

Fusarium spp. possess several virulence factors including the production of fumonisins, fusarins, moniliformin, and other mycotoxins. These factors can cause mycotoxicoses in humans due to suppression of humoral and cellular immunity. In addition, *Fusarium* spp. have the ability to produce proteases and collagenases, and also to adhere to prosthetic material such as catheters and contact lenses making biofilms (Dignani & Anaissie, 2004).

Release of toxins, enzyme production, and adherence to prosthetic materials, have all been postulated as virulence factors for *Fusarium* spp. infection. Specifically, mycotoxins produced by *Fusarium* spp. such as trichothecenes, deoxynivalenol (vomitoxin), and fumonisin affect the immune system by decreasing proliferation and function of lymphocytes, protein synthesis, and phagocytosis by macrophages and chemotaxis of neutrophils (Forsell et al., 1986; Nelson et al., 1994). In addition, these mycotoxins decrease the number of lymphocytes and MNCs, the levels of immunoglobulin M and immunoglobulin A and finally the phagocytic function of macrophages (Visconti et al., 1991; Qureshi & Hagler, 1992). Several mycotoxins induce leukopenia and marrow destruction and suppress platelet aggregation. Definitive evidence linking enzyme production to the virulence of *Fusarium* spp. is lacking. *In vitro* production of proteases by *F. solani* and of collagenases by *F. moniliforme* has been documented. However, whether these enzymes play a role in the pathogenesis of human fusariosis remains to be determined.

Polymorphism in the trichothecene mycotoxin gene cluster of *Fusarium* spp. has been found in phytopathogenic *Fusarium* spp. Phylogenetic analyses have demonstrated that polymorphism within these virulence-associated genes is specific and appears to have been maintained by balancing selection acting on chemotype differences that originated in the ancestor of *Fusarium* spp. In addition, chemotype-specific differences, evidence of adaptive evolution within trichothecene genes, have also been reported (Ward et al., 2002). Although data on polymorphism of human pathogenic *Fusarium* spp. are very limited, in an analysis of 33 *F. oxysporum* complex isolates coming from patients of the same US hospital, it was found that a recently dispersed, geographically widespread clonal lineage was responsible for over 70% of all clinical isolates investigated. Moreover, strains of the clonal lineage were conclusively shown to genetically match those isolated from the hospital water systems of three US hospitals, providing support for the hypothesis that hospitals may serve as a reservoir for nosocomial fusarial infections (O'Donnell et al., 2004).

The mode of *Scedosporium* spp. invasion and subsequent propagation to the host remains ambiguous. Clinical experience suggests the respiratory tract as the most probable portal of entry of *Scedosporium* spp. infection (Steinbach & Perfect, 2003). Consistently, lungs appear to be the most frequent sites of *S. apiospermum* infections (Salesa et al., 1993). However, spores may infect a patient through the gastrointestinal tract from ulcerative lesions or by direct inoculation from areas with trauma or a central venous catheter and eventually disseminate to multiple organs including brain, kidney or heart (Berenguer et al., 1997). Additionally, angiotropism and perineural invasion with subsequent dissemination along the nerve sheet may represent another means of spread of the disease as it has been suggested for other filamentous fungi (Frater et al., 2001).

Although the pathogenic mechanisms and virulence factors of *Scedosporium* spp. infection have not been fully elucidated, recent studies have described the role of different pathogenic determinants. Generally, *S. prolificans* isolates show comprehensive *in vitro* and *in vivo* resistance to currently used antifungals and therefore they are considered more virulent than *S. apiospermum* isolates (Hennequin et al., 1997; Ortoneda et al., 2002b; Carrillo & Guarro, 2001). One possible explanation of *S. prolificans* drug resistance is the presence of melanin (Ruiz-Diez & Martinez-Suarez, 2003). Melanins are dark brown or black pigments of high molecular weight formed by oxidative polymerization of phenolic compounds. Most fungal melanins are derived from the precursor molecule 1,8-dihydroxynaphthalene (DHN) through the polyketide biosynthetic pathway and reside in ascomycetes and related deuteromycetes (Jacobson, 2000). There are several hypotheses supporting the protective role of melanin against host defense mechanisms and environmental stress. A growing body of evidence supports melanin's function as antioxidant. Melanin may have a protective effect on the fungus by scavenging oxygen and nitrogen free radicals, produced by phagocytic cells during oxidative burst. Additional melanin pathogenetic mechanisms include sequestration of host defensive proteins, cross-linking or shielding cell wall constituents against hydrolytic enzymes, or conferring resistance to heat (Jacobson et al., 1995; Schnitzler et al., 1999). The role of melanin in virulence as protection for the pathogen against immunologically generated free radicals has been modelled on *Cryptococcus neoformans*. It has been shown that melanized cells survived approximately tenfold better than did non-melanized cells (Wang & Casadevall, 1994). Further studies have not demonstrated any apparent difference in terms of MIC of various antifungal agents, between the melanin-lacking mutants and the wild-type isolates of *S. prolificans* (Ruiz-Diez & Martinez-Suarez, 2003).

Larcher et al. have isolated and characterized a serine proteinase of the subtilisin family that is secreted by *S. apiospermum* (Larcher et al., 1996). From studies performed on cystic fibrosis patients it is known that proteinases, secreted by fungal pathogens like *A. fumigatus* or *S. apiospermum* contribute to pulmonary damage by degrading host proteins like fibrinogen and basement membrane laminin or indirectly by hypersensitivity mechanisms (Lake et al., 1990; Tronchin et al., 1993; Miller et al., 1993). Consequently, there is presumptive evidence, that serine proteinase from *S. apiospermum* may act like the alkaline proteinase of *A. fumigatus* by degrading human fibrinogen mediating, in this way, to the severe bronchopulmonary inflammation of the cystic fibrosis patients. In this regard, proteinase inhibitors may constitute an attractive future therapeutic alternative aiming to control the inflammation response (Larcher et al., 1996).

The predilection of *Scedosporium* spp. to CNS dissemination may be explained by the siderophore activity of these fungal strains (Panackal & Marr, 2004). Further biochemical and genetic studies are needed in order to determine the pathogenetic potential of *Scedosporium* spp.

Host Defences Against *Fusarium* and *Scedosporium* spp.

In vitro Studies

Most studies of host defences against filamentous fungi have focused to *A. fumigatus* and to *Rhizopus oryzae* (Diamond & Clark, 1982; Schaffner et al., 1982; Washburn et al., 1987). Knowledge of the host response against *Fusarium* and *Scedosporium* spp. is rather limited and less well understood.

Since *Fusarium* and *Scedosporium* spp. share some common features with *A. fumigatus*, such as utilizing the same portals of entry into human body as well as causing life-threatening and therapy-refractory infections especially in immuno-compromised patients, immune mechanisms similar to *A. fumigatus* probably apply to the two fungal genera. The main line of host defence against *A. fumigatus* consists of phagocytes, including circulating polymorphonuclear leukocytes (PMNs) and MNCs as well as monocyte-derived macrophages (MDMs). Although MNCs and macrophages can damage hyphae, the bulk of this role appears to fall upon the PMNs (Waldorf, 1989). Phagocytes are capable of damaging hyphal elements through oxygen-dependent and independent mechanisms. The oxygen-dependent mechanisms consist of a series of reactions starting with the production of superoxide anion (O_2^-), which is dismutated into hydrogen peroxide (H_2O_2). Myeloperoxidase (MPO) then catalyzes the conversion of H_2O_2 and halides to generate hypohalides and chloramines, which exert potent antifungal activities (Babior, 2000; Hampton et al., 1998). Cationic peptides (defensins and cathelici-dins) are part of the oxygen-independent pathway of phagocytic cells during the innate host response to fungi (De Lucca & Walsh, 1999; Ramanathan et al., 2002; Yang et al., 2002). A large body of evidence reveals that a variety of cytokines, chemokines and growth factors play an important role in the host response connect-ing the innate with the adaptive immunity and to the pathogenesis of filamentous fungal infections (Romani, 2004).

Recognition of fungal elements by macrophages occurs through Toll-like receptors and other surface receptors. These molecules transduce the signal of activation into the nucleus and induce expression of pro-inflammatory and other cytokines. However, no knowledge exists on the recognition activation of cells, intracellular transduction and gene expression in response to *Fusarium* and *Scedosporium* spp. Cytokines such as interleukin-6 (IL-6) and interleukin-8 (IL-8) but not tumor necrosis factor-α (TNF-α) are released by the phagocytes in response to *F. solani* and *S. prolificans*.

The immune response to *Fusarium* spp. may be species-dependent. Specifically *in vitro* studies have shown that human PMNs elicit higher oxidative burst in response to serum-opsonized hyphae of *F. solani* than of *F. oxysporum*. In addittion, PMNs induce much less damage to *F. solani* hyphae than to those of *F. oxysporum*. This finding indi-cates that *F. solani* hyphae are more resistant to the innate immune response than *F. oxysporum* hyphae. The increased incidence of invasive disease due to *F. solani* might be alternatively explained as arising from a higher level of environmental expo-sure of patients to this species than to *F. oxysporum*. However, it would seem extremely unlikely that there is no link between the apparently superior resistance of *F. solani* hyphae to MNC damage and the finding that this is the most frequently encountered of the two species in invasive infections of neutropenic patients (Winn et al., 2003).

Th1-type cytokines have exhibited certain enhancing activities on antifungal phagocytic responses. In this regard, effects of IL-15 on antifungal responses of human PMNs against *Fusarium* spp. have been studied. IL-15 did not affect PMN oxidative respiratory burst evaluated as O_2^- production in response to *F. solani* and *F. oxysporum*. However, IL-15 increased interleukin-8 (IL-8) release from PMNs challenged by *F. solani* hyphae, but not by *F. oxysporum* hyphae. Additionaly, release of TNF-α was not affected by the use of IL-15 (Winn et al., 2005).

Pulmonary alveolar macrophages (PAMs), the primary phagocytic cells of pulmonary host defence during neutropenia, exhibited fungicidal activity against conidia of *F. solani* and achieved a time-dependent increase in killing. In addition, when PAMs were incubated with DAMB or ABLC, it was found that ABLC and, to a lesser degree, DAMB additively augments the fungicidal activity of PAMs against conidia of *F. solani* (Roilides et al., 2006).

Different immunomodulatory effects of amphotericin B formulations (deoxycholate, liposomal, lipid complex and colloidal dispersion amphotericin B) on oxidative antifungal activities of human MNCs and PMNs against *F. solani* hyphae have been determined. Specifically, MNCs and PMNs pretreated with all amphotericin B formulations induce increased *F. solani* hyphal damage; while, ABLC appears most effective. Further, the effects of amphotericin B formulations on PMN-induced hyphal damage are significantly higher than those on MNC-induced damage. In contrast, O_2^- production by MNCs or PMNs upon hyphal challenge is not stimulated by amphotericin B formulations (Dotis, J., Simitsopoulou, M., Dalakiouridou, M., Konstantinou, T., Walsh, T.J., Roilides, E. Comparative study of the effects of amphotericin B formulations on antifungal activity of human phagocytes against *Aspergillus fumigatus* and *F. solani*, submitted).

In vitro studies have shown that *S. apiospermum* and *S. prolificans* conidia and hyphae are susceptible to phagocytes in a manner comparable to *A. fumigatus* with minor differences (Gil-Lamaignere et al., 2001, 2003). Specifically, MDMs are able to phagocytose *Scedosporium* conidia similarly to the phagocytosis of *Aspergillus*, despite the much bigger size of *S. prolificans* conidia. Additionally, MDMs inhibit germination of *S. prolificans* conidia less efficiently, as compared to *A. fumigatus* (Gil-Lamaignere et al., 2001). *In vitro* studies have demonstrated that phagocytes are capable of exhibiting sufficient oxidative burst to control *S. prolificans* strains in the presence of serum. In the absence of serum, however, the production of O_2^- appears to be lessened (Gil-Lamaignere et al., 2001). The way the opsonization status affects the oxidative burst in response to *S. prolificans* remains unclear and merits further investigation. Isolates of *S. prolificans* tested *in vitro* have been damaged in an effector cell-target ratio-dependent manner when challenged with both kinds of phagocytes. Moreover, phagocytes have tended to induce more damage to *S. prolificans* than to *A. fumigatus* (Gil-Lamaignere et al., 2001).

Further insight into the immunopathogenesis of *Scedosporium* spp. infection has been gained through *in vitro* studies of the phagocytic cell responses to amphotericin B-resistant and susceptible *S. apiospermum* isolates. Accordingly, it has been found that macrophages are able to phagocytose *S. apiospermum* conidia, damage hyphae in an effector cell-target ratio-dependent manner and release O_2^- in response to serum-opsonized hyphae. It has also been observed that hyphae of the two strains with different amphotericin B-susceptibility patterns have different levels of susceptibility to MPO products. This phenomenon, although not fully elucidated, may be related to the various levels of pathogenicity and antifungal drug resistance of *S. apiospermum* (Gil-Lamaignere et al., 2003).

ABLC has been reported that it displays a significant additive effect with PMNs against *S. prolificans* and *S. apiospermum in vitro* (Gil-Lamaignere et al., 2002a). Similarly, in another *in vitro* study, triazoles used in combination with PMNs have caused significant additive increase in the damage of the hyphae of *S. prolificans* and *S. apiospermum*. Furthermore, under certain conditions synergism has been noted between triazoles and PMNs against *S. prolificans* hyphae. Interestingly, the synergistic activity has been observed at the low concentrations of the antifungals used. This finding may be of particular importance especially in immunocompromised patients when a triazole reaches its trough level in plasma where such synergy may prevent fungal regrowth (Gil-Lamaignere et al., 2002b). Regardless of the mechanisms behind these collaborative effects, the findings from these studies would support the concomitant administration of antifungals and PMN transfusions to persistently neutropenic patients with invasive scedosporiosis. However, no cases of scedosporiosis have been reported that have been treated with PMN transfusions up to date.

Immunosuppression constitutes a significant risk factor for the surge of invasive fungal infections. In this regard, a number of studies have aimed to assess the immunomodulatory utility of cytokines in confronting emerging fungal pathogens (Steinbach & Perfect, 2003). *S. prolificans* has been shown to induce significantly more TNF and IL-6 release by human MNCs, as compared to *Aspergillus* spp. This could be attributed to the specific composition of the *S. prolificans* cell wall, although its exact composition is not known, which may yield more potent stimulatory molecules. Speculatively, this could be associated with the virulence of the specific fungus (Warris et al., 2005).

Besides it has been shown that the presence of IL-15 significantly enhances PMN-induced hyphal damage and oxidative respiratory burst of *S. prolificans* but not *S. apiospermum*. Additionally, IL-15 increases IL-8 release from PMNs challenged by *S. prolificans* whereas release of TNF is not affected. The inability of IL-15 to exhibit enhanced damage of *S. apiospermum* hyphae is in concordance with its greatest intrinsic virulence in humans. These findings suggest that IL-15 has species-specific enhancing effects on antifungal activities of PMNs against *Scedosporium* spp. Further, some of the cytokine-induced effects have been shown that are the result of direct actions on effector activities of PMNs while others, related to the increased release of IL-8, acting in an autocrine way on PMNs, result in enhanced indirect antifungal actions (Winn et al., 2005).

Among the cytokines studied that enhance PMN antifungal activity against *Scedosporium* spp. are interferon-γ (IFN-γ) and GM-CSF (Gil-Lamaignere et al., 2005). IFN-γ is produced endogenously primarily by T cells and is a potent activator of MNC-macrophages and PMNs (Gaviria et al., 1999b). This cytokine induces the Th1 response, which favours resistance to fungal disease, regulates the gene expression of NADPH oxidase subunits at the transcriptional level and potentiates the synthesis of antimicrobial peptides in macrophages (Chaves et al., 1996; Amezaga et al., 1992; Cassatella et al., 1990). GM-CSF acts on early as well as on late stages of haematopoiesis and increases the number of cells of the macrophage-monocyte system. This cytokine has been found to enhance phagocytosis, oxidative burst, increase the number and membrane binding of several classes of surface receptors on PMNs and inhibit PMN apoptosis (Rodriguez-Adrian et al., 1998; Giles, 1998;

Armitage, 1998). Treatment of PMNs with the combination of IFN and GM-CSF had broader effects on *Scedosporium* spp. enhancing PMN functions while cytokines alone had no effect. Despite the poor effect of either cytokine alone on the PMN oxidative burst after 22h, the combined treatment showed enhancement of oxidative burst in response to opsonized *S. apiospermum* hyphae. Similarly, after incubation with cytokines for 2h only the combination significantly enhanced the oxidative burst against serum-opsonized and non-opsonized hyphae of *Scedosporium* spp. Thus, in this study it has been demonstrated that IFN and GM-CSF exhibit a significant time- and species-dependent capability to enhance PMN activity against *Scedosporium* spp. (Gil-Lamaignere et al., 2005).

In vivo–ex vivo Studies

G-CSF is produced by macrophages, fibroblasts, and endothelial cells in virtually all organs in the body. It exerts its biological actions on precursors as well as on mature PMNs. In particular, G-CSF increases the number of circulating neutrophils, by stimulating the proliferation and differentiation of myeloid progenitor cells, and moreover enhances their phagocytic activity (Antachopoulos and Roilides, 2005). Administration of G-CSF in neutropenic animal models of invasive candidiasis, aspergillosis, or trichosporonosis was associated with faster recovery from neutropenia and improved survival (Hamood et al., 1994; Muranaka et al., 1997; Graybill et al., 1998). Consequently, theoretically, the administration of G-CSF would be beneficial in resolving scedosporiosis. No animal model studies have reported on the effects of G-CSF on the outcome of experiment fusariosis up to date. Ortoneda et al., in an immunosuppressed murine model of invasive infection by *S. prolificans*, demonstrated a modest efficacy of liposomal amphotericin B (LAMB) at 10 mg/kg/day combined with G-CSF (Ortoneda et al., 2002a). Subsequent studies showed that LAMB at very high doses (40 mg/kg/day) combined with G-CSF did not significantly improve the survival (Ortoneda et al., 2004). Interestingly, administration of G-CSF alone, was not more effective as compared to the control group (Ortoneda et al., 2004; Ortoneda et al., 2002a). In an immunocompetent murine model of disseminated *S. prolificans* infection it has been shown that posaconazole and GM-CSF have a combined effect in damaging *S. prolificans* hyphae *ex vivo*. However, when posaconazole and GM-CSF were administered to mice with invasive infection due to *S. prolificans*, they had selective beneficial effects on the burdens in certain organs but offered no additional benefit to survival (Simitsopoulou et al., 2004).

Clinical Data

Over the last decades, efforts to reconstitute host defence with granulocyte transfusion therapy have increased after the advances of the availability of recombinant hematopoietic growth factors and modern transfusion practices. In severely neutropenic patients suffering from fusariosis, treatment with G- or GM-CSF and granulocyte transfusions may also be considered (Dignani et al., 1997; Boutati & Anaissie, 1997; Price et al., 2000; Rutella et al., 2003; Durand-Joly et al., 2003) (Table 12.1). The beneficial effect of the transfusions seems to be enhanced by their administration to patients with good performance status as well as by their administration early during neutropenia and soon after onset of fusariosis.

The principle of therapy in fusariosis is early aggressive treatment with high doses of amphotericin B. When fusariosis is refractory to standard antifungal therapy or the patient does not tolerate therapy, amphotericin B can switch to a new triazole. In addition, G-CSF or GM-CSF plus granulocyte transfusion can be potential adjuvant therapeutic strategies combined with catheter removal (if the catheter is the source of fusariosis) and debridement of devitalized tissue (in localized disease). Long duration of antifungal treatment is crucial and has to be individualized for the better outcome of the patient.

The current clinical evidence on the role of antifungals and immunomodulatory factors in the host response against scedosporiosis is limited, involving mainly haematopoietic growth factors and IFN. Noteworthy, of the 39 cases reported in the literature with disseminated *Scedosporium* spp. infection, favourable outcome was reported only in four of them (Guarro et al., 2006). In these cases the positive outcome was attributed to the immunomodulatory factors administered in addition to antifungal drugs whereas the true role of antifungals was difficult to be established (Munoz et al., 2000; Girmenia et al., 1998; Ochiai et al., 2003; Phillips et al., 1991). Undoubtedly, the outcome of antifungal therapy

Table 12.1 Use of granulocyte transfusions in the treatment of fusariosis and scedosporiosis

Reference	Underlying condition	No of patients	Outcome
Minor et al. (1989)	Bone marrow transplantation	2	0/2 Responses
Barrios et al. (1990)	Bone marrow transplantation	1	1/1 Complete response
Helm et al. (1990)	Acute lymphocytic leukemia	1	1/1 Complete response
Spielberger et al. (1994)	Pancytopenia	1	1/1 Complete response
Dignani et al. (1997)	Neutropenia	3	2/3 Complete response
Boutati and Anaissie (1997)	Hematological malignancies	7[a]	3/7 Complete response
Girmenia et al. (1999)	Severe aplastic anemia	1[a]	0/1 Cesponse
Price et al. (2000)	Stem cell transplant	2	0/2 Responses
Rutella et al. (2003)	Hematological malignancies, neutropenia	1	1/1 Complete response
Durand-Joly et al. (2003)	B-acute leukemia	1[a]	1/1 Complete response
Total		19	10/19 Complete response

[a] Granulocyte transfusions after stimulation with G-CSF.

alone in scedosporiosis is poor with high overall mortality rates, and cases that have been initially treated with amphotericin B have been dismal (Guarro et al., 2006). Among the promising antifungals is voriconazole, though the results are still contradictory (Walsh et al., 2002; Perfect et al., 2003). Due to the multidrug resistance a crucial point in the management of scedosporiosis is immune function reconstitution. Characteristically, only 2 out of 16 patients infected with *S. prolificans* survived and their survival coincided with hematologic recovery (Berenguer et al., 1997). Similarily, in another review disseminated *S. prolificans* infection was fatal in all neutropenic patients (Revankar et al., 2002).

Conclusion

Although several non-dematiaceous molds can cause opportunistic fungal infections to humans, only *Fusarium* and *Scedosporium* spp. have been recognized as emerging pathogens with significant morbidity and mortality especially in immunocompromised patients. Our understanding of *Fusarium* and *Scedosporium* spp. has been advanced during the last decade. Currently, new molecular methods can be used additionally to conventional techniques to provide an earlier and better documented diagnosis even to the species level. In addition, new findings on the host immune response, based on PMN and MNC function, have been studied and well documented.

The current knowledge on the management of fusariosis and scedosporiosis primarily depends on *in vitro* and *in vivo* studies due to the low incidence of these mycoses and the subsequent difficulty in performing large clinical trials with sufficient predictive power. However, early aggressive treatment with high doses of standard antifungal drugs, combination treatment and newer antifungals agents, especially new azoles, seem to be very promising for the management of these difficult to treat infections. In addition, hemopoietic growth factors plus granulocyte transfusions can be potential adjuvant therapeutic strategies.

Future studies should aim to determine whether combination therapy with new agents, hemopoietic growth factors and cytokines, improves survival and treatment outcome in the most seriously debilitated patients who are afflicted with life-threatening infections due to *Fusarium* and *Scedosporium* spp.

References

Ajello, L. (1986). *Eur. J. Epidemiol.* 2:243–251.
Amezaga, M. A., Bazzoni, F., Sorio, C., Rossi, F., & Cassatella, M. A. (1992). *Blood*, 79:735–744.
Anaissie, E., Paetznick, V., Proffitt, R., Adler-Moore, J., & Bodey, G. P. (1991). *Eur. J. Clin. Microbiol. Infect. Dis.*, 10:665–668.
Anaissie, E. J., Kuchar, R. T., Rex, J. H., Francesconi, A., Kasai, M., Muller, F. M., Lozano-Chiu, M., Summerbell, R. C., Dignani, M. C., Chanock, S. J., & Walsh, T. J. (2001). *Clin. Infect. Dis.*, 33:1871–1878.
Antachopoulos, C. & Roilides, E. (2005). *Br. J. Haematol.*, 129:583–596.

Arikan, S., Lozano-Chiu, M., Paetznick, V., & Rex, J. H. (2001). *Antimicrob. Agents Chemother.*, 45:327–330.

Arikan, S., Lozano-Chiu, M., Paetznick, V., & Rex, J. H. (2002). *Antimicrob. Agents Chemother.*, 46:245–247.

Armitage, J. O. (1998). *Blood*, 92:4491–4508.

Babior, B. M. (2000). *Am. J. Med.*, 109:33–44.

Barrios, N. J., Kirkpatrick, D. V., Murciano, A., Stine, K., van Dyke, R. B., & Humbert, J. R. (1990). *Am. J. Pediatr. Hematol. Oncol.*, 12:319–324.

Bell, R. G. (1978). *Can. J. Microbiol.*, 24:856–863.

Berenguer, J., Rodriguez-Tudela, J. L., Richard, C., Alvarez, M., Sanz, M. A., Gaztelurrutia, L., Ayats, J., & Martinez-Suarez, J. V. (1997). *Medicine (Baltimore)*, 76:256–265.

Bodey, G. P., Boktour, M., Mays, S., Duvic, M., Kontoyiannis, D., Hachem, R., & Raad, I. (2002). *J. Am. Acad. Dermatol.* 47:659–666

Boutati, E. I. & Anaissie, E. J. (1997). *Blood* 90:999–1008.

Buzina, W., Feierl, G., Haas, D., Reinthaler, F. F., Holl, A., Kleinert, R., Reichenpfader, B., Roll, P., & Marth, E. (2006). *Med. Mycol.*, 44:473–477.

Carrillo, A. J. & Guarro, J. (2001). *Antimicrob. Agents Chemother.*, 45:2151–2153.

Cassatella, M. A., Bazzoni, F., Flynn, R. M., Dusi, S., Trinchieri, G., & Rossi, F. (1990). *J. Biol. Chem.*, 265:20241–20246.

Chaves, M. M., Silvestrini, A. A., Silva-Teixeira, D. N., & Nogueira-Machado, J. A. (1996). *Inflamm. Res.*, 45:313–315.

de Hoog, G. S., Marvin-Sikkema, F. D., Lahpoor, G. A., Gottschall, J. C., Prins, R. A., & Gueho, E. (1994). *Mycoses*, 37:71–78.

De Lucca, A. J. & Walsh, T. J. (1999). *Antimicrob. Agents Chemother.*, 43:1–11.

Defontaine, A., Zouhair, R., Cimon, B., Carrere, J., Bailly, E., Symoens, F., Diouri, M., Hallet, J. N., & Bouchara, J. P. (2002). *J. Clin. Microbiol.*, 40:2108–2114.

Diamond, R. D. & Clark, R. A. (1982). *Infect. Immun.*, 38:487–495.

Dignani, M. C. & Anaissie, E. J. (2004). *Clin. Microbiol. Infect.*, 10:67–75.

Dignani, M. C., Anaissie, E. J., Hester, J. P., O'Brien, S., Vartivarian, S. E., Rex, J. H., Kantarjian, H., Jendiroba, D. B., Lichtiger, B., Andersson, B. S., & Freireich, E. J. (1997). *Leukemia*, 11:1621–1630.

Durand-Joly, I., Alfandari, S., Benchikh, Z., Rodrigue, M., Espinel-Ingroff, A., Catteau, B., Cordevant, C., Camus, D., Dei-Cas, E., Bauters, F., Delhaes, L., & De Botton, S. (2003). *J. Clin. Microbiol.*, 41:4898–4900.

Dworzack, D. L., Clark, R. B., Borkowski, W. J., Jr., Smith, D. L., Dykstra, M., Pugsley, M. P., Horowitz, E. A., Connolly, T. L., McKinney, D. L., Hostetler, M. K., et al. (1989). *Medicine (Baltimore)*, 68:218–224.

Espinel-Ingroff, A. (1998). *J. Clin. Microbiol.*, 36:198–202.

Espinel-Ingroff, A. (2001). *J. Clin. Microbiol.*, 39:954–958.

Espinel-Ingroff, A. (2003). *Rev. Iberoam. Micol.*, 20:121–136.

Espinel-Ingroff, A., Boye, K., & Sheehan, D. J. (2001). *Mycopathologia*, 150:101–115.

Forsell, J. H., Witt, M. F., Tai, J. H., Jensen, R., & Pestka, J. J. (1986). *Food Chem. Toxicol.*, 24:213–219.

Frater, J. L., Hall, G. S., & Procop, G. W. (2001). *Arch. Pathol. Lab. Med.*, 125:375–378.

Gaviria, J. M., van Burik, J. A., Dale, D. C., Root, R. K., & Liles, W. C. (1999a) *J. Infect. Dis.*, 179:1038–1041.

Gaviria, J. M., van Burik, J. A., Dale, D. C., Root, R. K., & Liles, W. C. (1999b). *J. Infect. Dis.*, 179:1301–1304.

Ghannoum, M. A. & Kuhn, D. M. (2002). *Eur. J. Med. Res.*, 7:242–256.

Gil-Lamaignere, C., Maloukou, A., Rodriguez-Tudela, J. L., & Roilides, E. (2001). *Med. Mycol.*, 39:169–175.

Gil-Lamaignere, C., Roilides, E., Lyman, C. A., Simitsopoulou, M., Stergiopoulou, T., Maloukou, A., & Walsh, T. J. (2003). *Infect. Immun.*, 71:6472–6478.

Gil-Lamaignere, C., Roilides, E., Maloukou, A., Georgopoulou, I., Petrikkos, G., & Walsh, T. J. (2002a). *J. Antimicrob. Chemother.*, 50:1027–1030.

Gil-Lamaignere, C., Roilides, E., Mosquera, J., Maloukou, A., & Walsh, T. J. (2002b). *Antimicrob. Agents Chemother.*, 46:2234–2237.

Gil-Lamaignere, C., Winn, R. M., Simitsopoulou, M., Maloukou, A., Walsh, T. J., & Roilides, E. (2005). *Med. Mycol.*, 43:253–260.

Giles, F. J. (1998). *Clin. Infect. Dis.*, 26:1282–1289.

Gilgado, F., Cano, J., Gene, J., & Guarro, J. (2005). *J. Clin. Microbiol.*, 43:4930–4942.

Gilgado, F., Serena, C., Cano, J., Gene, J., & Guarro, J. (2006). *Antimicrob. Agents Chemother.*, 50:4211–4213.

Girmenia, C., Iori, A. P., Boecklin, F., Torosantucci, A., Chiani, P., De Fabritiis, P., Taglietti, F., Cassone, A., & Martino, P. (1999). *Haematologica*, 84:114–118.

Girmenia, C., Luzi, G., Monaco, M., & Martino, P. (1998). *J. Clin. Microbiol.*, 36:1436–1438.

Gopinathan, U., Garg, P., Fernandes, M., Sharma, S., Athmanathan, S., & Rao, G. N. (2002). *Cornea*, 21:555–559.

Graybill, J. R., Bocanegra, R., Najvar, L. K., Loebenberg, D., & Luther, M. F. (1998). *Antimicrob. Agents Chemother.*, 42:2467–2473.

Guarro, J., Gaztelurrutia, L., Marin, J., & Barcena, J. (1991). *Enferm. Infecc. Microbiol. Clin.*, 9:557–560.

Guarro, J. & Gene, J. (1995). *Eur. J. Clin. Microbiol. Infect. Dis.*, 14:741–754.

Guarro, J., Kantarcioglu, A. S., Horre, R., Luis Rodriguez-Tudela, J., Cuenca Estrella, M., Berenguer, J., & Sybren De Hoog, G. (2006). *Med. Mycol.*, 44:295–327.

Guarro, J., Pujol, I., & Mayayo, E. (1999). *Antimicrob. Agents Chemother.* 43:1256–1257.

Hamood, M., Bluche, P. F., De Vroey, C., Corazza, F., Bujan, W., & Fondu, P. (1994). *Mycoses*, 37:93–99.

Hampton, M. B., Kettle, A. J., & Winterbourn, C. C. (1998). *Blood*, 92:3007–3017.

Hayden, R. T., Isotalo, P. A., Parrett, T., Wolk, D. M., Qian, X., Roberts, G. D., & Lloyd, R. V. (2003). *Diagnostic Mol. Pathol.*, 12:21–26.

Healy, M., Reece, K., Walton, D., Huong, J., Shah, K., & Kontoyiannis, D. P. (2004). *J. Clin. Microbiol.* 42:4016–4024.

Helm, T. N., Longworth, D. L., Hall, G. S., Bolwell, B. J., Fernandez, B., & Tomecki, K. J. (1990). *J. Am. Acad. Dermatol.*, 23:393–398.

Hennequin, C., Benailly, N., Silly, C., Sorin, M., Scheinmann, P., Lenoir, G., Gaillard, J. L., & Berche, P. (1997). *Antimicrob. Agents Chemother.*, 41:2064–2066.

Hopfer, R. L., Walden, P., Setterquist, S., & Highsmith, W. E. (1993). *J. Med. Vet. Mycol.*, 31:65–75.

Hubel, K., Dale, D. C., & Liles, W. C. (2002). *J. Infect. Dis.*, 185:1490–1501.

Issakainen, J., Jalava, J., Eerola, E., & Campbell, C. K. (1997). *J. Med. Vet. Mycol.*, 35:389–398.

Jacobson, E. S. (2000). *Clin. Microbiol. Rev.*, 13:708–717.

Jacobson, E. S., Hove, E., & Emery, H. S. (1995). *Infect. Immun.*, 63:4944–4945.

Jensen, T. G., Gahrn-Hansen, B., Arendrup, M., & Bruun, B. (2004). *Clin. Microbiol. Infect.*, 10:499–501.

Katragkou, A., Dotis, J., Kotsiou, M., Tamiolaki, M., & Roilides, E. (2007). *Mycoses* (accepted after modification).

Kaufman, L., Standard, P. G., Jalbert, M., & Kraft, D. E. (1997). *J. Clin. Microbiol.*, 35:2206–2209.

Lake, F. R., Tribe, A. E., McAleer, R., Froudist, J., & Thompson, P. J. (1990). *Thorax*, 45:489–491.

Larcher, G., Cimon, B., Symoens, F., Tronchin, G., Chabasse, D., & Bouchara, J. P. (1996). *Biochem. J.*, 315(Pt 1):119–126.

Lewis, R. E., Wiederhold, N. P., & Klepser, M. E. (2005). *Antimicrob. Agents Chemother.*, 49:945–951.

Marin, J., Sanz, M. A., Sanz, G. F., Guarro, J., Martinez, M. L., Prieto, M., Gueho, E., & Menezo, J. L. (1991). *Eur. J. Clin. Microbiol. Infect. Dis.*, 10:759–761.

Marr, K. A., Carter, R. A., Crippa, F., Wald, A., & Corey, L. (2002). *Clin. Infect. Dis.*, 34:909–917.

Martino, P., Gastaldi, R., Raccah, R., & Girmenia, C. (1994). *J. Infect.*, 28:7–15.

Miller, M. A., Greenberger, P. A., Amerian, R., Toogood, J. H., Noskin, G. A., Roberts, M., & Patterson, R. (1993). *Am. Rev. Respir. Dis.*, 148:810–812.

Minor, R. L., Pfaller, M. A., Gingrich, R. D., & Burns, L. J. (1989). *Bone Marrow Transplant.*, 4:653–658.

Munoz, P., Marin, M., Tornero, P., Martin Rabadan, P., Rodriguez-Creixems, M., & Bouza, E. (2000). *Clin. Infect. Dis.*, 31:1499–1501.

Muranaka, H., Suga, M., Nakagawa, K., Sato, K., Gushima, Y., & Ando, M. (1997). *Infect. Immun.*, 65:3422–3429.

Mursch, K., Trnovec, S., Ratz, H., Hammer, D., Horre, R., Klinghammer, A., de Hoog, S., & Behnke-Mursch, J. (2006). *Childs Nerv. Syst.*, 22:189–192.

Nelson, P. E., Dignani, C. & Anaissie, E. J. (1994). *Clin. Microbiol. Rev.*, 7:479–504.

Nucci, M. & Anaissie, E. (2002). *Clin. Infect. Dis*, 35:909–920.

Nucci, M., Marr, K. A., Queiroz-Telles, F., Martins, C. A., Trabasso, P., Costa, S., Voltarelli, J. C., Colombo, A. L., Imhof, A., Pasquini, R., Maiolino, A., Souza, C. A., & Anaissie, E. (2004) *Clin. Infect. Dis.*, 38:1237–1242.

O'Donnell, K., Sutton, D. A., Rinaldi, M. G., Magnon, K. C., Cox, P. A., Revankar, S. G., Sanche, S., Geiser, D. M., Juba, J. H., van Burik, J. H., Padhye, A., Anaissie, E. J., Francesconi, A., Walsh, T. J., & Robinson, J. S. (2004). *J. Clin. Microbiol.*, 42:5109–5120.

Ochiai, N., Shimazaki, C., Uchida, R., Fuchida, S., Okano, A., Ashihara, E., Inaba, T., Fujita, N., & Nakagawa, M. (2003). *Leuk. Lymphoma*, 44:369–372.

Ortoneda, M., Capilla, J., Pastor, F. J., Serena, C., & Guarro, J. (2004). *Diagn. Microbiol. Infect. Dis.*, 50:247–251.

Ortoneda, M., Capilla, J., Pujol, I., Pastor, F. J., Mayayo, E., Fernandez-Ballart, J., & Guarro, J. (2002a) *J. Antimicrob. Chemother.*, 49:525–529.

Ortoneda, M., Pastor, F. J., Mayayo, E., & Guarro, J. (2002b) *J. Med. Microbiol.*, 51:924–928.

Ouellet, T. & Seifert, K. A. (1993). *Phytopathologia*, 8:1003–1007.

Panackal, A. A. & Marr, K. A. (2004). *Semin. Respir. Crit. Care. Med.*, 25:171–181.

Paphitou, N. I., Ostrosky-Zeichner, L., Paetznick, V. L., Rodriguez, J. R., Chen, E., & Rex, J. H. (2002). *Antimicrob. Agents Chemother.* 46:3298–3300.

Perfect, J. R., Marr, K. A., Walsh, T. J., Greenberg, R. N., DuPont, B., de la Torre-Cisneros, J., Just-Nubling, G., Schlamm, H. T., Lutsar, I., Espinel-Ingroff, A., & Johnson, E. (2003). *Clin. Infect. Dis.*, 36:1122–1131.

Pfaller, M. A., Messer, S. A., Hollis, R. J., & Jones, R. N. (2002). *Antimicrob. Agents Chemother.*, 46:1032–1037.

Phillips, P., Forbes, J. C., & Speert, D. P. (1991). *Pediatr. Infect. Dis. J.*, 10:536–539.

Price, T. H., Bowden, R. A., Boeckh, M., Bux, J., Nelson, K., Liles, W. C., & Dale, D. C. (2000). *Blood*, 95:3302–3309.

Pujol, I., Guarro, J., Gene, J., & Sala, J. (1997). *J. Antimicrob. Chemother.*, 39:163–167.

Qureshi, M. A. & Hagler, W. M., Jr. (1992). *Poult. Sci.* 71:104–112.

Raad, I., Tarrand, J., Hanna, H., Albitar, M., Janssen, E., Boktour, M., Bodey, G., Mardani, M., Hachem, R., Kontoyiannis, D., Whimbey, E., & Rolston, K. (2002). *Infect. Control Hosp. Epidemiol.*, 23:532–537.

Rainer, J. & De Hoog, G. S. (2006). *Mycol. Res.*, 110:151–160.

Rainer, J., de Hoog, G. S., Wedde, M., Graser, Y., & Gilges, S. (2000). *J. Clin. Microbiol.*, 38:3267–3273.

Ramanathan, B., Davis, E. G., Ross, C. R., & Blecha, F. (2002). *Microbes Infect.*, 4:361–372.

Revankar, S. G., Patterson, J. E., Sutton, D. A., Pullen, R., & Rinaldi, M. G. (2002). *Clin. Infect. Dis.*, 34:467–476.

Rippon, J. W., Larson, R. A., Rosenthal, D. M., & Clayman, J. (1988). *Mycopathologia*, 101:105–111.

Rodriguez-Adrian, L. J., Grazziutti, M. L., Rex, J. H., & Anaissie, E. J. (1998). *Clin. Infect. Dis.*, 26:1270–1278.

Romani, L. (2004). *Nat. Rev. Immunol.*, 4:1–23.

Ruiz-Diez, B. & Martinez-Suarez, J. V. (2003). *Curr. Microbiol.*, 46:228–232.

Rutella, S., Pierelli, L., Sica, S., Serafini, R., Chiusolo, P., Paladini, U., Leone, F., Zini, G., D'Onofrio, G., Leone, G., & Piccirillo, N. (2003). *Cytotherapy*, 5:19–30.

Salesa, R., Burgos, A., Ondiviela, R., Richard, C., Quindos, G., & Ponton, J. (1993). *Scand. J. Infect. Dis.*, 25:389–393.

Salkin, I. F., McGinnis, M. R., Dykstra, M. J., & Rinaldi, M. G. (1988). *J. Clin. Microbiol.*, 26:498–503.

Sampathkumar, P. & Paya, C. V. (2001). *Clin. Infectious Dis.*, 32:1237–1240.

Schaffner, A., Douglas, H., & Braude, A. (1982). *J. Clin. Invest.*, 69:617–631.

Schnitzler, N., Peltroche-Llacsahuanga, H., Bestier, N., Zundorf, J., Lutticken, R., & Haase, G. (1999). *Infect. Immun.*, 67:94–101.

Simitsopoulou, M., Gil-Lamaignere, C., Avramidis, N., Maloukou, A., Lekkas, S., Havlova, E., Kourounaki, L., Loebenberg, D., & Roilides, E. (2004). *Antimicrob. Agents Chemother.*, 48:3801–3805.

Spielberger, R. T., Falleroni, M. J., Coene, A. J., & Larson, R. A. (1994). *Clin. Infect. Dis.*, 18:490–491.

Steinbach, W. J. & Perfect, J. R. (2003). *J. Chemother.*, 15(Suppl. 2):16 27.

Tadros, T. S., Workowski, K. A., Siegel, R. J., Hunter, S., & Schwartz, D. A. (1998). *Hum Pathol*, 29:1266–1272.

Torres, H. A., Hachem, R. Y., Chemaly, R. F., Kontoyiannis, D. P., & Raad, I. I. (2005). *Lancet Infect. Dis.*, 5:775–785.

Tosti, A., Piraccini, B. M., & Lorenzi, S. (2000). *J. Am. Acad. Dermatol.*, 42:217–224.

Tronchin, G., Bouchara, J. P., Larcher, G., Lissitzky, J. C., & Chabasse, D. (1993). *Biol. Cell*, 77:201–208.

Visconti, A., Minervini, F., Lucivero, G., & Gambatesa, V. (1991). *Mycopathologia*, 113:181–186.

Waldorf, A. R. (1989). *Immunol. Ser.*, 47:243–271.

Walsh, T. J., Goodman, J. L., Pappas, P., Bekersky, I., Buell, D. N., Roden, M., Barret, J., & Anaissie, E. J. (1999). *Antimicrob. Agents Chemother.*, 45:3487–3496.

Walsh, T. J., Groll, A., Hiemenz, J., Fleming, R., Roilides, E., & Anaissie, E. (2004). *Clin. Microbiol. Infect.*, 10(Suppl. 1):48–66.

Walsh, T. J. & Groll, A. H. (1999). *Transpl. Infect. Dis.*, 1:247–261.

Walsh, T. J., Hiemenz, J. W., Seibel, N. L., Perfect, J. R., Horwith, G., Lee, L., Silber, J. L., DiNubile, M. J., Reboli, A., Bow, E., Lister, J., & Anaissie, E. J. (1998). *Clin. Infect. Dis.*, 26:1383–1396.

Walsh, T. J., Lutsar, I., Driscoll, T., Dupont, B., Roden, M., Ghahramani, P., Hodges, M., Groll, A. H., & Perfect, J. R. (2002). *Pediatr. Infect. Dis. J.*, 21:240–248.

Walsh, T. J., Peter, J., McGough, D. A., Fothergill, A. W., Rinaldi, M. G., & Pizzo, P. A. (1995). *Antimicrob. Agents Chemother.*, 39:1361–1364.

Wang, Y. & Casadevall, A. (1994). *Infect. Immun.*, 62:3004–3007.

Ward, T. J., Bielawski, J. P., Kistler, H. C., Sullivan, E., & O'Donnell, K. (2002). *Proc. Natl Acad. Sci.*, 99:9278–9283.

Warris, A., Netea, M. G., Verweij, P. E., Gaustad, P., Kullberg, B. J., Weemaes, C. M., & Abrahamsen, T. G. (2005). *Med. Mycol.*, 43:613–621.

Washburn, R. G., Gallin, J. I., & Bennett, J. E. (1987). *Infect. Immun.*, 55:2088–2092.

Wilson, C. M., O'Rourke, E. J., McGinnis, M. R., & Salkin, I. F. (1990). *J. Infect. Dis.*, 161:102–107.

Winn, R. M., Gil-Lamaignere, C., Roilides, E., Simitsopoulou, M., Lyman, C. A., Maloukou, A., & Walsh, T. J. (2005). *Cytokine*, 31:1–8.

Wood, G. M., McCormack, J. G., Muir, D. B., Ellis, D. H., Ridley, M. F., Pritchard, R., & Harrison, M. (1992). *Clin. Infect. Dis.*, 14:1027–1033.

Yang, D., Biragyn, A., Kwak, L. W., & Oppenheim, J. J. (2002). *Trends Immunol*, 23:291–296.

Index